This item must be returned or renewed by the last date shown below. The loan period may be shortened if it is reserved by another reader. A fine will be due if it is not returned on time.

DATE OF RETURN

Biomolecular Sensors

Biomolecular Sensors

Edited by
Electra Gizeli and Christopher R. Lowe

London and New York

First published 2002
by Taylor & Francis
11 New Fetter Lane, London EC4P 4EE

Simultaneously published in the USA and Canada
by Taylor & Francis Inc,
29 West 35th Street, New York, NY 10001

Taylor & Francis is an imprint of the Taylor & Francis Group

© 2002 Electra Gizeli, Christopher R. Lowe

Typeset in Times by Wearset Ltd, Boldon, Tyne and Wear
Printed and bound in Great Britain by TJ International Ltd, Padstow,
Cornwall

Every effort has been made to ensure that the advice and information in
this book is true and accurate at the time of going to press. However,
neither the publisher nor the authors can accept any legal responsibility or
liability for any errors or omissions that may be made. In the case of drug
administration, any medical procedure or the use of technical equipment
mentioned within this book, you are strongly advised to consult the
manufacturer's guidelines.

British Library Cataloguing in Publication Data
A catalogue record for this book is available from the British Library

Library of Congress Cataloging in Publication Data
Biomolecular sensors / edited by Electra Gizeli, Christopher R. Lowe.
    p. cm.
Includes bibliographical references and index.
1. Biosensors I. Gizeli, Electra, 1964– II. Lowe, C. R. (Christopher R.)

R857.B54 B5335 2002
610'.28–dc21

           2001052294

ISBN 0-7484-0791-X (alk. paper)

Contents

Notes on contributors   ix
Preface   xi
Acknowledgements   xiii

PART I
Biological recognition   **1**

1   Principles of antigen–antibody recognition   **3**
M.H.V. VAN REGENMORTEL AND D. ALTSCHUH
Introduction   3
Structure of immunoglobulins   4
Structure of paratopes   5
Structure of epitopes   6
Antigen–antibody interactions   9
The specificity of immunological reagents   13
Structure–function relationships in immunochemistry   14
References   15

2   Protein–protein interactions   **19**
GIDEON SCHREIBER
Characterization of protein–protein interfaces   19
Double mutant cycle analysis of a protein–protein interface   23
*Mutational induced structural rearrangement in protein
    interfaces   25*
Kinetic pathway of protein–protein association   25
The mechanism of association of a protein complex   26
*Measuring protein–protein interactions in homogeneous and
    heterogeneous phase   27*
Affinity versus activity   29
Summary   29
References   30

vi   *Contents*

3   DNA interactions   **33**

BRIAN E. CATHERS AND MICHAEL J. WARING

Analogue recognition   36

Middle of the spectrum   37

Digital recognition   39

References   42

PART II
Immobilisation of biomolecules   **47**

4   Immobilisation chemistry of biological recognition molecules   **49**

ANDREW G. MAYES

Introduction   49

Creation of the molecular recognition interface on the transducer   49

The transducer surface   50

Adsorption to the transducer surface   60

Entrapment methods   60

Covalent coupling chemistry   64

*Techniques giving some control over the orientation of the immobilised
    antibody   74*

Spatial control of surface immobilisation   78

Conclusion   83

References   83

5   Binding isotherms and kinetics of immobilized biological systems   **87**

CLAUS DUSCHL

Introduction   87

The determination of binding constants and of kinetic rate constants   89

Conclusions   118

References   118

PART III
Transducer technology   **121**

6   Optical transducers   **123**

MARTHA LILEY

Introduction   123

Optical transducers: general considerations   124

Direct detection   133

Fluorescence transducers   158

Conclusions and outlook   170

Acknowledgements   172

References   172

7 Acoustic transducers **176**

ELECTRA GIZELI

Introduction 176
Elastic waves in solids 176
Acoustic wave devices 180
Acoustic wave sensors for studying biomolecular interactions 183
Comparison of acoustic sensors 200
Conclusions 202
Acknowledgements 202
References 202

8 Immunoassays using enzymatic amplification electrodes **207**

FRIEDER W. SCHELLER, CHRISTIAN G. BAUER, ALEXANDER MAKOWER, ULLA
WOLLENBERGER, AXEL WARSINKE AND FRANK F. BIER

Introduction 207
Coupling of immunoassays with enzymatic recycling electrodes 207
Conclusions 230
Acknowledgement 232
References 232

PART IV
Applications **239**

**9 Surface plasmon resonance: development and use of BIACORE instruments
for biomolecular interaction analysis** **241**

BENGT IVARSSON AND MAGNUS MALMQVIST

Introduction 241
Application demands 244
SPR-refractometer instrumental configurations 245
Sensor surface chemistry 256
Biomolecular interaction analysis-BIA 259
Marketing 263
References 265

10 IAsys: the resonant mirror biosensor **269**

R.J. DAVIES AND P.R. EDWARDS

Introduction 269
Modus operandi – *light and surfaces 269*
Developments in kinetics 271
Recent applications 274
Conclusion 287
References 287

11   Commercial quartz crystal microbalances: theory and applications   **291**

C.K. O'SULLIVAN AND G.G. GUILBAULT

Introduction   291
Quartz crystal microbalance – theory   291
Commercial systems   295
Applications   295
Conclusion and future directions   299
Bibliography   300

12   The quartz crystal microbalance with dissipation monitoring (QCM-D)   **304**

MICHAEL RODAHL, PATRIK DAHLQVIST, FREDRIK HÖÖK AND
BENGT KASEMO

Introduction   304
The QCM beyond the Sauerbrey regime   305
Application examples   307
Example 1: DNA   307
Example 2: Mussel adhesive protein   309
Concluding remarks   312
Appendix   313
References   314

Index   317

Contributors

Dr C.G. Bauer, Analytical Biochemistry, Institute of Biochemistry and Biology, University of Potsdam, Karl-Liebnecht-Str. 24–25, Haus 25, 14476, Golm, Germany.

Dr F. Bier, Analytical Biochemistry, Institute of Biochemistry and Biology, University of Potsdam, Karl-Liebnecht-Str. 24–25, Haus 25, 14476, Golm, Germany.

Dr Brian Cathers, New Biotics, Inc., 11760 E. Sorrento Valley Road, San Diego, CA 92121.

Mr P. Dahlqvist, Q-Sense AB, Holtermansgatan 1B, SE-412 92 Göteburg, Sweden.

Dr R. Davies, Unipath Ltd, Priory Business Park, MK44 3UP.

Dr Claus Duschl, Nanotype GmbH-Lochhamer Schlag 12, 082166 Gräfelfing, Germany.

Dr P. Edwards, Affinity Sensors Ltd, Saxon Way, Cambridge, UK, CB3 8SL.

Dr Electra Gizeli, University of Cambridge, Institute of Biotechnology, Tennis Court Rd, Cambridge, UK, CB2 1QT.

Prof. G. Guilbaut, Laboratory of Sensor Development, Department of Chemistry, National University of Ireland, Cork, Ireland.

Dr F. Hook, Department of Applied Physics, Chalmers University of Technology, S-412 96, Gothenburg, Sweden.

Mr Bengt Ivarsson, Biacore AB, S-75450, Uppsala, Sweden.

Prof. B. Kasemo, Department of Applied Physics, Chalmers University of Technology, S-412 96, Gothenburg, Sweden.

Dr Martha Liley, Advanced Microsystems, CSEM, SA, Jacquet-Droz 1, CH-2007 Neuchâtel.

Dr A. Makower, Analytical Biochemistry, Institute of Biochemistry and Biology, University of Potsdam, Karl-Liebnecht-Str. 24–25, Haus 25, 14476, Golm, Germany.

Dr Magnus Malmqvist, Sweden.

Dr Andrew G. Mayes, School of Chemical Sciences, University of East Anglia, Norwich, UK, NR4 7TJ.

Dr Ciara K. O'Sullivan, Bioengineering and Bioelectrochemistry Group, Department of Chemical Engineering, Universtat Rovira i Virgili, Avinguda Paisos, Catalans, Tarragona, Spain.

Dr M. Rodahl, Q-Sense AB, Holtermansgatan 1B, SE-412 92 Göteburg, Sweden.

Prof. F.W. Scheller, Analytical Biochemistry, Institute of Biochemistry and Biology, University of Potsdam, Karl-Liebnecht-Str. 24–25, Haus 25, 14476, Golm, Germany.

Dr Gideon Schreiber, Department of Biological Chemistry, Weizmann Institute of Science, Rehovot 76100, Israel.

Prof. M.H.V. Van Regenmortel, Laboratoire d'Immunochemie, Institut de Biologie Moleculaire et Cellulaire, 15 rue R. Descartes, 67084, Strasbourg Cedex, France. UMR 7100, Ecole Supérieure de Biotechnologie de Strasbourg, Parc d'Innovation, BD Sébastien Brandt, F.67400, Illkirch, France.

Prof. M. Waring, University of Cambridge, Department of Pharmacology, Tennis Court Rd, Cambridge, UK, CB2 1QJ.

Dr A. Warsinke, Analytical Biochemistry, Institute of Biochemistry and Biology, University of Potsdam, Karl-Liebnecht-Str. 24–25, Haus 25, 14476, Golm, Germany.

Dr U. Wollenberger, Analytical Biochemistry, Institute of Biochemistry and Biology, University of Potsdam, Karl-Liebnecht-Str. 24–25, Haus 25, 14476, Golm, Germany.

Preface

The recent substantial advances in the understanding of the genomes of pro- and eukaryotic organisms which underpin growth, expression, differentiation and productivity have far reaching consequences in pure, strategic and applied science. The "-omics" triad of genomics, proteomics and metabolomics is beginning to enable scientists to establish outline mechanistic descriptions of the key bioinformatic drivers, the metabolic pathways, and the accumulation and/or assembly of components leading to all classes of end-product.

The development of new therapeutic drugs ranks among the most laborious and capital intensive of all mankind's activities. Recent estimates of the number of individual genes in the human genome (~30,000) and the number of unique chemical structures theoretically attainable using existing chemistries (~108) suggest that up to 1012 assays would be required to map the structure–activity space for all potential therapeutic targets. However, this need for specific, well-validated molecular targets and the unwieldy scale and duration of drug development are now being addressed by these new "-omics" disciplines. Together, they have the capacity to revolutionise the discovery and development of drugs by allowing the molecular components of cells to be subjected to comprehensive high-throughput screening and analysis. The goal of genomics is to sequence all the genes in the human genome and to specify the patterns of gene expression associated with particular cellular states, whilst proteomics identifies and documents their proteomes, or protein profiles. However, many key aspects of proteomics are not encoded at the genetic level and can only be clarified by analysis at the protein level. For example, factors such as concentration, transcriptional alteration, post-translational modification (proteolysis, phosphorylation and glycosylation), intermittent or permanent formation of complexes with other proteins or cellular components, compartmentation within the cell and modulation of biological activity with a plethora of small molecules, influence the function of the protein, but are not genetically encoded. These modifications often play a crucial role in the activity, localisation and turnover of an individual protein.

The inability of genomics to address the protein level in sufficient detail is a crucial shortcoming, since it is at this level that the molecular life of the cell is transacted and at which most disease processes develop. Proteomics is widely expected to supersede genomics as the most productive application of biotechnology in the next few years. The field of proteomics will require the development of new sets of analytical tools that allow the characterisation of complex interactions between proteins and other biological molecules, and how these biorecognition phenomena are translated into dynamic and macroscopic events. This new armamentarium of tools will exploit technologies and systems being developed outside the biological sciences and particularly within the physical sciences and engineering. The area of biomolecular sensors, in which biology and the physical sciences come together, and which encapsulates a broader definition than the more conventional analytical devices based on biosensors, is now just beginning to materialise as a force to be reckoned with as an aid to understanding the underlying biophysical principles of molecular recognition. This scenario represents a very healthy encroachment of the physical sciences into biology and may be expected to lead to new understanding of both biology and the physical sciences.

This book comprises contributed chapters prepared by experts directly involved in biophysical research and development. However, whilst the biomolecular sensor literature abounds with many combinations of biological recognition systems and transducers, and this book covers most of the principal types, it does concentrate on those that have enjoyed commercial success. The first parts cover the basic principles of biological recognition, particularly with respect to protein-protein and antibody-antigen interactions, how these molecules may be immobilised to transducers and the nature of the binding kinetics and isotherms. These sections are designed to acquaint all readers, irrespective of their background discipline, with the fundamental biological wherewithal to tackle Part III on the transducer technology. This section deals with the principles of how the biorecognition process is converted into a usable and processible signal with optical, acoustic, electrical and other approaches. Finally, and particularly in view of the comments made above, the last four chapters show how these fundamental principles have been commercialised into instruments which exploit the technology in areas as diverse as biological research, medicine, plant science and the environment.

This book should remind all readers of two important lessons: first, that the recent history of biomolecular sensors suggests that mistakes have been made, both in underestimating the complexity of the technology involved in such a facile concept, and in the much slower than expected penetration of established diagnostics markets, where the "ten times cheaper or ten times better" rule applies. The second lesson is that whilst taking biology to physics seemed a good idea twenty years ago, it now seems much more sensible to take advanced physics to biology in order to unravel the mysteries of life itself. Viewed this way, biomolecular sensors are much more likely to have a significant impact in the future, particularly in sectors such as basic biology, genomics, proteomics and high throughput screening.

Christopher R. Lowe
Electra Gizeli
Cambridge, 2001

Acknowledgements

Figure 5.2 reproduced from C. Duschl *et al.* (1996), *Biophysics Journal*, 75, 583–594 with permission from the Biophysical Society; Figure 5.3 reproduced from J. Spinke *et al.* (1993), *Journal of Chemical Physics*, 99, 7012–7019 with permission of the American Institute of Physics; Figure 5.5 from H.C. Berg, *Random Walks in Biology* © 1993 Reprinted by permission of Princeton University Press; Figure 5.7 reproduced from D.G. Myszka *et al.* (1996), *Biophysics Journal*, 70, 1965–1995 with permission from the Biophysical Society; Figure 5.8 reproduced from D.G. Myszka *et al.* (1996), *Protein Science*, 5, 2468–2478 with permission of Cambridge University Press; Figure 5.9 reprinted from J. Piehler *et al.* (1997), *Journal of Immunological Methods*, 201, 189–206 with permission from Elsevier Science; Figure 6.9 reproduced from Cush *et al.* (1993), *Biosensors and Bioelectronics*, 8, 347–354 with permission of Elsevier Science; Figure 6.10 reproduced from Spinke *et al.* (1997), *Sensors and Actuators B*, 38–39, 256–260 with permission of Elsevier Science; Figure 6.18 reproduced from Deacon *et al.* (1991), *Biosensors and Bioelectronics*, 6, 193–199 with permission of Elsevier Science; Figure 6.19 reproduced from A.P. Duveneck *et al.* (1999), *Proc SPIE: Advanced Materials and Optical Systems for Chemical and Biological Detection*, volume 3858 with permission; Figure 6.22 reproduced from Anderson *et al.* (1993), *Biosensors and Bioelectronics*, 8, 249–256 with permission of Elsevier Science; Figure 7.8a and Figure 7.8b reprinted with permission from K. Ijiro *et al.* (1998), *Langmuir*, 14, 2796–2800 © 1998 American Chemical Society; Figure 7.9 reprinted with permission from S. Ghafouri and M. Thompson (1999), *Langmuir*, 15, 564–572. © 1999 American Chemical Society; Figure 7.10 reprinted with permission from Y. Okahata and H. Ebato (1989), *Analytical Chemistry*, 61, 2185–2188. © 1989 American Chemical Society; Figure 7.11 reprinted with permission from R. Dahint *et al.* (1999), *Analytical Chemistry*, 71, 3150–3156. © 1999 American Chemical Society; Figure 7.12 reproduced from J.C. Andle *et al* (1995), *Sensors and Actuators B*, 24–25, 129–133 with permission of Elsevier Science; Figure 7.13 reprinted with permission from K. Melzak *et al.* (2001), *Langmuir*, 17, 1594–1598; Figure 7.14 reproduced from G.L. Harding *et al.* (1997), *Sensors and Actuators A*, 61, 279–286 with permission of Elsevier Science; Figure 8.2, Table 8.1 and Table 8.2 reproduced from U. Wollenberger *et al.* (1997) *Frontiers in Biosensorics, Band II, Practical Applications*, Birkhäuser Verlag, Basel, 45–70 with permission; Figures 8.11, 8.12 and 8.13 (1997), Anal Chim Acta 344, 122; Figure 10.2 reproduced with permission from D-B. Borza and W.T. Morgan (1998), *The Journal of Biological Chemistry*, 273, 5493–5499 © The American Society for Biochemistry & Molecular Biology; Figure 10.3 reproduced from Hirmo *et al.* (1998), *Analytical Biochemistry*, 257, 63–66 with permissions from Academic Press; Figure 10.4 reproduced with permission from N. Athanassopoulou *et al.* (1999), *Biochemical Society Transactions*, 27:2, 340–344 © Biochemical Society; Figure 10.5 reproduced with permission from A. Meyboom *et al.* (1998), *The Journal of Biological Chemistry*, 272, 14600–14605 © The American Society for Biochemistry & Molecular Biology; Figure 10.6 reproduced from Torigoe *et al.* (1997), *Nucleic Acids Symposium Series*, 37, 441–448 with permission of Oxford University Press; Figure 10.7 reprinted from D. Bracewell *et al.* (1998), *Biosensors and Bioelectronics*, 13, 847–853 with permission from Elsevier Science.

Part I
Biological recognition

1 Principles of antigen–antibody recognition

M.H.V. Van Regenmortel and D. Altschuh

Introduction

The term 'antigen' refers to any entity, whether a cell, a macromolecular assembly or a single molecule, which can elicit an immune response in a competent, vertebrate host and be recognized specifically by the products of that immune response. The ability of antigens to react specifically with complementary antibodies is known as 'antigenic reactivity' or 'antigenicity', while their capacity to generate an immune response is called 'immunogenicity'. It should be noted that immunogenicity is not an intrinsic property of the antigen but a relational property that depends on the gene repertoire and regulatory mechanisms of the host being immunized and which has no meaning outside the context of the host (Berzofsky 1985). For instance, mouse serum albumin is an antigenic protein that is immunogenic in the rabbit but not normally in the mouse because of the regulatory mechanism known as immunological tolerance.

The distinction between antigenicity and immunogenicity is clearly revealed in the case of small antigens, i.e. molecules with a relative molecular mass smaller than about 2000. Such molecules, which are called haptens, can bind to antibodies and are thus antigenic but they are not immunogenic on their own since they acquire immunogenicity only after being conjugated to a carrier molecule (Morel-Montero and Delaage 1994).

The antigenic reactivity of a protein is located in restricted parts of the molecule known as epitopes, that make direct contact with complementary antibody molecules. Each epitope corresponds to a discrete patch of amino acid residues which binds specifically to the binding site or paratope on an antibody. Antibody molecules belong to the family of proteins known as immunoglobulins (Ig) which are subdivided into five classes: IgG, IgM, IgA, IgE and IgD. The basic architecture of each immunoglobulin class is the same and consists of a four chain structure. The N-terminal portion of the four chains shows a wide variability in sequence concentrated in six hypervariable regions known as complementarity-determining regions (CDRs). These hypervariable regions form six loops comprising a total of about fifty residues which participate in the formation of the paratope of the immunoglobulin molecule. The most common type of immunoglobulin, IgG, contains two identical paratopes located at distal ends on the surface of the molecule.

The antibody nature of an immunoglobulin can be identified only by its complementary antigen, and reciprocally the epitope nature of a small region of the antigen can be recognized only by means of an antibody molecule. Epitopes and paratopes are thus relational entities that can be defined only by their mutual complementarity and not by any intrinsic feature of each partner existing independently of the relational nexus. Since the occurrence of a binding reaction is required for identifying the interacting partners, it is the functional activity during the binding process that makes it possible to define each of the two partners and not a structure identifiable before the interaction has taken place. Defining epitopes on their own would be like defining husbands in a world devoid of wives. Attempts to define a paratope on its own without reference to its complementary epitope would be equally futile. Since the structure and activity of a binding site cannot be dissociated, they should be viewed in an integrated manner as a four-dimensional structure-functioning complex (Van Regenmortel 1996).

Structure of immunoglobulins

Immunoglobulin molecules are heterodimers consisting of four polypeptide chains linked by disulfide bridges (Nezlin 1994). Two identical heavy (H) chains of 450–600 amino acid residues are linked to two light (L) chains of about 220 residues. The sequence of the N-terminal domains of both H and L-chains vary in different antibodies and are called variable (V) regions while the remaining domains in each chain are invariant and are called constant (C) regions. Within each V region, three segments which exhibit sequence hypervariability form the CDRs. In the heavy chain of IgG, there are three constant domains called CH1, CH2 and CH3. Between the CH1 and CH2 domains, there is an additional segment called the 'hinge' which contains the disulfide bridges that connect two heavy chains and which confers a certain amount of flexibility on the molecule. The light chain contains a single constant domain called 'CL'.

In mammals there are five immunoglobulin classes or isotypes that differ in the sequence and carbohydrate content of their heavy chains. These five classes known as IgG, IgM, IgA, IgD and IgE contain heavy chains that are called γ, μ, α, δ and ϵ respectively (Table 1.1). In contrast, the L-chains are the same for all immunoglobu-

Table 1.1 Properties of immunoglobulin classes

Properties	IgG	IgM	IgA	IgD	IgE
H-chain type	γ	μ	α	δ	ϵ
L-chain type	κ or λ	κ or λ	κ or λ	κ or λ	κ or λ
Structure	H_2L_2	$(H_2L_2)_5$ + J chain	H_2L_2 and $(H_2L_2)_2$ + SC + J chain	H_2L_2	H_2L_2
Molecular weight	150×10^3	950×10^3	160×10^3	180×10^3	190×10^3
Sedimentation coefficient	7 S	19 S	7 and 11 S	7 S	8 S
Carbohydrate (%)	3	12	8	10	12
Approximate concentration in serum (mg/ml)	10	1	2	0.04	0.0003
Ability to fix complement	+	+	−	−	−

lin classes, although there are two types of them, i.e. lambda (λ) or kappa (κ). The lambda and kappa L-chains have different sequences and are usually free of carbohydrate. In mammals there are several IgG subclasses which differ in the structure of their H-chains and in the length of the hinge and number of disulfide bridges linking the two H-chains. In humans, the subclasses are called IgG1, IgG2, IgG3 and IgG4 while in mice, they are called IgG1, IgG2a, IgG2b and IgG3. In IgM and IgE, the hinge region of the H-chain is replaced by an extra domain pair, which holds the H-chains together.

Immunoglobulins can be cleaved at the middle of their H-chains by various proteases (Nezlin 1994). Papain cleaves the γ-chains of IgG at the N-terminal side of the disulfide bridges that keep the H-chains together, thereby generating two Fab (fragment antigen binding) fragments and one Fc (fragment crystallizable) fragment. Each Fab fragment contains at its tip one of the two identical paratopes of the immunoglobulin constituted by the three CDRs of the H-chain and the three CDRs of the L-chain.

The sensitivity of various IgG subclasses to proteolytic cleavage varies and depends on the length and sequence of their hinge regions. In exceptional cases, it is possible to obtain by pepsin digestion of immunoglobulins a minimal antigen-binding fragment called Fv. The Fv fragment consists of only the variable regions of the H and L-chains and retains the antigen-binding capacity of the parent immunoglobulin. Recombinant DNA technology now makes it possible to obtain Fv fragments in a routine manner (Plückthun 1994).

IgG is the most abundant class of immunoglobulin in the serum of mammals and is synthesized predominantly after secondary exposure to antigen. IgM molecules are the predominant class of immunoglobulin produced during early phases of the immune response. IgM molecules consist of five subunits, which resemble IgG molecules and are linked together by disulfide bridges (Table 1.1). Each pentameric IgM molecule is composed of ten μ chains, ten light chains and one joining (J) chain of Mr 15×10^3 which participates in the polymerization process. The pentameric IgM molecules have ten antigen-combining sites.

Structure of paratopes

Immunoglobulin domains fold into two β-pleated sheets of strands arranged in antiparallel direction and linked by a disulfide bridge. This characteristic folding pattern is known as the immunoglobulin fold. The VL and VH domains associate non-covalently to form a β-barrel structure which places the six CDR loops close to each other. The CDR loops vary not only in sequence but also in length from one antibody to the next (Lesk and Tramontano 1993). For five of the hypervariable loops, there is only a limited number of main-chain conformations which depend upon the nature of a few key residues. These commonly occurring main-chain conformations have been called canonical structures (Chothia *et al.* 1989). The H3 loop is different from the other five loops in that it is usually quite long (up to twenty-five residues) and adopts a greater variety of structures. The range of length in residues of the other loops is: L1: 10–17; L2: 7; L3: 7–11; H1: 5–7 and H2: 9–12 (Rees *et al.* 1994). The existence of a limited number of canonical structures has made it possible to develop homology-based modelling methods for predicting the structure of antibody combining sites from primary structure data. These methods succeed in modelling the main-chain conformations of the combining sites to about 1 Å resolution, but not the side chains.

Although the six CDR loops of an antibody together usually comprise about fifty residues, each individual paratope is made up of atoms from not more than fifteen to twenty residues. This means that about two-thirds of the CDR residues of the immunoglobulin molecule do not participate directly in the interaction with an individual complementary epitope. These non-interacting residues are capable of binding to other epitopes that bear no structural resemblance to the first epitope. Although the same immunoglobulin molecule may thus harbour two totally independent paratopes (Bhattacharjee and Glaudemans 1978), it is more commonly found that several sub-sites in the immunoglobulin binding pocket partly overlap. In this case binding of an epitope to one paratope prevents a second epitope from being accommodated at another sub-site of the immunoglobulin molecule. The ability of a first epitope to inhibit the binding of a second epitope to the same immunoglobulin therefore does not necessarily mean that the two epitopes are structurally related.

Since the relationship between an antibody and its antigen is never of an exclusive nature, immunoglobulins are always multi-specific. In addition to recognizing the epitope against which it was elicited, an antibody is always capable of binding to a variety of related or unrelated epitopes. Usually the antibody will bind to its homologous antigen with higher affinity than to most heterologous antigens. However, it is not uncommon to find heterologous antigens to which the antibody binds more strongly than to the homologous antigen, a phenomenon known as heterospecific or heteroclitic binding (Underwood 1985; Harper *et al.* 1987). Heterospecificity is not unexpected since the clonal selection of a B-cell and its subsequent differentiation into an antibody producing plasmocyte can be triggered by an immunogen endowed with only moderate affinity for the B-cell receptor. The highest possible degree of fit between epitope and paratope is not required for initiating B-cell differentiation and the low affinity antibodies that are produced by the differentiated plasmocyte may show a higher affinity for other epitopes that possess a superior degree of complementarity with the antibody.

Structure of epitopes

Protein epitopes are usually classified as either continuous or discontinuous, depending on whether the amino acid residues in the epitope are contiguous in the polypeptide chain or not. Continuous epitopes correspond to short peptide fragments of a protein that are able to bind to antibodies raised against the intact protein. Because the peptide fragment usually does not retain the conformation present in the folded protein and mostly represents only a portion of a more complex epitope, it tends to react only weakly with the anti-protein antibodies.

Discontinuous epitopes represent the vast majority of protein epitopes. They are made up of residues that are not contiguous in the sequence but are brought close together by the folding of the peptide chain. Although this classification is widely used, there is in fact no clear-cut borderline between continuous and discontinuous epitopes (Van Regenmortel 1994). In general, discontinuous epitopes contain several segments of a few contiguous residues and some of these may be able on their own to bind to the paratope. In such a case, the segments may be given the status of continuous epitopes in spite of being part of a larger discontinuous epitope. Conversely, so-called continuous epitopes usually contain a number of residues that are not implicated in the binding reaction and can be replaced by any of the other nineteen possible amino acids without impairing the antigenic reactivity of the peptide. This makes the continuous epitope, functionally speaking, discontinuous. In practice, whenever a peptide of up to about ten residues is found to bind to an antibody, it is called a continuous epitope.

The same problem of definition exists with paratopes. Since paratopes are usually constituted of six hypervariable loops, they clearly are discontinuous structures. However, it has been shown that peptides corresponding to individual CDR sequences are able to bind the antigen with the same specificity as the intact antibody, although with lower affinity (Laune *et al.* 1997). This situation is analogous to that observed with continuous epitopes, and such CDR peptides that are able to bind the antigen can be given the status of continuous paratopes (Van Regenmortel 1998a).

Regarding the number of epitopes likely to be found on a protein molecule, it is accepted that the entire accessible surface of a protein consists of a large number of overlapping epitopes (Benjamin *et al.* 1984). In view of the relational nature of epitopes, the number of epitopes in a protein can be estimated from the total number of different monoclonal antibodies (Mabs) that can be raised against the protein. On this basis, it has been estimated that the insulin molecule possesses at least 115 epitopes (Schroer *et al.* 1983).

It must be stressed that it is impossible to draw sharp boundaries between the many individual overlapping epitopes which collectively can be viewed as forming an antigenic site at the surface of the protein. The same residues of a protein antigen may be involved in a number of different epitopes built up from individual atomic groups that interact in a unique manner with completely unrelated paratopes (Lescar *et al.* 1995). Conversely, the same antibody may sometimes recognize two epitopes devoid of any sequence similarity. Such findings are due to the fact that the recognition between epitope and paratope does not take place at the level of whole residues but at the level of individual atoms (Van Regenmortel 1999a). The methods used for localizing epitopes in proteins are listed in Table 1.2. These methods have been described in detail elsewhere (Van Regenmortel 1994; Morris 1996).

When a protein is cleaved into fragments, the residues that make up the discontinuous epitopes are mostly scattered on individual peptides. Since most protein epitopes are discontinuous, it is to be expected that antibodies raised against intact proteins generally will not react with peptide fragments derived from the protein. Only in rare cases will the linear peptide fragment harbour a sufficient number of residues from the original epitope to enable the peptide to bind to antibodies raised against the intact antigen.

Another difficulty in defining protein epitopes arises from the fact that the antigenic reactivity of a protein is not the same in the native and denatured forms of the molecule. It is thus important to specify whether the epitopes being studied pertain to the native protein or not. This problem has become particularly relevant because of the popularity of solid-phase immunoassays. When proteins are adsorbed to a layer of plastic in such assays, they tend to undergo some physical distortion or denaturation and in many cases the epitopes corresponding to the native state are

Table 1.2 Methods used for localizing epitopes in proteins

Method	Number of residues in epitope	Criterion for residue allocation
Crystallographic analysis of antigen–antibody complexes	15–22	Contact in epitope–paratope interface

Synthetic peptides as antigenic probes in binding studies
   Free peptides, or
   peptides adsorbed to solid-phase,
   conjugated to a carrier
   or attached to support used
   for synthesis
	5–8	Cross-reactive binding with antiprotein antibodies

Identification of critical residues in peptides by systematic replacement of residues	3–5	Decrease in cross-reactive binding
Peptide sequences inserted in recombinant proteins	5–8	Cross-reactive binding with antiprotein antibodies
Binding of antipeptide antibodies to protein	5–8	Induction of cross-reactive antibodies
Binding of antiprotein antibodies to chemically modified or mutated proteins	3–5	Decrease in binding compared with unmodified protein
Protection of antibody-bound residues of epitope to chemical attack	5–10	Chemical reactivity of individual residues
Competitive topographic mapping	Only relative position of epitope is defined	Not applicable

not preserved (Darst *et al.* 1988). In order to ascertain whether antibodies truly recognize the 'native' form of the protein, it is necessary to test these antibodies in a liquid-phase type of assay that preserves the original conformation of the protein antigen (Spangler 1991).

X-ray crystallographic studies of antigen–antibody complexes lead to the characterization of structural epitopes defined by the set of residues or atoms considered to be in contact with atoms of the antibody. According to this criterion, the size of epitopes is usually about fifteen to twenty-two residues.

The contribution of individual amino acid residues of the epitope to its antigenic activity can be established by measuring the change in binding affinity resulting from single residue substitutions (Benjamin and Perdue 1996; Kelley 1996). Alanine scanning and site-directed mutagenesis studies have shown that only three to five residues of the epitope defined on a structural basis contribute significantly to the binding energy of the antigen–antibody interaction. Such studies lead to the definition of a so-called energetic epitope which is much smaller than the structural epitope.

The contribution of individual residues of a peptide fragment to its antigenicity can be established by replacement studies in which each residue of a synthetic peptide is, in turn, replaced by the other nineteen amino acids (Geysen *et al.* 1988). When all the analogues are tested for their binding capacity, it is usually found that only a few residues are critical for binding and cannot be replaced by any residue without impairing binding. Other residues can be replaced by any of the amino acids without affecting antigenic reactivity. It is usually assumed that these indifferent replaceable residues are not contact residues interacting with the antibody, although it cannot be excluded that their main chain atoms interact with the paratope.

Continuous epitopes of proteins are usually identified by testing the antigenic reactivity of sets of overlapping synthetic peptides encompassing the entire sequence of the protein. This approach requires prior knowledge of the protein sequence. It is also possible to test the reactivity of sets of peptides that do not correspond to parts of the protein antigen sequence. Following the development of random combinatorial peptide libraries obtainable by chemical synthesis (Lam *et al.* 1996) or phage display (Daniels and Lane 1996), it is relatively straightforward to obtain the 6.4×10^7 hexapeptides or 2.5×10^{10} octapeptides that can be assembled using the twenty amino acids. When these peptides are tested for their ability to react with antiprotein antibodies, some antigenically active peptides may correspond very closely to part of the sequence of the protein immunogen used to raise the antibodies. In that case, the active peptide will be considered a continuous epitope of the protein. If the sequence of the antigenically active peptide is not present in the immunogen, the peptide will be considered to correspond to a mimotope, i.e. a structure believed to mimic a discontinuous epitope of the protein. The term mimotope was coined by Geysen *et al.* (1986) and was originally defined as a peptide able to bind to an antibody but unrelated in sequence to the epitope used to induce the antibody. At present, the term mimotope is used in a broader sense and is applied to any mimic of a protein epitope, irrespective of the degree of sequence similarity with the epitope (Van Regenmortel 1998a).

It should be emphasized that in order to qualify as a mimotope, a peptide must not only be able to bind to the antibody but it must also be able to elicit antibodies that react with the epitope that is being mimicked. This additional requirement stems from the fact that a single immunoglobulin molecule is multi-specific and may be able to bind to totally unrelated epitopes. When a peptide is labelled a mimotope of epitope A solely on the basis of its capacity to bind to an anti-A antibody, it cannot be excluded that the so-called mimotope binds to a different sub-site from the one that interacts with epitope A. It is therefore essential to establish that the putative mimotope is able to induce antibodies that cross-react with epitope A. Only in this case can the antigenically active peptide be labelled a mimotope of epitope A.

It is not uncommon for peptides showing little sequence similarity to cross-react with the same antibody. This is due to the fact that only some atoms of each residue of the epitope are involved in the binding interaction and that these essential atoms may originate from more than one type of residue (Van Regenmortel 1999a).

Antigen–antibody interactions

Our knowledge of the structural basis of antigen–antibody interaction has been mainly derived from X-ray crystallographic studies of complexes of monoclonal antibody fragments with their protein antigens (Padlan 1996). Usually, an area of about

700–$900\,\text{Å}^2$ of the antigen surface, comprising between fifteen and twenty-two amino acid residues, is found to be in contact with about the same number of residues of the antibody. The paratopes comprise residues originating from most or all CDRs of molecules as well as from some unvariant framework residues. In general, about half of the CDR residues implicated in the paratope are aromatic residues. All protein epitopes studied so far by X-ray crystallography are discontinuous and are made up of residues from between two and five separate regions of the antigen polypeptide chain. Most haptens because of their small size make fewer contacts with CDR residues, although this does not necessarily result in a lower binding affinity, provided the size of the hapten is not smaller than $Mr = 400$ (Chappey *et al.* 1994).

The binding forces that contribute to the energy of interaction of protein–antibody complexes are listed in Table 1.3. The binding is mediated by many intermolecular hydrogen bonds and by a small number of salt bridges. More than 90 per cent of the hydrogen bonds involve side chains, while very few involve main chain atoms. In the three anti-lysozyme antibodies of known structure, only about a quarter of the surface formed by the CDRs is in contact with the antigen. In these three antibodies, tyrosine and tryptophan residues contribute 155 of the 302 interatomic contacts between the antibodies and the antigen. All paratopes contain a much greater frequency of aromatic residues than is usual at the surface of proteins. These aromatic side chains can make large rotations with little entropic cost and they contribute significantly to the binding energy.

It should be noted that many available crystal structures of immune complexes have been determined to a resolution of only about $3.0\,\text{Å}$, which is insufficient to reveal the precise nature of the atomic interactions. In a study of the lysozyme – D1.3 antibody complex at a resolution of $1.8\,\text{Å}$, it was found that forty-eight water molecules were present at the antigen–antibody interface (Bhat *et al.* 1994). A considerable number of water molecules have also been found at the interface in several other antigen–antibody complexes (Malby *et al.* 1994; Chacko *et al.* 1995). These water molecules are not expelled from the interface during the interaction because of the imperfect complementarity between the two surfaces, and they contribute hydrogen bonds to the antigen–antibody interaction. The presence of many water molecules at the interface is at odds with the earlier generalization, drawn from crystallographic data obtained at lower resolution, that water is extruded from epitope–paratope interfaces (Davies *et al.* 1990).

When suspended in aqueous solution, protein molecules tend to repel each other because of the hydrophilic nature of their surfaces. This repulsion has to be overcome before specific binding can occur between complementary epitopes and paratopes (Van Oss 1995). The major forces acting in the primary attraction between epitopes and paratopes are hydrophobic and electrostatic interactions. In spite of their name, hydrophilic molecules bind water molecules at an appreciable level, albeit less strongly than hydrophobic compounds. The attraction between a hydrophilic and a hydrophobic entity is often considerable and is always stronger than between two repulsive hydrophilic groups. The strong hydrophobic nature of the paratope is thus well adapted to maximise binding with the predominantly hydrophilic epitopes. Only after the primary binding has occurred through hydrophobic and electrostatic interactions are the epitope and paratope brought sufficiently close together to allow van der Waals interactions and hydrogen bonds to become operative.

Table 1.3 Examples of epitope–paratope interactions[1]

Antibody (Fab)	Surface buried ($\mathring{A}$)	Ligand	Relative molecular mass (M_r)	Surface buried ($\mathring{A}$)	van der Waals contacts (side chains)	van der Waals contacts (main chains)	Hydrogen bonds (side chains)	Hydrogen bonds (main chains)	Ion pairs	K_a (M^{-1})
McPC603	151	Phosphocholine	169	138	54	4	3	–	2	1.7×10^5
4-4-20	338	Fluorescein	320	247	107	36	3	–	0	3.4×10^{10}
B13I2	503	Myohemerythrin peptide	818	439	113	27	13	2	2	–
17/9	488	Hemagglutinin peptide	1055	418	143	36	14	1	1	5.0×10^7
D1.3	537	Lysozyme	14000	541	110	18	14	2	0	2×10^8
HyHEL-5	744	Lysozyme	14000	741	145	25	13	0	3	2×10^{10}
HyHEL-10	716	Lysozyme	14000	759	180	21	19	1	0	5×10^9
NC41	822	Neuraminidase	50000	838	154	20	13	1	0	–

Note

1 For references, see Padlan 1994

The interaction between antigen and antibody at equilibrium may be expressed as:

$$A + B \underset{k_d}{\overset{k_a}{\rightleftharpoons}} AB \tag{1.1}$$

Where A represents free antigen, B free antibody, AB the antigen–antibody complex, k_a ($M^{-1}sec^{-1}$) the association rate constant and k_d (sec^{-1}) the dissociation rate constant, respectively.

At equilibrium, the affinity constant can be expressed as an association constant

$$K_A = \frac{k_a}{k_d} = \frac{[AB]}{[A][B]} \quad \text{(units } M^{-1}\text{)} \tag{1.2}$$

or as a dissociation constant

$$K_D = \frac{k_d}{k_a} = \frac{[A][B]}{[AB]} \quad \text{(units M)} \tag{1.3}$$

Whereas the quantity of complex formed depends on the equilibrium affinity constant, the time necessary to reach equilibrium depends on the kinetic rate constants. It must be emphasized that all antigen–antibody interactions are reversible and that the magnitude of the kinetic rate constants directly influences the outcome of an immunoassay. The relationship between k_d values and the half-life of an interaction is shown in Table 1.4. It can be concluded, for instance, that if a monoclonal antibody has a k_d value of $10^{-2}sec^{-1}$, it will be of little use in a solid-phase immunoassay because the antibody will become dissociated during the washing step of the assay. Kinetic rate constants of antigen–antibody interactions can be readily determined with optical biosensors based on surface plasmon resonance (Morton *et al.* 1995; Karlsson and Roos 1997; Myszka 1997).

Two competing explanatory models have been used for many years to describe the mechanism of antigen–antibody recognition: the lock-and-key and induced fit models. According to the lock-and-key model, the two partners interact as rigid bodies and the recognition depends on very precise steric and chemical complementarity without any deformation in either of the two partners. The alternative induced fit model is based on the occurrence of conformational rearrangements of main and

Table 1.4 Relationship between dissociation rate constant (k_d) and half-life of an interaction

k_d (sec^{-1})	$t\,1/2$
10^{-6}	8 days
5×10^{-6}	1.6 days
10^{-5}	19.2 hours
5×10^{-5}	3.8 hours
10^{-4}	1.9 hours
5×10^{-4}	23 min
10^{-3}	11.5 min
5×10^{-3}	2.3 min
10^{-2}	70 sec
10^{-1}	7 sec

side chain atoms in one or both reactants which allows them to improve their complementarity and form more stable complexes. Since most crystallographic studies of antigen–antibody complexes performed in recent years show the presence of at least some degree of induced fit (Wilson and Stanfield 1994), the lock-and-key model appears today less satisfactory for describing immunochemical interactions.

The specificity of immunological reagents

The term specificity is derived from the word species and it describes what pertains to and characterizes a species. Immunological specificity has its historical roots in bacteriology and in the original belief that distinct bacterial species could be differentiated by their reaction with antisera that were absolutely specific and did not cross-react with other bacterial species. Subsequently, it was recognized that serological cross-reactions between different bacterial species were very common and specificity became a matter of degree rather than an all or none reaction (for a review, see Van Regenmortel 1998b).

Berzofsky and Schechter (1981) pointed out that the concept of specificity of antibodies is complementary to the concept of cross-reactivity of antigens. One type of cross-reactivity more properly called shared reactivity, arises when a given antibody recognizes the same epitope in two different proteins. This situation is illustrated in Figure 1.1 where the monoclonal antibody (Mab) anti-a recognizes the same epitope a in the two antigens 1 and 2. Another type of cross-reactivity called true cross-reactivity by Berzofksy and Schechter (1981) arises when a Mab (anti-c in Figure 1.1) recognizes two different but related epitopes, c and e, in the two antigens.

An examination of Figure 1.1 provides an answer to the question of whether a Mab or a polyclonal antiserum should be considered the more specific reagent. The question is in fact meaningless unless it includes a reference to what is to be differentiated. Antibody specificity is a ternary relational property that acquires a

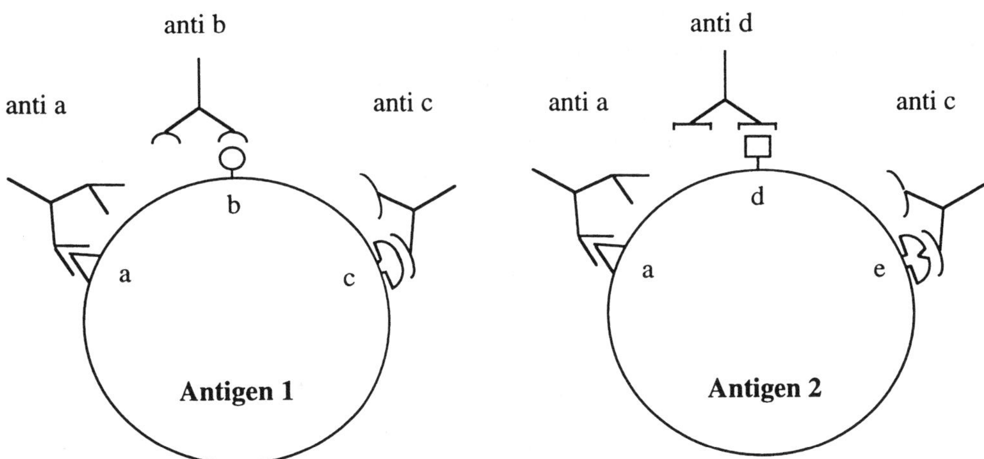

Figure 1.1 Schematic representation of cross-reactivity of (i) an antibody (anti-a) recognizing the same epitope (a) in two different antigens and (ii) an antibody (anti-e) recognizing two different but related epitopes (c and e) in two different antigens

meaning only with respect to the antibody's capacity to react differently with two or more antigens and thereby to discriminate between them (Van Regenmortel 1998b). If the aim is to distinguish between antigens 1 and 2 (Figure 1.1), Mab anti-a would be considered non-specific while Mabs anti-b and anti-d would be considered highly specific since they do not cross-react at all. For discriminating between the two antigens, a polyclonal antiserum containing a mixture of antibodies anti-a, b and c would be more specific than Mab anti-a but less specific than Mab anti-b. In fact, it is actually misleading to refer to a Mab as being specific for a protein antigen or a macromolecular assembly since an antibody can only be specific for an epitope. Part of the confusion surrounding the concept of specificity arises because of the failure to distinguish a single epitope from a multi-epitopic antigen.

In practice it is the need of the investigator to differentiate between two antigens that makes it possible to say whether a particular reagent is specific or not. An antibody will be called specific if it achieves the level of discrimination that is required in any particular case. The same antibody may thus be considered specific or non-specific depending on the task at hand and it may be more meaningful to speak of discrimination potential rather than of specificity.

It is often assumed that antibodies of greater specificity will also show greater affinity, because high affinity antibodies should possess a better stereochemical complementarity with their antigens than antibodies of low affinity. However, there is no necessary link between affinity and specificity. For instance, an antibody of low affinity may discriminate better between two antigens because its reactivity with the heterologous antigen is below the detection level of a particular assay. In general, antibodies of high affinity can be expected to retain more residual binding activity with heterologous antigens, and this tends to increase their cross-reactive potential.

Structure–function relationships in immunochemistry

The nature of the relationship between the structure of a protein and its biological function is a central question in molecular biology which has not yet been answered satisfactorily. Most molecular biologists adhere to the reductionist claim that 'structure determines function' and they believe that the behaviour of complex biological systems can be fully described and understood in terms of the physico-chemical properties of their constituent parts. Explanations in molecular biology often take the form of an apparently causal, linear pathway that links the nucleotide sequence to the protein sequence, then to the protein tertiary structure, to specific binding activity and finally to biological function. Such a causal pathway seems to possess explanatory power because it is presented as if the context and cellular environment did not contribute to the occurrence of each successive step. This confusion arises from the failure to appreciate that causality is a relation between successive events and not between material objects. A cause is an event that is a necessary and sufficient condition for the occurrence of a later event, called the effect. Since any biological event is always simultaneously influenced by a large number of factors, many authors consider that simple, linear causality is not an appropriate concept for analyzing biological phenomena (Sattler 1986; Mahner and Bunge 1997). A gene certainly does not 'cause' a phenotypic trait independently of the cellular context or machinery. Similarly the sequence of a protein does not 'cause' its conformation irrespective of the physico-chemical environment; in many cases chaperones are also required for achieving the correct folding. A biological event such as a binding interaction between a receptor and its ligand also cannot be 'caused' by something that is not an event, i.e. the structure of the ligand. Furthermore, the structure of one of the interacting molecules on its own, i.e. without considering a particular relation with an identified partner, cannot be used to deduce binding activity. Since a binding site cannot be visualized in the absence of a specific interaction, structure and activity must be viewed in an integrated manner in a four-dimensional space–time framework (Van Regenmortel 1996 and 1999b).

The final hypothetical causal link between the protein sequence and biological function is actually the most nebulous one because of the ambiguity of the term function. As discussed by Kitcher (1998), the function of an entity is what it is designed to do. In biology the only source of design is the action of natural selection which retains a useful function during evolution. The selection mechanism is blind to structure: it selects for effects and cannot discriminate between different structures with similar effects (Rosenberg 1994). A functional explanation for the presently observed structure is thus given in terms of superior fitness in the past, rather than the usual structural explanation for an observed function. Since the term function implies some advantage to the biological system as a whole and involves teleological considerations, it is preferable to refer to structure–activity relationships rather than to structure–function relationships. In the absence of causal links in structure–activity relationships, correlations rather than causal relations should be investigated.

Since the primary requirement for biological activity is binding, the advent of biosensor instruments that measure binding kinetics with considerable precision without labeling the reactants has greatly simplified the search for structure–activity correlations (Karlsson and Roos 1997; Myszka 1997). Biosensors make it possible to find out which molecules interact, how many of them interact, how quickly and how strongly they interact, what the binding stoichiometry is and whether conformational changes occur upon binding.

About a third of all current biosensor studies deal with antigen–antibody interactions (Van Regenmortel *et al.* 1998; Huber *et al.* 1999; Myszka 1999; Rich and Myszka 2000). Biosensors are particularly useful for selecting immunological reagents suitable as diagnostic probes and for the mapping of epitopes and paratopes (Daiss and Scalice 1994; Saunal and Van Regenmortel 1995). Site-directed mutagenesis and alanine-scanning mutagenesis studies make it possible to measure the change in binding affinity resulting from each substitution and thus to evaluate the energetic contribution of individual side-chains to the interaction (Kelley 1996). Since biosensors allow very small differences in rate constants to be detected (Rauffer-Bruyère *et al.* 1997), they provide the most reliable information for correlating the atomic coordinates of an antigen–antibody complex with its binding characteristics (Van Regenmortel 2000).

References

Benjamin, D.C., Berzofsky, J.A., East, I.J., Gurd, F.R.N., Hannum, C., Leach, S.J., Margoliash, E., Michael, J.G., Miller, A., Prager, E.M., Reichlin, M., Sercaz, E.E., Smith-Gill, S.J., Todd, P.E. and Wilson, A.C. (1984) The antigenic structure of proteins: a reappraisal. *Ann. Rev. Immunol.* 2, 67–101.

Benjamin, D.C. and Perdue, S.S. (1996) An analysis of site-directed mutagenesis in epitope mapping. *Methods: A Companion to Meth. Enzymol.* 9, 508–515.

Berzofsky, J.A. (1985) Intrinsic and extrinsic factors in protein antigenic structure. *Science,* 229, 932–940.

Berzofsky, J.A. and Schechter, A.N. (1981) The concepts of cross-reactivity and specificity in immunology. *Molec. Immunol.* 18, 751–763.

Bhat, T.N., Bentley, G.A., Boulot, G., Greene, M.I., Tello, D., Dallacqua, W., Souchon, H., Schwarz, F.P., Mariuzza, R.A. and Poljak, R.J. (1994) Bound water molecules and conformational stabilisation help mediate an antigen–antibody association. *Proc. Natl. Acad. Sci. USA,* 91, 1089–1093.

Bhattacharjee, A.K. and Glaudemans, C.P.J. (1978) Dual binding specificities in MOPC 384 and 870 murine myeloma immunoglobulins. *J. Immunol.* 120, 411–413.

Chacko, S., Silverton, E., Kam-Morgan, L., Smith-Gill, S., Cohen, G. and Davies, D. (1995) Structure of an antibody–lysozyme complex; unexpected effect of a conservative mutation. *J. Mol. Biol.* 245, 261–274.

Chappey, O., Debray, M., Niel, E. and Scherrmann, J.M. (1994) Association constants of monoclonal antibodies for hapten: Heterogeneity of frequency distribution and possible relationship with hapten molecular weight. *J. Immunol. Meth.* 172, 219–225.

Chothia, C., Lesk, A.M., Tramontano, A., Levitt, M., Smith-Gill, S.J., Air, G., Sheriff, S., Padan, E.A., Davies, D., Tulip, W.R., Colman, P.M., Spinelli, S., Alzari, P.M. and Poljak, R.J. (1989) Conformations of immunoglobulin hypervariable regions. *Nature* 342, 877–883.

Daiss, J.L. and Scalice, E.R. (1994) Epitope mapping on BIAcore: theoretical and practical considerations. *Methods: A Companion to Methods in Enzymol.* 6, 143–156.

Daniels, D.A. and Lane, D.P. (1996) Phage peptide Librairies. *Methods,* 9, 494–507.

Darst, S.A., Robertson, C.R. and Berzofsky, J.A. (1988) Adsorption of the protein antigen myoglobin affects the binding of conformation-specific monoclonal antibodies. *Biophys. J.* 53, 533–539.

Davies, D.R., Padlan, E.A. and Sheriff, S. (1990) Antigen–antibody complexes. *Ann. Rev. Biochem.* 59, 439–473.

Geysen, H.M., Rodda, S.J. and Mason, T.J. (1986) A priori delineation of a peptide which mimics a discontinuous antigenic determinant. *Mol. Immunol.* 23, 709–715.

Geysen, H.M., Mason, T.J. and Rodda, S.J. (1988) Cognitive features of continuous antigenic determinants. *J. Molec. Recogn.* 1, 32–41.

Harper, M., Lema, F., Boulot, G. and Poljak, R.J. (1987) Antigen specificity and cross-reactivity of monoclonal anti-lysozyme antibodies. *Molec. Immunol.* 24, 97–108.

Huber, A., Demartis, S. and Neri, D. (1999) The use of biosensor technology for the engineering of antibodies and enzymes. *J. Mol. Recogn.* 12, 155–168.

Karlsson, R. and Roos, H. (1997) Reaction kinetics. In: *Principles and Practice of Immunoassay,* 2nd Edition, Price, C.P. and Newman, D.J. (eds), London, MacMillan, pp. 99–122.

Kelley, R.F. (1996) Thermodynamics of antigen binding and evaluation of humanized antibodies. In: *Structure of Antigens* Vol. 3, Van Regenmortel, M.H.V. (ed.), CRC Press, Boca Raton, pp. 1–20.

Kitcher, P. (1998) Function and design. In: *The Philosophy of Biology,* Hull, D.I. and Ruse M. (eds), Oxford University Press, New York, pp. 258–279.

Lam, K.S., Lake, D., Salmon, S.E., Smith, J., Chen, M-L., Wade, S., Abdul-Latif, F., Leblova, Z., Ferguson, R.D., Krchnak, V., Sepetov, N.F. and Lebl, M. (1996) Application of a one-bead one-peptide combinatorial library method for B-cell epitope mapping. *Methods,* 9, 482–493.

Laune, D., Molina, F., Ferrières, G., Mani, J.C., Cohen, P., Simon, D., Bernard, T., Piechaczyk, M., Pau, B. and Granier, C. (1997) Systematic exploration of the antigen binding activity of synthetic peptides isolated from the variable regions of immunoglobulins. *J. Biol. Chem.* 272, 30937–30944.

Lescar, J., Pelligrini, M., Souchon, H., Tello, D., Poljak, R.J., Peterson, N., Greene, M. and Alzari, M. (1995) Crystal structure of a cross-reaction complex between Fab F9.13.7 and guinea fowl lysozyme. *J. Biol. Chem.* 270, 18067–18076.

Lesk, A.M. and Tramontano, A. (1993) An atlas of antibody combining sites. In: *Structure of Antigens* Vol. 2, Van Regenmortel, M.H.V. (ed.), CRC Press, Boca Raton, pp. 1–29.

Mahner, M. and Bunge, M. (1997) *Foundations in Biophilosophy*. Springer, Berlin.

Malby, R.L., Tulip, W.R., Harley, V.R., McKimm-Breschkin, J.L., Laver, W.G., Webster, R.G. and Colman, P.M. (1994) The structure of complex between the NC10 antibody and influenza virus neuramidase and comparison with the overlapping binding site of the NC41 antibody. *Structure*, 2, 733–746.

Morel-Montero, A. and Delaage, M. (1994) Immunochemistry of pharmacological substances. In: *Immunochemistry*, Van Oss, C.J. and Van Regenmortel, M.H.V. (eds), Marcel Dekker, New York, pp. 357–372.

Morris, G.E. (ed.) (1996) *Epitope Mapping Protocols*. Humana Press, Totowa.

Morton, T.A., Myszka, D.G. and Chaiken, I.M. (1995) Interpreting complex binding kinetics from optical biosensors: a comparison of analysis by linearization, the integrated rate equations, and numerical integration. *Anal. Biochem.* 227, 176–185.

Myszka, D.G. (1997) Kinetic analysis of macromolecular interactions using surface plasmon resonance. *Curr. Opin. Biotech.* 8, 50–57.

Myszka, D.G. (1999) Survey of the 1998 Optical Biosensor Literature. *J. Mol. Recogn.* 12, 309–408.

Nezlin, R. (1994) Immunoglobulin structure and function. In: *Immunochemistry*, Van Oss, C.J. and Van Regenmortel, M.H.V. (eds), Marcel Dekker, New York, pp. 3–45.

Padlan, E.A. (1996) X-ray crystallography of antibodies. *Adv. Protein Chem.* 49, 57–133.

Plückthun, A. (1994) Recombinant antibodies. In: *Immunochemistry*, Van Oss, C.J. and Van Regenmortel, M.H.V. (eds), Marcel Dekker, New York, pp. 201–236.

Rauffer-Bruyère, N., Chatellier, J., Weiss, E., Van Regenmortel, M.H.V. and Altschuh, D. (1997) Cooperative effects of mutations in a recombinant Fab on the kinetic of antigen binding. *Molec. Immunol.* 34, 165–173.

Rees, A.R., Pedersen, J.T. Searle, S.M.J., Henry, A.H. and Webster, D.M. (1994) Antibody structure from X-ray crystallography and molecular modeling. In: *Immunochemistry*, Van Oss, C.J. and Van Regenmortel, M.H.V. (eds), Marcel Dekker, New York, pp. 615–650.

Rich, R.L. and Myszka, D.G. (2000) Survey of the 1999 surface plasma resonance biosensor literature. *J. Mol. Recogn.* 13, 388–407.

Rosenberg, A. (1994) *Instrumental Biology of the Disunity of Science*. University of Chicago Press, Chicago.

Sattler, R. (1986) *Bio-philosophy. Analytic and Holistic Perspectives*. Springer Verlag, Berlin.

Saunal, H. and Van Regenmortel, M.H.V. (1995) Kinetic and functional mapping of viral epitopes using biosensor technology. *Virology*, 213, 462–471.

Schroer, J.A., Bender, T., Feldmann, R.J. and Kim, K.J. (1983) Mapping epitopes on the insulin molecule using monoclonal antibodies. *Eur. J. Immunol.* 13, 693–700.

Spangler, B.D. (1991) Binding to native proteins by antipeptide monoclonal studies. *J. Immunol.* 146, 1591–1595.

Underwood, P.A. (1985) Theoretical considerations of the ability of monoclonal antibodies to detect antigenic differences between closely related variants, with particular reference to heterospecific reactions. *J. Immunol. Meth.* 85, 295–307.

Van Oss, C.J. (1995) Hydrophobic, hydrophilic and other interactions in epitope–paratope binding. *Molec. Immunol.* 32, 199–211.

Van Regenmortel, M.H.V. (1994) The recognition of proteins and peptides by antibodies. In: *Immunochemistry*, Van Oss, C.J. and Van Regenmortel, M.H.V. (eds), Marcel Dekker, New York, pp. 277–301.

Van Regenmortel, M.H.V. (1996) Mapping epitope structure and activity: from one-dimensional prediction to four-dimensional description of antigenic specificity. *Methods*, 9, 465–472.

Van Regenmortel, M.H.V. (1998a) Mimotopes, continuous paratopes and hydropathic complementarity: Novel approximations in the description of immunochemical specificity. *J. Dispersion Sci. & Technol.* 19, 1199–1219.

Van Regenmortel, M.H.V. (1998b) From absolute to exquisite specificity. Reflections on the fuzzy nature of species, specificity and antigenic sites. *J. Immun. Meth.* 216, 37–48.

Van Regenmortel, M.H.V. (1999a) Molecular dissection of protein antigens and the prediction of epitopes. In: *Synthetic Peptides as Antigens*, Van Regenmortel, M.H.V. and Muller, S. (eds), Elsevier, pp. 1–77.

Van Regenmortel, M.H.V. (1999b) Biosensors and the search for structure–activity correlations. *J. Mol. Recogn.* 12, 277–278.

Van Regenmortel, M.H.V. (2001) Analysing structure–function relationships with biosensors. *Cell. Mol. Life Sci.* 58, 794–800.

Van Regenmortel, M.H.V., Altschuh, D., Chatelier, J., Christensen, L., Rauffer-Bruyère, N., Richalet-Secordel, P., Witz, J. and Zeder-Lutz, G. (1998) Measurement of antigen–antibody interactions with biosensors. *J. Mol. Recogn.* 11, 163–167.

Wilson, I.A. and Stanfield, R.L. (1994) Antibody–antigen interactions: new structures and new conformational changes. *Curr. Opin. Struct. Biol.* 4, 857–867.

2 Protein–protein interactions

Gideon Schreiber

Living organisms generate their complex repertoire of proteins by the stepwise assembly of small chemical entities (amino acids). After a folding process and/or further enzymatic transformations these complicated molecules attain their final three-dimensional structures, which determine their biological activities through the specific interactions with their ligands. An infinite variety of different molecular structures can be realized by varying the length of the chains and the sequence of the building units. Specific protein–protein interactions are used by all living organisms for a large variety of functions, including regulation, building of super molecular structures, the immune response, cell recognition and more. Those who study protein–protein interactions hope to resolve the high resolution structure of the complex, and indeed, more and more of these are being solved. However, whether or not this goal is reached, biophysical data must be collected in order to gain insight into the affinity, kinetics and thermodynamics of the interaction. Methods for measuring rate constants and affinities of interaction can be grouped according to those which measure the interaction of proteins in solution in homogeneous phase and those where the interaction of a protein in solution is monitored through binding to a second, surface-bound protein (heterogeneous phase measurements). Probing the interaction in solution involves techniques such as optical spectroscopy, measurements of activity (or inhibition), analytical ultracentrifugation, isothermal titration calorimetry and equilibrium dialysis. Heterogeneous phase measurements are done using ELISA assays or a variety of transducer based techniques, which are the focus of this book. Homogeneous and heterogeneous phase measurements mimic protein interactions found in nature. For example the interaction between proteases and their protein inhibitors occur between two proteins in homogeneous phase, while the interaction between a ligand and its receptor is in the heterogeneous phase.

Characterization of protein–protein interfaces

Structural characterization

Some fundamental properties characterizing a protein interface have been determined from the atomic structures of protein complexes, or by comparing the bound with the unbound structures of these proteins (see Table 2.1).

Table 2.1 Structural, kinetic and thermodynamic interaction parameters for a range of protein complexes

		Pdb entry	Interface area ($\mathring{A}^2$)	Number of H-bonds	ΔG of binding (kcal/mol)	k_{on} ($M^{-1}s^{-1}$)	k_{off} (M^{-1})	ΔH (kcal/mol)
Protease Subtilisin	*Inhibitor* CI2	2sni	1630	10	−16	6×10^6	1×10^{-5}	
Antibody Fv	*Antigen* D1.3-Lysozyme	1vfb	1400	9	−11	8×10^4	7×10^{-4}	−20
Enzyme barnase	*Inhibitor* barstar	jtg	1570	13	−19	4×10^8	4×10^{-6}	−19
TEM1	BLIP		2560	11	−12.5	2×10^5	1×10^{-4}	
Signal transduction Ras-GTP	cRaf1	1gua	1310	12	−9.2	4×10^7	7	
Hormone hGH (G120R)	*Receptor* hGH-R (1:1)	1a22	1350		−12	2×10^6	3×10^{-4}	−9.4

Size and shape

The size of the buried surface area of a protein interface varies between as little as $1100\,\text{Å}^2$ to over $4000\,\text{Å}^2$, with the average size being $1600\,\text{Å}^2$ (Jones and Thornton 1996). Plasticity upon complex formation varies, from almost rigid docking as found in the case of barnase-barstar or TEM-1β-lactamase and its protein inhibitor-BLIP (Buckle *et al.* 1994; Strynadka *et al.* 1996), to large structural rearrangement upon association as in the case of hirudin-thrombin and AChE-fasciculin (Rydel *et al.* 1990; Harel *et al.* 1995).

Chemical character of the interface

On average, the fraction of non-polar, polar and charged residues occurring at the interface is similar to that found on the surface of proteins. However, the spread around the average values is large. All the twenty amino acids can be found in protein interfaces; however, some are more abundant than others. Most noteworthy are the aromatic amino acids, which constitute about 12 percent of protein surfaces, but contribute about 19 percent to the buried surface area (Jones and Thornton 1996). Charged groups are less abundant in interfaces, constituting 28 percent of buried surface area, compared to 38 percent of protein surface area covered by them. The same non-covalent interactions which stabilize protein structures (hydrophobic, Van der Waals, hydrogen bonds and salt bridges) are also found to stabilize protein complexes. The relative importance of each type of interaction varies between different complexes analyzed. For example, at the interface of Human-Growth-Hormone with its receptors, most energy of interaction is generated by a small number of aromatic and hydrophobic interactions (Clackson and Wells 1995). On the other hand, in the interaction between barnase and barstar, a number of salt bridges seem to generate most of the binding energy for the high affinity of this interaction (Buckle *et al.* 1994; Schreiber and Fersht 1995).

Surface complimentarity

It is often assumed that two proteins fit one another like a hand and glove. But in reality, close to 30 percent of a protein–protein interface is filled with water molecules, which bridge the gaps between the proteins (Lawrence and Colman 1993). Thus, the fit between proteins is far from perfect. This is demonstrated readily by computational docking. Many of the docking algorithms search for surface complimentarity. Yet, the algorithms do not always recognize the precise site of interaction, even if one starts the computation with the known structure of the complex. This means that, in addition to a reasonably good fit of the surfaces, chemical complimentarity has an important role in determining binding specificity.

Secondary structure of the binding site

No strong preference was observed for any particular type of secondary structure or structural motif in protein–protein interfaces. Moreover, interaction surfaces consist of amino acids which are located on different parts of the primary amino acid chain. Nevertheless, some protein families share canonical binding structures. Best known are the immunoglobulin-like β-structures found in antibodies, surface receptors, etc., where the ligand binding site is located at the loops connecting the β-sheets.

Conformational changes upon complex formation

From the structural analysis of a large number of protein complexes one can distinguish different levels of structural rearrangement upon complexation. In some cases the interaction can be described as rigid docking, which is accompanied only by rearrangement of some side-chain conformations. In other cases segment movements involving the main chain can be found. In yet other interactions, gross domain movements have been observed upon complexation. Protein complexes of all these three groups have been found, with no obvious difference in affinity or binding kinetics between them. For a detailed list of structural responses upon complexation of a large number of protein complexes, see Conte *et al.* 1999.

Thermodynamic characterization

A detailed study of a protein–protein interaction includes the kinetic and thermodynamic characterization of the interaction, as well as mapping the contribution of specific residues toward binding. Optical transducer devices have improved the speed, precision and accuracy with which binding constants can be determined. The most common approach to determine which residues contribute to binding is to perform an "Ala scan" using site directed mutagenesis on the proposed binding site. The rationale behind Ala-scanning mutagenesis is that by mutation to Ala all interactions of a side chain, except for the C_β atom, are eliminated. Ala substitutions eliminate interactions without introducing new interactions, or changing the backbone conformation (except for mutations of Gly and Pro (introducing C_β for Gly and perturbing the protein backbone for Gly and Pro)).

The use of mutagenesis to probe protein–protein interactions has become a standard method, producing a large data set of proteins where both structure and detailed energetics of the binding sites are available. It is generally found that only a limited number of amino acids are hot spots of binding. Only 5 percent of the interface residues contribute to the free energy of binding more than 2 kcal/mol (Bogan and Thorn 1998). Most residues are either neutral, or contribute little toward the stability of the complex. Understanding why some residues have a large contribution toward the stability of a complex, while the majority of residues have only a small influence on affinity, has been the focus of a number of studies in recent years. Just by analyzing the structure of a complex it is currently impossible to identify hot spots of binding, or to design residues to function as hot spots. However it has been noted that hot spots of binding are located near the center of the buried surface area, and that these residues are relatively protected from bulk solvent. Still, the majority of residues which meet the above criteria are not hot spots of binding (Bogan and Thorn 1998).

The free energy of binding (ΔG) is a function of the enthalpy (ΔH) and entropy (ΔS) according to $\Delta G = \Delta H - T\Delta G$. The enthalpy of binding may be determined directly using isothermal titration calorimetry, or indirectly from the temperature dependence of ΔG (vant Hoff). Enthalpies and entropies of binding were measured for a number of protein–protein interactions, including hGH–hGH–R and barnase–barstar (Pearce *et al.* 1996; Frisch *et al.* 1997). It was anticipated that a better understanding of the energy landscape of the interaction interface could be extracted from separately evaluating the contributions of enthalpy and entropy. Unfortunately, the use of enthalpies for the analysis of structure–thermodynamic relationships appears to be complicated by enthalpy–entropy compensation of weak intermolecular interactions. These tend to cancel out in measurements of free energy, which is thus the preferred quantity for simple analysis of interactions. Even the use of ΔC_p as a measure of hydrophobic and hydrophilic buried surface area upon complexation proved to be fruitless, with poor correlation found between these parameters for the interaction of barnase and barstar (Frisch *et al.* 1997) as well as other protein–protein and protein–peptide interactions (Stites 1997).

Double mutant cycle analysis of a protein–protein interface

The currently-popular reductionist approach to study the energetical contribution of amino acids toward stability or binding is by constructing single mutations. However, the interpretation of such experiments is severely limited by the fact that the properties of a residue are a function of the entire system, and not necessary the sum of its parts. As a result, the energetic importance of a particular residue to binding cannot currently be explained by a physical analysis of the binding site. A powerful experimental and theoretical approach to study protein–protein complexes is the use of double and higher-order mutant cycles, where interactions between amino acids are treated within their native context (Figure 2.1). Such cycles reveal

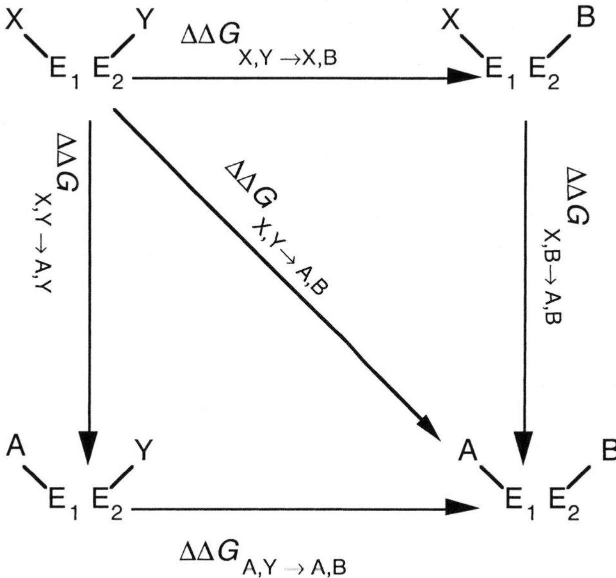

Figure 2.1 Scheme for a double mutant cycle where E_1 and E_2 represent two proteins, and X and Y are two residues which are mutated to A and B on E_1 and E_2 respectively. The coupling energy between X and Y is calculated from the difference between the pairwise mutation (A, B) and the two single mutations:
$$\Delta\Delta G_{int} = \Delta\Delta G_{X \to A, Y \to B} - \Delta\Delta G_{X \to A} - \Delta\Delta G_{Y \to B}$$

whether the contributions from a pair of residues are additive or whether the effects of mutations are coupled (Horovitz and Fersht 1990). We recently performed a study which attempted to determine which of the potential pairwise interactions actually contribute toward the stability of the complex between barnase and barstar (Schreiber and Fersht 1995). We decided on a strategy in which pairwise interaction energies were determined between most residues in the interface, irrespective of the distance between them. Double mutant cycles were constructed between a subset of five barnase and seven barstar residues, which were identified from structural and mutagenesis studies to be important in stabilizing the complex. The advantage of such a global approach is that it allows us to identify long range interactions (which may not be direct) as well as standard short range interactions, and to evaluate their contribution to the free energy of binding. In general, the coupling energy between pairs of residues was found to decrease as the distance between them increases (Schreiber and Fersht 1995), and residues separated by less than 7 Å interact. From the analysis of the dependence of $\Delta\Delta G_{int}$ on the distance between two residues for a number of protein systems, it appears that values of $\Delta\Delta G_{int}$ of >0.6 kcal/mol are found for pairs of residues in close contact (distance up to 5 Å – see Figure 2.2). Not all interactions are direct, as the contribution of one residue to binding is composed of its direct interactions, and of the indirect contribution toward the interactions of

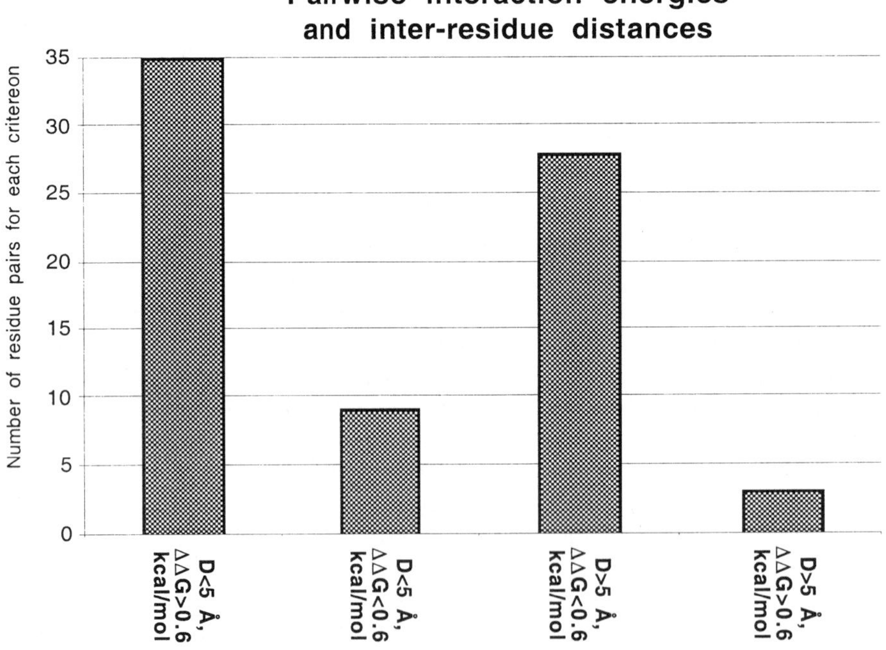

Figure 2.2 Free energies of interactions ($\Delta\Delta G_{int}$) were related to the distance (D) between the two residues, as measured from the crystal structures of the complexes between barnase with barstar, FvD1.3 with FvE5.3, D1.3 with HEL and TEM1-β-lactamase with BLIP. Residues were divided into four groups according to the distance and $\Delta\Delta G_{int}$ found for each pair of residues. The figure demonstrates that interaction energies of more than 0.6 kcal/mol were determined in thirty-five out of thirty-eight cases for residues where the distance between them is up to 5 Å

its closest neighbors with the complementary protein. This point was emphasized by the apparent interaction energy between two negatively charged residues in the barnase–barstar binding interface (Glu 73 on barnase and Asp 39 on barstar) which are located 4.5 Å one from the other. A more detailed analysis of this interaction using higher order multiple mutant cycles has suggested this interaction to be mediated through a number of positively charged residues in close proximity to Glu 73 and Asp 39 (Schreiber *et al.* 1997). At separations greater than 7 Å, the effects of mutations are mostly additive. In other words the two residues affect binding independently of each other.

Mutational induced structural rearrangement in protein interfaces

Quantitative mutational analysis relies on the assumption that structural rearrangement is small, and limited to the close surroundings of the mutated residue. This issue was examined experimentally, by solving the X-ray crystal structure of mutant complexes, mostly substitutions to Ala. Indeed, in most cases the mean RMS (root mean square) deviation of C^α atoms of mutant structures is in the range found between several different structure determinations of the same protein. Thus, the effects of mutation on backbone conformations are no greater than effects caused by crystallization conditions and packing. On the other hand, at the vicinity of the mutated residue, side-chain and small backbone rearrangements are observed, causing some closure of the gap opened by the mutation. The cavity is usually filled with water molecules, which form hydrogen bonds similar to the mutated residue (Ysern *et al.* 1994; Pearce *et al.* 1996; Vaughan *et al.* 1999).

Kinetic pathway of protein–protein association

For a simple two state diffusion controlled system, the binding affinity (K_d) equals to the ratio between the dissociation rate constant (k_{off}) and the association rate constant (k_{on}). Dissociation rate constants for heterodimers can be as slow as $10^{-6}\,\mathrm{s}^{-1}$ (where the half life of the complex is measured in days) as in the case of barnase-barstar and colicin E9 interacting with a DNase (Schreiber and Fersht 1993; Wallis *et al.* 1995). However the range of k_{off} measured for most interactions is 10^{0}–$10^{-5}\,\mathrm{s}^{-1}$. Single mutations of "hot spot" amino acids can increase rates of dissociation by up to 10,000 fold. Rates of association are in the range of 10^3 to $10^9\,\mathrm{M}^{-1}\mathrm{s}^{-1}$, with most complexes associating at a rate of 10^5 to $10^6\,\mathrm{M}^{-1}\mathrm{s}^{-1}$. This rate of association reflects the diffusion limited motion, which predicts rates of collision of up to $10^{10}\,\mathrm{M}^{-1}\mathrm{s}^{-1}$, and the strict orientational constraints required for specific binding. The effect of mutations on the rate of association is relatively small, with mutations of charged residues altering the rate of association by up to only ten-fold. The dynamic measuring range of currently available optical transducer devices is sufficient to measure in real time a large portion of the protein–protein interactions of interest. These devices can measure association rate constants of up to $10^7\,\mathrm{M}^{-1}\mathrm{s}^{-1}$, and dissociation rate constants in the range of 10^{-1}–$10^{-5}\,\mathrm{s}^{-1}$.

The mechanism of association of a protein complex

The association reaction of two proteins, A and B, to form the complex AB can be described by the following two-step kinetic scheme:

$$A + B \underset{k_{-1}}{\overset{k_1}{\longleftrightarrow}} A:B \overset{k_2}{\longrightarrow} AB \tag{2.1}$$

where k_1 and k_{-1} are the rate constants for the association and dissociation of a low affinity loose complex, referred to as the "encounter complex" ($A{:}B$), which is held together by long-range electrostatic interactions. The second step is the docking of the two proteins at a rate k_2, to give the final high affinity complex (AB). This requires the formation of short range interactions between specific amino acids, expulsion of water from the interface, and structural rearrangement (Schreiber and Fersht 1996; Gabdoulline and Wade 1997; Vijayakumar *et al.* 1998; Selzer and Schreiber 1999).

The structure of the transition state of association, which seems to be similar to that of the encounter complex (Vijayakumar *et al.* 1998; Selzer and Schreiber 1999) but can be determined experimentally, was analyzed for the interaction between barnase and barstar (Schreiber and Fersht 1996). A double mutant cycle analysis of activation energies for a large set of residues showed that pairs of charged residues located up to $10\,\text{Å}$ from each other interact during the transition state for association, while remote charges and non-charged residues, even if closed, do not interact (Schreiber and Fersht 1996). Moreover, the only residues which affect the rate of association of the complex are charged residues mutated to neutral ones, while mutations of non-charged residues had no effect on the rate of association. Masking the electrostatic effect, using salt, reduces the rate of association to a basal level, which is independent of mutation (Schreiber and Fersht 1996; Selzer and Schreiber 1999). A thermodynamic analysis of the transition state, comparing changes in $\Delta H^{\ddagger}$ and $\Delta S^{\ddagger}$ and $\Delta G^{\ddagger}$ upon mutation has shown that changes in activation enthalpy and free energy correlate, with the activation entropy being around zero. These data conform to the view that the transition state is an early event during association, which occurs prior to desolvation of the interface and the formation of short range interactions between amino acids. The activation enthalpy and entropy of this early transition-state are small, as only some rotational and translational freedom is lost, and only a small amount of binding enthalpy is gained. The formation of the final complex is accompanied by a loss of translational and rotational degrees of freedom of the individual proteins as well as that of residues involved in binding, and the partial dehydration of the binding sites of the proteins involved. The rate of formation of the final complex k_2, is a first order rate constant (as the local protein concentrations are high, and not dependent on the global protein concentrations), which, for diffusion controlled reactions, is faster than the observed second order rate constant, k_1, at least at low protein concentrations. As a result, a plot of protein concentration versus the observed rate of association is linear, which can be interpreted as k_{on} being dependent on the diffusion limited rate of formation of the encounter complex and not on k_2.

As electrostatic forces are the dominant factor in determining the rate of association it is possible to account for the rate increase using electrostatic energy calculations:

$$\ln k_{\rm on} = \ln k_{\rm on}^0 - \frac{U}{RT}\left(\frac{1}{1+\kappa a}\right) \tag{2.2}$$

where $\ln k_{\rm on}^0$ is the basal rate of association in the absence of electrostatic forces, κ is the Debye-Hückel screening parameter which relates to the ionic strength of the solution, a being the minimal distance of approach between the molecules which can be set to 6 Å and U is the calculated electrostatic interaction energy between the proteins. U can be calculated either by electrostatic energy calculations (using *DelPhi* software) or from the salt dependence of the association rate constant (Selzer and Schreiber 1999). The validity of this simple equation was demonstrated on a number of protein complexes (Selzer and Schreiber 1999).

While the observed rates of association of protein complexes can be explained in simple biophysical terms, the picture concerning rates of dissociation is more complex. The reason for this is clear, as rates of association are mostly dictated by long-range interactions, with the exact structure being of minimal importance. The magnitude of the dissociation rate constant is sensitive to the exact nature of the interacting side chains and relies primarily on the formation or breakage of short-range non-covalent bonds.

Measuring protein–protein interactions in homogeneous and heterogeneous phase

A now common method for heterogeneous phase measurements is the use of biosensor technology, which monitors the binding of a protein in solution to an immobilized partner; best known is the BIAcore, which probes changes in surface plasmon resonance. It provides direct access to both association and dissociation rate constants, low sample consumption and, due to its high sensitivity, it covers a broad range of measurable affinity constants. Nevertheless, there is increasing evidence that there may be large deviations in the affinities and the kinetic rate constants determined from standard binding measurements using surface-plasmon-resonance (SPR) (Nieba *et al.* 1996). A number of factors have been named to contribute to the potential discrepancies. (i) Mass transport limitations – when reaction takes place in homogeneous solution it proceeds at the same rate uniformly throughout the medium. On the other hand, if reaction takes place at a surface the concentrations of reactants and products will change locally at that surface. This leads to substantial differences in concentration over distances, which make the transport from areas of high concentration to areas of low concentration an important factor in the reaction sequence. This problem is especially severe for rapid interactions (Eddowes 1990). (ii) Accessibility and heterogeneity of the ligand binding sites. This is caused by the random attachment during the covalent coupling process of protein to the carbocymethyldextran chain. (iii) Rebinding during the dissociation phase. The local concentrations of immobilized protein on the dextran layer is in the 100 μM range (depending on the MW, and surface density of the immobilized protein). As more of the covalently bound proteins are available for ligand binding, the probability of rebinding increases. This problem is even more severe for smaller peptides, as rates of association increase (because of less steric constraints) (Nieba *et al.* 1996; O'Shannessy and Winzor 1996). (iv) Most transducer surfaces are composed of a layer of negatively charged dextran, upon which proteins are immobilized. Many proteins have a strong net electrostatic charge, which will interact (either by binding or repulsion) with the surface. This phenomenon causes a negatively charged protein in the soluble fraction to associate more slowly and to dissociate faster from its surface bound counterpart. This was observed independently for the interaction of TEM1-β-lactamase (net charge -7) binding to immobilized BLIP (Albeck and Schreiber 1999), and for ifnar2 (net charge -13) binding to immobilized interferon (Piehler and Schreiber 1999). Usually, biosensor binding experiments are done in the presence of salt at physiological concentrations. This amount of salt is enough to inhibit non-specific binding caused by electrostatic forces, yet long range electrostatic forces are still important. The Debye length for a single charge at this ionic strength is about 8 Å. As a result it was found that the salt dependence of an association reaction differs depending on whether the reaction rate is monitored with one of the components immobilized to a charged dextran layer, or in homogeneous phase in solution.

Because of the reasons listed above it is expected that absolute values of binding may vary whether measured in homogeneous or heterogeneous phase. Moreover, variations in absolute affinity and rate constants are also observed for a protein interaction, depending on which of the proteins is immobilized to the surface. Still, the question remains whether these variations are a consequence of global effects of the environment on the protein, or of local effects related to a small number of surface residues. This question can be addressed by analyzing the effects of site specific mutations on the binding kinetics and affinity. Figure 2.3 shows a comparison of

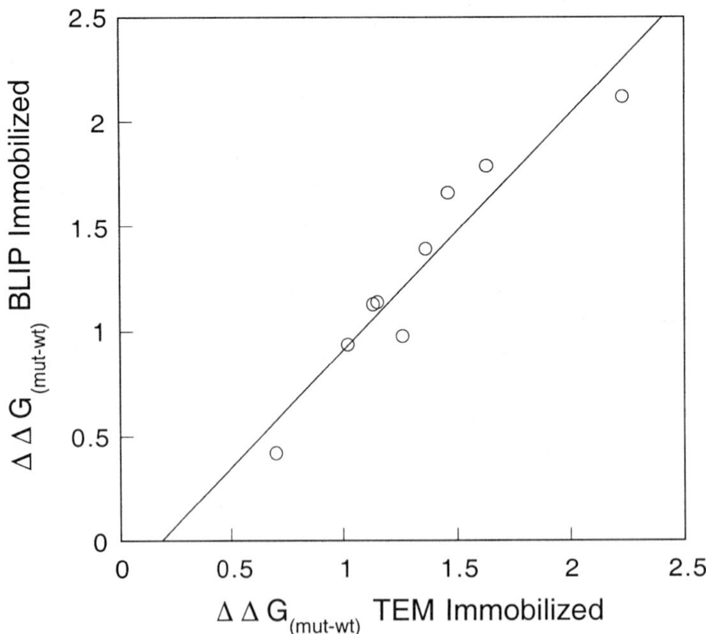

Figure 2.3 Experimentally determined values of $\Delta\Delta G_{(mut\text{-}wt)}$ for the interaction between TEM1-β-lactamase and BLIP measured with a BIAcore 2000 with either one of the two proteins immobilized to a CM5 sensor chip (Albeck and Schreiber 1999). Each circle represents a particular mutation. Values are in kcal/mol

binding affinities measured for a number of mutants of TEM-1 interacting with BLIP, with either TEM or BLIP bound to the surface of a CM5 chip using a BIAcore 2000 for measurement (Albeck and Schreiber 1999). While the absolute values of ΔG of binding differ, values of $\Delta\Delta G_{(mut-wt)}$ were the same, independent of which protein was immobilized to the surface. The same values of $\Delta\Delta G_{(mut-wt)}$ were also obtained from a titration experiment done with both proteins in homogeneous phase. A good correlation between values of $\Delta\Delta G_{(mut-wt)}$ calculated from BIAcore data versus those calculated from radio-immuno assays was also observed for the interaction of human-growth-hormone with its receptor (Cunningham and Wells 1993), despite differences in absolute values. The underlying reason for the similarity of $\Delta\Delta G_{(mut-wt)}$ values obtained from different methods of measurements is that many of the system-specific perturbations, which affect the magnitude of absolute values of ΔG, cancel out when values of $\Delta\Delta G_{(mut-wt)}$ are calculated. This suggests that many of the experimental artifacts affect the system as a whole, and that single mutations have only a limited potential to alter these global effects. Since most of the time we are more interested in determining relative effects of mutations, which tell us about the importance of specific residues to binding, this observation is comforting.

Affinity versus activity

The reason for studying interaction affinities between proteins is often to relate those to function. Thus, we have to assume that binding affinities obtained from cell free measurements (either in homogeneous or heterogeneous phase) and the biological activity of the ligand are related. In the simplest case, where the protein–protein interaction of interest occurs between two soluble proteins, as for example between an Ab and an Ag, or between a protease and its inhibitor, affinities were found to directly relate to activity. The question is more difficult to answer when the protein–protein interaction is part of a cascade of a regulatory event. A comparison of the binding affinity of Ras toward Raf (which is followed by specific phosphorylation of Raf) to the ability of phosphorylated Raf to activate transcription has shown a very good correlation for some of the mutants studied, but not for all (Block *et al.* 1996). Similarly, for the interaction between interferon and its receptors, a very good correlation was found between biological anti-viral activity and *in vitro* binding affinity (measured on RIfS) for a large set of mutations of interferon (Figure 2.4). On the other hand, human-growth-hormone (hGH) binding affinity to its receptor (hGH-R, measured on a BIAcore) exceeds the requirements for cellular activity. A systematic study of the affinity and activity of a series of mutants, in which the binding affinity was reduced up to 500-fold (from the highest affinity mutant) showed that the EC_{50} value for cell proliferation was unaffected until affinity was reduced about thirty-fold from wild-type hGH. Thus, the affinity of native hGH-receptor is much higher than is needed for maximal activity (Pearce *et al.* 1999).

Summary

With the complexity of protein–protein interactions being revealed at the molecular level, it is becoming more and more important to analyze these interactions

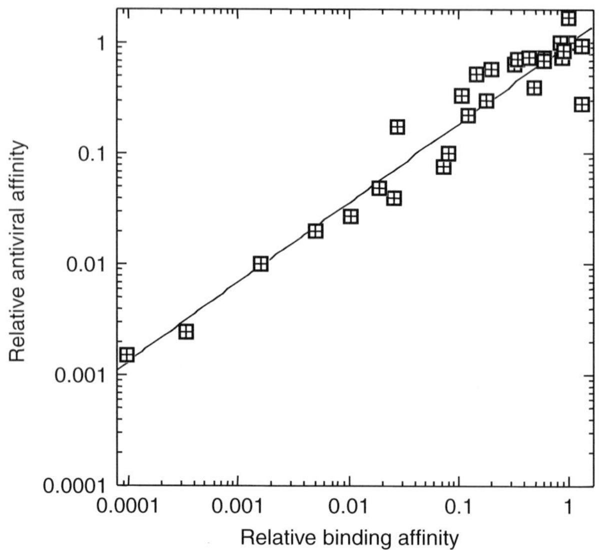

Figure 2.4 Correlation between anti-viral activity of mutant IFNα2 proteins and their binding affinities to ifnar2, as measured using a biosensor with ifnar2 bound to the surface. Anti-viral activity was measured by a cytopathic effect inhibition assay in human WISH cells for the *E. coli* produced IFNα2 proteins, using NIH IFNβ as a standard

quantitatively. The development of real-time measuring devices, and the improvements made in heterologous protein expression and purification techniques have made this task more feasible. In this chapter I have tried to give some background on the nature of protein–protein interactions, and of methodologies and problems in determining their kinetic and thermodynamic parameters.

References

Albeck, S. and Schreiber, G. (1999) Biophysical characterization of the interaction of the β-lactamase TEM-1 with its protein inhibitor BLIP. *Biochemistry* 38, 11–21.
Block, C., Janknecht, R., Herrmann, C., Nassar, N. and Wittinghofer, A. (1996) Quantitative structure–activity analysis correlating Ras/Raf interaction in vitro to Raf activation in vivo. *Nat. Struct. Biol.* 3, 244–51.
Bogan, A.A. and Thorn, K.S. (1998) Anatomy of hot spots in protein interfaces. *J. Mol. Biol.* 280, 1–9.
Buckle, M., Schreiber, G. and Fersht, A.R. (1994) Protein–protein recognition: crystal structural analysis of a barnase–barstar complex at 2.0 Å resolution. *Biochemistry* 33, 8878–89.
Clackson, T. and Wells, J.A. (1995) A hot spot of binding energy in a hormone–receptor interface. *Science* 267, 383–6.
Conte, L.L., Chothia, C. and Janin, J. (1999) The atomic structure of protein–protein recognition sites. *J. Mol. Biol.* 285, 2177–98.

Cunningham, B.C. and Wells, J.A. (1993) Comparison of a structural and a functional epitope. *J. Mol. Biol.* 234, 554–63.

Eddowes, M.J. (1990) *Biosensors: A Practical Approach.* IRC Press, Oxford.

Frisch, C., Schreiber, G., Jonson, C.M. and Fersht, A.R. (1997) Thermodynamics of the interaction of barnase and barstar: changes in free energy versus changes in enthalpy on mutation. *J. Mol. Biol.* 267, 696–706.

Gabdoulline, R.R. and Wade, R.C. (1997) Simulation of the diffusional association of barnase and barstar. *Biophysical J.* 72, 1917–29.

Harel, M., Kleywegt, G.J., Ravelli, R.B., Silman, I. and Sussman, J.L. (1995) Crystal structure of an acetylcholinesterase–fasciculin complex: interaction of a three-fingered toxin from snake venom with its target. *Structure* 3, 1355–66.

Horovitz, A. and Fersht, A.R. (1990) Strategy for analysing the cooperativity of intramolecular interactions in peptides and proteins. *J. Mol. Biol.* 214, 613–17.

Jones, S. and Thornton, J.M. (1996) Principles of protein–protein interactions. *Proc. Natl. Acad. Sci. U.S.A.* 93, 13–20.

Lawrence, M.C. and Colman, P.M. (1993) Shape complementarity at protein/protein interfaces. *J. Mol. Biol.* 234, 946–50.

Nieba, L., Krebber, A. and Pluckthun, A. (1996) Competition BIAcore for measuring true affinities: large differences from values determined from binding kinetics. *Anal. Biochem.* 234, 155–65.

O'Shannessy, D.J. and Winzor, D.J. (1996) Interpretation of deviations from pseudo-first-order kinetic behavior in the characterization of ligand binding by biosensor technology. *Anal. Biochem.* 236, 275–83.

Pearce, K.H., Cunningham, B.C., Fuh, G., Teeri, T. and Wells, J.A. (1999) Growth hormone binding affinity for its receptor surpasses the requirements for cellular activity. *Biochemistry* 38, 81–9.

Pearce, K.H., Ultsch, M.H., Kelley, R.F., de Vos, A.M. and Wells, J.A. (1996) Structural and mutational analysis of affinity-inert contact residues at the growth hormone-receptor interface. *Biochemistry.* 35, 10300–7.

Piehler, J. and Schreiber, G. (1999) Biophysical analysis of the interaction of human ifnar2 expressed in *E. coli* with IFNα2. *J. Mol. Biol.* 289, 57–67.

Rydel, T.J., Ravichandran, K.G., Tullinsky, A., Bode, W., Huber, R., Roitsch, C. and Fenton, J.W. (1990) The structure of a complex of recombinant hirudin and human α-thrombin. *Science* 245, 277–80.

Schreiber, G. and Fersht, A.R. (1993) The interaction of barnase with its polypeptide inhibitor barstar studied by protein engineering. *Biochemistry* 32, 5145–50.

Schreiber, G. and Fersht, A.R. (1995) Energetics of protein–protein interactions: analysis of the barnase–barstar interface by single mutations and double mutant cycles. *J. Mol. Biol.* 248, 478–86.

Schreiber, G. and Fersht, A.R. (1996) Rapid, electrostatic assisted, association of proteins. *Nature Struct. Biol.* 3, 427–31.

Schreiber, G., Frisch, C. and Fersht, A.R. (1997) The role of Glu73 of barnase in catalysis and the binding of barstar. *J. Mol. Biol.* 270, 111–22.

Selzer, T. and Schreiber, G. (1999) Predicting the rate enhancement of protein complex formation from the electrostatic energy of interaction. *J. Mol. Biol.* 287, 409–19.

Stites, W.S. (1997) Protein–protein interactions: interface structure, binding thermodynamics, and mutational analysis. *Chem. Rev.* 97, 1233–50.

Strynadka, N.C., Jensen, S.E., Alzari, P.M. and James, M.N. (1996) A potent new mode of beta-lactamase inhibition revealed by the 1.7 Å X-ray crystallographic structure of the TEM-1-BLIP complex. *Nat. Struct. Biol.* 3, 290–7.

Vaughan, C.K., Buckle, A.M. and Fersht, A.R. (1999) Structural response to mutation at a protein–protein interface. *J. Mol. Biol.* 286, 1487–506.

Vijayakumar, M., Wong, K.Y., Schreiber, G., Fersht, A.R., Szabo, A. and Zhou, H.Z. (1998) Electrostatic enhancement of diffusion-controlled protein–protein association: comparison of theory and experiment on barnase and barstar. *J. Mol. Biol.* 278, 1015–24.

Wallis, R., Moore, G.R., James, R. and Kleanthous, C. (1995) Protein–protein interactions in colicin E9 DNase-immunity protein complexes. 1. Diffusion-controlled association and femtomolar binding for the cognate complex. *Biochemistry* 34, 13743–50.

Ysern, X., Fields, B.A., Bhat, T.N., Goldbaum, F.A., Dall'Acqua, W., Schwarz, F.P., Poljak, R.J. and Mariuzza, R.A. (1994) Solvent rearrangement in an antigen–antibody interface introduced by site-directed mutagenesis of the antibody combining site. *J. Mol. Biol.* 238, 496–500.

3   DNA interactions

Brian E. Cathers and Michael J. Waring

The DNA molecule, that icon of molecular biology which has been adopted to serve almost as an icon of modern biology in general, is a very special structure from the standpoint of molecular recognition. Because it is the ultimate repository of genetic information in living things, unicellular as well as multicellular organisms, its essentially linear array of genes encoded in an immensely long sequence of nucleotides represents a unique vehicle for chemical recognition to modulate biological function and affords a vital target for medical intervention. To put the issue in a nutshell, interactions with DNA which are directed at recognising specific portions of its sequence of base pairs have the potential to extract from DNA its *raison d'être* ultimately eliciting a biological response. Elsewhere in this volume will be found numerous examples of biotechnological applications and goals where the capability to identify, react with, or even modify the nucleotide sequence of DNA furnishes the fundamental chemistry to achieve something useful. Here we confine ourselves to a consideration of the principles upon which sequence recognition may be based, illustrated with examples which serve to focus discussion of issues that will receive more substantial treatment later. In this chapter, we propose to adopt a special framework upon which to hang our discussion of structural aspects of the DNA recognition process. We will view DNA interactions along a spectrum from recognition by shape and other secondary structural features (analogue), to direct sequence readout (digital). The distinction between these two extreme modalities of perceiving sequence information is valid for other macromolecules too, of course, but the elegance of the concept is particularly striking as it applies to DNA. While the ligands referred to are by no means an exhaustive compilation, most examples selected to illustrate the various modes of interaction with DNA will be chosen from a diverse set of small drug-like molecules, though we do not hesitate to enlist large macromolecules when appropriate. But first, to understand DNA interactions and the fundamental basis for recognition processes, it behoves us to look at the DNA receptor itself in order to visualise those features of its structure that allow for recognition.

In the middle part of the twentieth century there were many misconceptions about the structure of DNA, and even the earliest models depicting DNA as a long helical polymer placed the bases pointing outwards with the pentose–phosphate backbone in the centre. Certain critical observations were made in the early 1950s by Wilkins and Franklin using X-ray diffraction techniques, namely that the diffraction pattern of a moist sample of DNA changed significantly from low humidity (the A-form) to high humidity (the B-form). Their findings fired the enthusiasm of

Watson and Crick to piece together a picture for the secondary structure of DNA based upon a helical model (Watson and Crick 1953). Using Chargaff's 1:1 base pairing rules (in isolated DNA there is always an equal number of As and Ts, and an equal number of Gs and Cs, but the AT:GC ratio is different from species to species) and the keto form of the bases, there quickly emerged a pattern of recognition. If a purine is always paired with a pyrimidine, then the symmetry of the helix could be maintained. This led to what is now known as the Watson–Crick form of DNA (Figure 3.1), and in 1962 Watson, Crick and Wilkins were awarded the Nobel Prize for solving the structure and elucidating the pairing rules for duplex DNA. Today we take for granted the complementarity of the anti-parallel strands in duplex DNA, and in everyday laboratory practice we make use of the now familiar base pairing rules. But what structural features of duplex DNA underlie recognition by proteins and small molecules?

There are several forms of duplex DNA, and as mentioned above Franklin and Wilkins very early on recognised that the helical form of DNA changed from what was called A-form to B-form at different levels of humidity. Another commonly recognised isomer of DNA is Z-form. As it is unclear what role Z-DNA may play in biological function we will limit the discussion of DNA recognition to the A- and B-forms; however, good descriptions of Z-DNA can be found elsewhere (Dickerson *et al.* 1982; Rich *et al.* 1984). The canonical B-form of DNA depicted in Figure 3.1 is characterised by the existence of two grooves of unequal width arising from the right-handed twisting of the helix. The wider of the two grooves, called the major groove, is 12 Å across and 8 to 9 Å deep. The smaller or minor groove is nearly as deep (7 to 8 Å) but is only 6 Å across. The shorter and bulkier A-form of DNA is also right-handed, but its minor groove becomes very wide and shallow and almost disappears. The major groove narrows but is still about as deep as it is in B-form DNA. The walls of both grooves are formed by the pentose–phosphate backbones

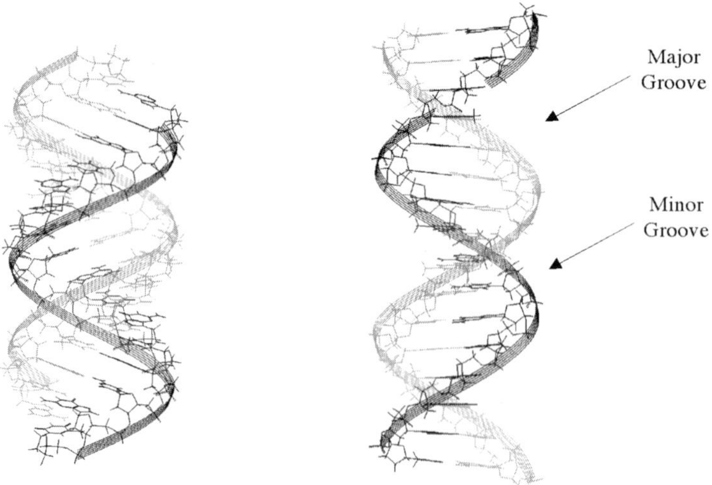

Figure 3.1 Canonical A-form (left) and B-form (right) duplex DNA of identical sequence: 5′-(ACGT)₄. Structures produced using Insight II (MSI, San Diego, CA)

with their connecting phosphodiester groups located on the outermost, solvent accessible edges of the walls. The floor of each well in B-DNA runs at a large angle to the plane of the bases and is formed by the edges of the stacked base pairs (in A-DNA the bases are tilted about 19° relative to the helix axis). It is on the floor of the grooves that specific hydrogen bonding interactions with other molecules can take place. Figure 3.2 depicts the arrangement of functional groups exposed within the floor of each groove.

While the variety of functional groups is small, it is the pattern, shape, and disposition of these hydrogen bond donor and acceptor groups reflecting the primary nucleotide sequence that gives rise to unique recognition elements within DNA. For example, one of the major structural characteristics of the minor groove that enables GC-rich sequences to be targeted preferentially over AT-rich sequences is the exocyclic 2-amino group of guanine. This hydrogen bond donating group protrudes into the cavity of the minor groove and can provoke steric clashes with potential minor groove binding elements. The effects of exocyclic amino groups in the minor groove have been extensively studied through substitution of diaminopurine for adenine in AT base pairs or of inosine for guanosine in GC base pairs (Marchand *et al.* 1992; Waring and Bailly 1994; Bailly and Waring 1995; Bailly *et al.* 1995a; Bailly *et al.* 1995b; Bailly *et al.* 1996; Mollegaard *et al.* 1997; Waring and Bailly 1997; Bailly *et al.* 1998). Work along these lines has demonstrated the profound effect such a small group within the minor groove can have on both the sequence selectivity of molecules that bind in the minor groove and the binding affinity of proteins that recognise elements in the major groove but depend upon the flexibility of DNA for

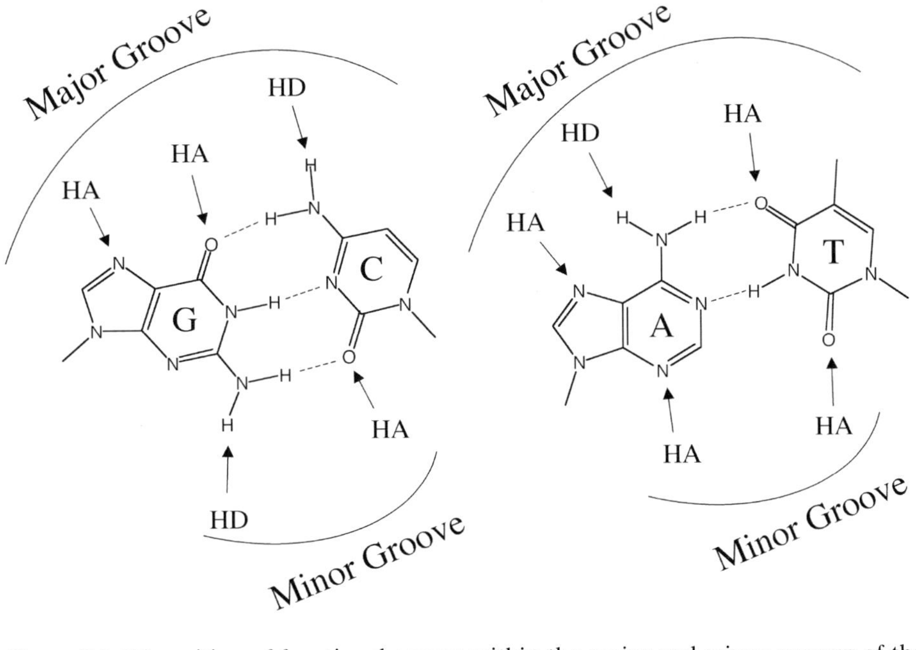

Figure 3.2 Disposition of functional groups within the major and minor grooves of the DNA double helix as represented by a G:C and an A:T base pair. The functional groups are labelled as either hydrogen bond donating (HD) or hydrogen bond accepting (HA)

binding. This last concept of flexibility or bendability is a contributory aspect of DNA recognition that is largely lost with the cartoon representations of canonical A-form and B-form DNA. Natural or induced perturbations in the secondary structure of DNA play a large part in the specificity of analogue recognition processes. One recurring theme in DNA bending is its association with A-tracts, whereby runs of five or more consecutive adenine nucleotides cause a considerable curvature of the helical axis towards the minor groove. Different models describing DNA bending can be found in an excellent review by Crothers *et al.* (1990) and the references therein.

Analogue recognition

While most small molecules tend to bind in the tight confines of the minor groove, proteins more often take advantage of the expanse and greater number of hydrogen bond contacts available within the major groove. Protein binding specificity does not usually come from a direct read-out of the primary DNA sequence, but rather from recognition of multiple sequence motifs. It is the combination of local variations in the secondary structure of DNA (determined by a specific DNA sequence or family of sequences) and the relative position of multiple recognition elements within the primary sequence that gives rise to the specificity of protein–DNA interactions. For example, proteins like the oestrogen receptor bind as dimers to palindromic DNA sequences (Anderson *et al.* 1998). While not a direct sequence readout, this type of recognition can be just as specific.

Proteins utilise two main motifs of secondary structure for binding to DNA: α-helices and β-sheets. For a review of the structural families and recognition principles involved in protein–DNA interactions, see Pabo and Sauer (1992) and Harrison (1991) and references therein. The α-helix secondary structural element is by far the most commonly employed for DNA recognition. The positively charged amino terminal end of the helix is typically directed into the major groove, gaining access to the backbone phosphate groups through ionic interactions. Once inside the groove both hydrogen bonding and van der Waals interactions between amino acid side chains and the nucleobase functional groups that make up the floor of the groove stabilise the complex.

Proteins stabilise and position the α-helix in several different ways. The most thoroughly studied example is the helix-turn-helix motif. This is commonly employed by both prokaryotic and eukaryotic regulatory proteins. The helix-turn-helix motif is generally twenty residues long and is characterised by two adjacent α-helices that cross each other at an angle of about 120°. This small motif is not usually stable as a stand-alone peptide and requires other secondary structural features to provide a hydrophobic pocket for proper formation. The helix-turn-helix motif is therefore often found embedded within other domains. A good example of this type of protein–DNA interaction is provided by the FIS protein (Yuan *et al.* 1991; Kostrewa *et al.* 1992).

Helical elements of protein secondary structure stabilised by a specific interaction with zinc (II) make up another large class of α-helical DNA binding motifs. Loop-helix domains resemble the prokaryotic helix-turn-helix domains both in regard to how they interact with DNA and in the formation of protein dimers. Two α-helices are connected by a short loop with a zinc ion bound to stabilise the tertiary structure. The first helix is involved in DNA recognition within the major groove while the second helix provides contacts for protein dimerisation. A zinc-stabilised β-turn-α-helix motif is called a zinc finger. The first crystal structure of a protein–DNA complex was of the zinc finger protein zif268 (Pavletich and Pabo 1991). By analysing zif268 and many other related zinc finger protein–DNA interactions, a recognition code that can be loosely applied to other α-helical protein–DNA interactions has been devised (Jacobs 1992; Desjarlais and Berg 1993; Fairall *et al.* 1993; Choo and Klug 1995; Choo and Klug 1997).

The eukaryotic transcription factor called TATA binding protein (TBP) employs β-sheets to recognise DNA nucleotide sequences through binding in the minor groove. It was originally thought that since there were fewer recognition elements within the minor groove, TBP binding would be rather promiscuous. However, it was subsequently shown that TBP is actually quite selective for closely related TATA sequences (Wong and Bateman 1994), and it was proposed that this specificity is derived from the ability of certain sequences to adopt different conformations. TBP has been shown to dramatically bend and unwind the TATA box upon binding (Kim *et al.* 1993a; Kim *et al.* 1993b). Therefore, the flexibility of DNA may predispose certain sequences to TBP binding. While contributions to TBP recognition from small or subtle differences in structure between closely related TATA sequences can explain in part the selectivity of TBP, the idea of DNA flexibility provides a much more palatable explanation.

Middle of the spectrum

Of the DNA interactions that fall in the middle of the analogue to digital spectrum, many involve small drug-like molecules. Some of the most interesting examples employ not only the subtle elements of hydrogen bonding and hydrophobic interactions described above but also lead to covalent reaction with the receptor, thereby highlighting another facet of DNA, namely the nucleophilicity of the nucleobases themselves. There are numerous DNA-alkylating agents, some with absolutely no sequence selectivity and others with very well-defined targets (Waring 1981). Given the variety of nucleophiles to be found in DNA, it is perhaps hard to imagine how even a modestly potent electrophile could discriminate based upon primary sequence. How then do these reactive molecules find specific target sites on DNA? Part of the answer comes from the flexibility and microscopic heterogeneity in secondary structure characteristic of large molecules of DNA. Many small molecule–DNA interactions derive specificity from induced or natural perturbations in its helical secondary structure. One well-known group of compounds in this category, characterised over many years by work devoted to the mechanism of sequence recognition and selectivity, will serve to illustrate not only DNA alkylation but also the radical changes that can take place in the DNA secondary structure upon ligand binding.

Cyclopropanepyrrololindole- (CPI-) containing molecules were originally synthesised by scientists at Pharmacia and Upjohn. One of the most thoroughly studied compounds in this class, the natural product (+)-CC-1065 (Hanka *et al.* 1978), led to the clinically interesting DNA-alkylators Adozelesin and Carzelesin (Li 1991; Bhuyan *et al.* 1992; Li 1992), then to the crosslinking bis-alkylator Bizelesin (Mitchell *et al.* 1991). Bizelesin was constructed from two CPI subunits connected via a rigid linker (Figure 3.3). It was believed that the sequence selectivity of single CPI-containing molecules could be increased by linking the two warheads and allowing them to cooperate in directing the molecule to the site of highest reaction potential. The increased selectivity of Bizelesin bore out this expectation to some degree. Even though the preferred site for Bizelesin binding was identified as the palindromic hexamer TAATTA (Sun and Hurley 1993), which is two CC-1065 sites placed end to end on opposite strands, the sequence specificity did not simply arise from the combined effects of individual CPI monomers. It turned out to be much more complicated (Seaman and Hurley 1996). Whereas CC-1065 induces a bent structure upon binding, Bizelesin crosslinking actually straightens normally bent A-tract DNA (Thompson and Hurley 1995). To explain this apparent anomaly one has to look carefully at the final structure. The central base pair either rearranges to form a Hoogsteen pair or is completely disrupted into an open, unpaired state (Seaman and Hurley 1993). It seems that the increased sequence selectivity of Bize-

A

(+)-CC-1065

B

Bizelesin-DNA Adduct

C

Figure 3.3 Sequence selective DNA alkylators. A. Structure of the natural product CC-1065 highlighting the cyclopropanepyrrololindole- (CPI-) moiety. B. Structure of Bizelesin as a DNA adduct. C. Schematic representation of a highly reactive crosslinking site for Bizelesin with the adenine nucleotides susceptible to alkylation indicated by pointers

lesin is not due to something peculiar about the structure of the ligand itself, but rather to the inherent flexibility of certain DNA sequences. The initial mono-adduct formed with Bizelesin is reversible, so only those DNA sequences that are capable of undergoing the dramatic conformational changes necessary for crosslinking will be trapped by Bizelesin.

Another area where perturbation of DNA structure due to its flexibility plays an important role in recognition and binding is intercalation. First characterised in the 1960s and historically the earliest well-defined mode of drug binding to DNA (Lerman 1961; Fuller and Waring 1964; Gale *et al.* 1981), intercalation is the process whereby a planar aromatic ring system becomes inserted into DNA between the stacked base pairs of the double helix. It was initially postulated as a model for acridine binding to DNA (inspired by efforts to explain frameshift mutagenesis by these compounds), then refined by studies on ethidium bromide, and later shown to apply to the interaction between DNA and many other drugs – particularly antitumour and antimicrobial agents (Gale *et al.* 1981). Intercalation involves extension, stiffening and local unwinding of the double helix. It is the latter feature which is most characteristic of intercalating agents, and consequently ligand-induced removal and reversal of the supercoiling of closed circular duplex DNA furnishes strong evidence for intercalative binding to DNA (Waring 1970). Intercalators are frequently sequence-neutral with regards to nucleotide sequence selectivity, though some antibiotics such as actinomycin can recognise GC-rich binding sites – in sharp contrast to minor groove-binding ligands like distamycin and netropsin which are characteristically AT-specific (Van Dyke *et al.* 1982; Lane *et al.* 1983; Fox and Waring 1984). The discovery of bis-intercalation by the antibiotic echinomycin, which recognises CpG sequences preferentially flanked by AT base pairs (Waring and Wakelin 1974; Low *et al.* 1984; Van Dyke and Dervan 1984), sparked off a burst of activity to make synthetic bis-intercalators which produce major distortion of the secondary structure of DNA in the hope that such molecules might be endowed with enhanced sequence-recognition properties and improved biological activity. Progress along these lines has been reviewed by Wakelin and Waring (1990).

Digital recognition

While proteins and most small molecules derive DNA-binding specificity through analogue recognition processes, there are examples of man-made molecules capable of directly reading out the primary sequence information within the macromolecule, sometimes with very little effect on its double helical secondary structure. The idea of absolute sequence-specific recognition of DNA by non-protein and non-nucleic acid substances has only recently come to fruition in the form of two very interesting classes of compounds. The first, devised by Ole Buchardt and Peter Nielsen, is actually an enzymically and chemically stable analogue of a polynucleotide that takes ultimate advantage of the DNA digital recognition motif. Buchardt and Nielsen's concept was to utilise the same purine and pyrimidine bases found in nucleic acids but to eliminate the normal electrostatic repulsion found in duplex DNA by replacing the pentose–phosphate backbone with an uncharged aminoethyl–glycine backbone (Figure 3.4a) (Nielsen *et al.* 1994). The resulting peptide nucleic acids (PNAs) are both nuclease and protease resistant and can form heteroduplexes with DNA (and RNA) that are several-fold more stable than analogous DNA–DNA duplexes

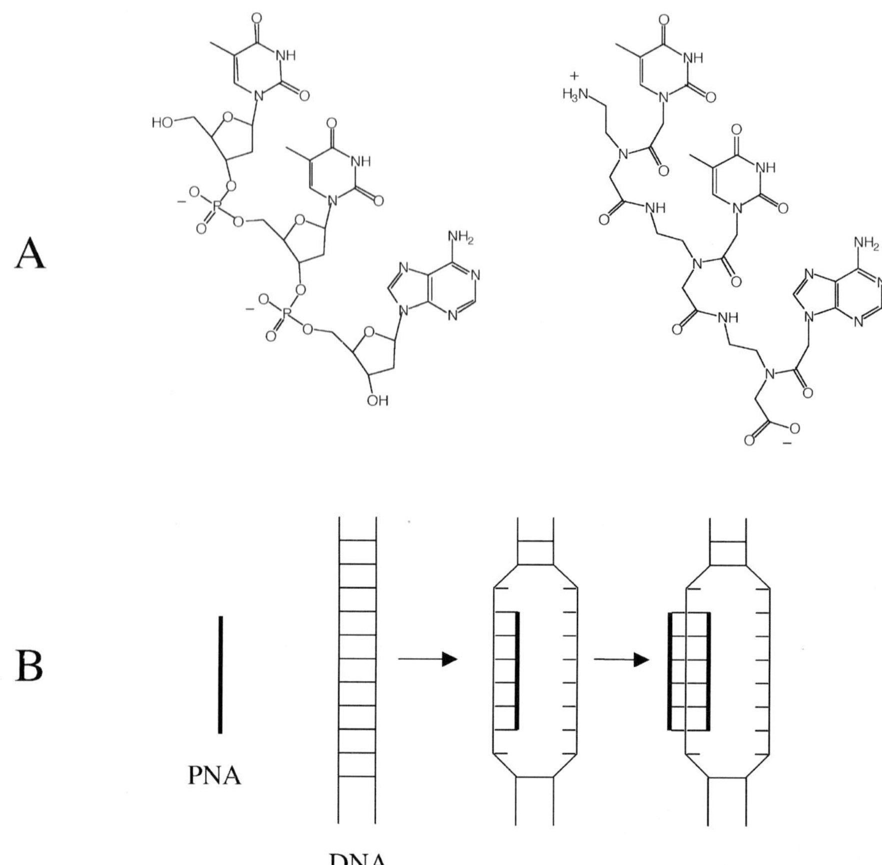

Figure 3.4 Digital recognition of DNA by a peptide nucleic acid (PNA) A. Structure depicting the different backbone chemistry of DNA (left) and PNA (right). B. Schematic representation of DNA duplex invasion by a small peptide nucleic acid (PNA) molecule and the triplex and single stranded DNA loop that result

(Demidov *et al.* 1995; Gangamani *et al.* 1997; Jensen *et al.* 1997; Leijon *et al.* 1997). The stable PNA molecules are excellent anti-sense agents *in vitro*, binding with very high specificity and avidity to mRNA, ultimately inhibiting translation (Nielsen 1997; Good and Nielsen 1998). PNAs can also invade duplex DNA (Nielsen *et al.* 1991; Demidov *et al.* 1993; Kurakin *et al.* 1998).

Pyrimidine-containing PNA molecules bind to complementary duplex targets by displacing the homopyrimidine DNA strand and forming a ternary complex containing *two* PNA molecules associated with the remaining DNA strand. One PNA strand forms Watson–Crick base pairs while the other forms Hoogsteen hydrogen bonds with the captured DNA strand. The homopyrimidine DNA strand is essentially left out as a single stranded loop (Figure 3.4b) (Demidov *et al.* 1993). This property is potentially useful for gene-directed therapy whereby binding to a specific site within a gene can inhibit transcription or conversely may even serve as an artificial promoter to switch on transcription from the displaced DNA strand (Good and Nielsen 1997). However, the rate of association of PNA with DNA is dramatically slowed at physiological salt concentration even though preformed PNA_2-DNA complexes are quite stable at high ionic strength. Due to poor cellular uptake of PNA molecules little progress has been made in the clinic; however, PNAs and PNA conjugates have become increasingly important in the laboratory as probes for monitoring DNA-related phenomena (Gangamani *et al.* 1997; Jensen *et al.* 1997; Kozlov *et al.* 1998). There are obvious parallels between the behaviour of PNAs and that of triplex-forming oligonucleotides which bind to duplex DNA in the major groove, predominantly at homopurine–homopyrimidine sequences, to produce an anti-gene effect formally analogous to the anti-sense effect they exert at the level of mRNA and translation. Space does not permit a detailed discussion of these interesting molecules which afford another very valuable approach to gene targeting. Their chemistry and biochemistry have been excellently reviewed elsewhere (Helene 1991; Frank-Kamenetskii and Mirkin 1995).

The second class of compounds endowed with the ability to recognise DNA digitally in a sequence-specific manner do not actually pair with its bases in a Watson–Crick (or Hoogsteen) manner at all; on the contrary, unlike PNAs these substances fail to disrupt the normal Watson–Crick base pairing of duplex DNA when they bind to it (Figure 3.5a). Tracing their lineage back to the antibiotics netropsin and distamycin, Peter Dervan's pyrrole–imidazole polyamides bind in the minor groove and recognise short, mixed-base pair sequences of DNA. Unlike the natural products, which have long been known to be highly AT-selective, these

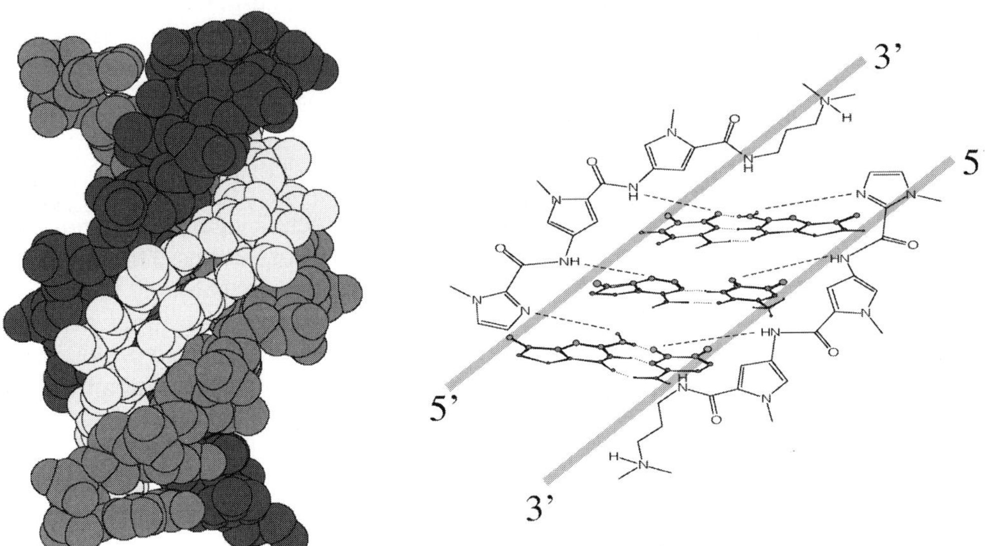

Figure 3.5 Digital recognition of DNA through minor groove contacts. A. Space filling model of a minor groove binding hairpin polyamide (lightest shade of grey) demonstrating that the DNA secondary structure is well preserved. Structure taken from the Brookhaven Protein Data Bank (408D, Kielkopf *et al.* 1998b). B. Structure depicting the hydrogen bonds between the polyamide and functional groups in the minor groove that give rise to the sequence selectivity

polyamides were designed to utilise the exocyclic amino group of guanine to direct specific binding to regions containing G-C base pairs. After recognising the potential for side-by-side binding within the minor groove (Mrksich *et al.* 1992), first described for distamycin (Pelton and Wemmer 1989; Pelton and Wemmer 1990), Dervan developed solid phase synthetic methods for preparing many different polyamide chains (Baird and Dervan 1996). The pairing code for sequence specific recognition by imidazole (Im) and pyrrole (Py) containing polyamides was deduced as follows (Pelton and Wemmer 1989; Pelton and Wemmer 1990; Mrksich *et al.* 1992; Wade *et al.* 1992; Wade *et al.* 1993; Trauger *et al.* 1996; White *et al.* 1997). An Im/Py pair distinguishes a GC base pair from CG and from both of AT and TA, while the opposite Py/Im pair recognises a CG base pair but is disfavoured at GC, AT, or TA sites (the very interesting fact that the stereochemistry or directionality of the hydrogen bond from the exocyclic amino group of guanine is so critical for the discrimination between GC and CG base pairs is wonderfully depicted elsewhere (White *et al.* 1997; Kielkopf *et al.* 1998a)). The Py/Py pair recognises TA or AT equally well but is disfavoured at either GC or CG sites, while the fourth possible pair, Im/Im, is disfavoured by all the base pairs. Figure 3.5b depicts the hydrogen bonding interactions that give rise to the specificity at the GC, AT, or CG sites. Recently, by adding a third amino acid (3-hydroxypyrrole, Hp) to the repertoire, discrimination between AT and TA sites has been made possible. A complete pairing code now exists for minor groove recognition by three-component polyamides (see Table 2 in White *et al.* 1998). By constructing hairpin polyamides where the requisite Im, Py or Hp pairings are kept in place by a judiciously constructed linker, high sequence-specificity and tight binding to the targeted region within a test DNA molecule can be assured. With good cellular permeability and the ability to target specific DNA sequences, hopes are high that these compounds will become clinically relevant as anti-gene therapeutics.

One of the purposes of this chapter has been to emphasise that the DNA molecule does not behave like the simple, rigid canonical B-form that is most often portrayed in classes and in textbooks. By its very nature and purpose this highly adaptable and flexible molecule must be recognised in an exquisitely sequence-specific manner to allow for the exact timing and production of mRNA, and ultimately protein, that is so critical to the survival of cells and organisms. The carrier of genetic information encoded with only four repeating units contains within its sequence all the elements necessary for highly specific recognition either through analogue or digital means. The boom in biotechnology now allows us to probe interactions with DNA in real time, which in turn provides kinetic information to go along with our static, structural snapshots of DNA interactions.

References

Anderson, I., Bartley, C.R., Lerch, R.A., Gray, W.G.N., Friesen, P.D. and Gorski, J. (1998) Estrogen receptor requires no accessory factors for high-affinity binding to a consensus response element. *Biochemistry* 37, 17287–17298.

Bailly, C., Mollegaard, N.E., Nielsen, P.E. and Waring, M.J. (1995a) The influence of the 2-amino group of guanine on DNA conformation. Uranyl. *EMBO Journal* 14, 2121–2131.

Bailly, C., Payet, D., Travers, A.A. and Waring, M.J. (1996) PCR-based development of DNA substrates containing modified bases: An efficient system for investigating the role of the exocyclic groups in chemical and structural recognition by minor groove binding drugs and proteins. *Proceedings of the National Academy of Science, USA* 93, 13623–13628.

Bailly, C., Suh, D., Waring, M.J. and Chaires, J.B. (1998) Binding of daunomycin to diaminopurine- and/or inosine-substituted DNA. *Biochemistry* 37, 1033–1045.

Bailly, C. and Waring, M.J. (1995) Transferring the purine 2-amino group from guanines to adenines in DNA changes the sequence-specific binding of antibiotics. *Nucleic Acids Research* 23, 885–892.

Bailly, C., Waring, M.J. and Travers, A.A. (1995b) Effects of base substitutions on the binding of a DNA-bending protein. *Journal of Molecular Biology* 253, 1–7.

Baird, E.E. and Dervan, P.B. (1996) Solid phase synthesis of polyamides containing imadazole and pyrrole amino acids. *Journal of the American Chemical Society* 118, 6141–6146.

Bhuyan, B.K., Smith, K.S., Adams, E.G., Petzold, G.L., McGovren, J.P. (1992) Lethality, DNA alkylation, and cell cycle effects of adozelesin (U-73975) on rodent and human cells. *Cancer Research* 52, 5687–5692.

Choo, Y. and Klug, A. (1995) Designing DNA-binding proteins on the surface of filamentous phage. *Current Opinion in Biotechnology* 6, 431–436.

Choo, Y. and Klug, A. (1997) Physical basis of a protein-DNA recognition code. *Current Opinion in Structural Biology* 7, 117–125.

Crothers, D.M., Haran, T.E. and Nadeau, J.G. (1990) Intrinsically bent DNA. *Journal of Biological Chemistry* 265, 7093–7096.

Demidov, V., Frank-Kamenetskii, M.D., Egholm, M., Buchardt, O. and Nielsen, P.E. (1993) Sequence selective double strand DNA cleavage by peptide nucleic acid (PNA) targeting using nuclease S1. *Nucleic Acids Research* 21, 2103–2107.

Demidov, V.V., Yavnilovich, M.V., Belotserkovskii, B.P., Frank-Kamenetskii, M.D. and Nielsen, P.E. (1995) Kinetics and mechanism of polyamide ("peptide") nucleic acid binding to duplex DNA. *Proceedings of the National Academy of Science, USA* 92, 2637–2641.

Desjarlais, J.R. and Berg, J.M. (1993) Use of a zinc-finger consensus sequence framework and specificity rules to design specific DNA binding proteins. *Proceedings of the National Academy of Science, USA* 90, 2256–2260.

Dickerson, R.E., Drew, H.R., Conner, B.N., Wing, R.M., Fratini, A.V. and Kopka, M.L. (1982) The anatomy of A-, B-, and Z-DNA. *Science* 216, 475–485.

Fairall, L., Schwabe, J.W.R., Chapman, L., Finch, J.T. and Rhodes, D. (1993) The crystal structure of a two zinc-finger peptide reveals an extension to the rules for zinc-finger/DNA recognition. *Nature (London)* 366, 483–487.

Fox, K.R. and Waring, M.J. (1984) *Nucleic Acids Res.* 12, 9271–9285.

Frank-Kamenetskii, M.D. and Mirkin, S.M. (1995) Triplex DNA structures. *Annual Reviews of Biochemistry* 64, 65–95.

Fuller, W. and Waring, M.J. (1964) *Ber. Bunsenges. Physik. Chem.* 68, 805–808.

Gale, E.F., Cundliffe, E. Reynolds, P.E., Richmond, M.H. and Waring, M.J. (1981) *The Molecular Basis of Antibiotic Action*, 2nd edn, Wiley, London.

Gangamani, B.P., Kumar, V.A. and Ganesh, K.N. (1997) Spermine conjugated peptide nucleic acids (spPNA): UV and fluorescence studies of PNA-DNA hybrids with improved stability. *Biochemical & Biophysical Research Communications* 240, 778–782.

Good, L. and Nielsen, P.E. (1997) Progress in developing PNA as a gene-targeted drug. *Antisense and Nucleic Acid Drug Development* 7, 431–437.

Good, L. and Nielsen, P.E. (1998) Antisense inhibition of gene expression in bacteria by PNA targeted to mRNA. *Nature Biotechnology* 16, 355–358.

Hanka, L.J., Dietz, A., Gerpheide, S.A., Kuentzel, S.L. and Martin, D.G. (1978) CC-1065 (NSC-298223), a new antitumor antibiotic. Production, in vitro biological activity, microbiological assays and taxonomy of the producing microorganism. *Journal of Antibiotics* 31, 1211–1217.

Harrison, S.C. (1991) A structural taxonomy of DNA-binding domains. *Nature (London)* 353, 715–719.

Helene, C. (1991) The anti-gene strategy: control of gene expression by triplex-forming-oligonucleotides. *Anticancer Drug Design* 6, 569–584.

Jacobs, G.H. (1992) Determination of the base recognition positions of zinc fingers from sequence analysis. *The EMBO Journal* 11, 4507–4517.

Jensen, K.K., Orum, H., Nielsen, P.E. and Norden, B. (1997) Kinetics for hybridization of peptide nucleic acids (PNA) with DNA and RNA studied with the BIAcore technique. *Biochemistry* 36, 5072–5077.

Kielkopf, C.L., Baird, E.E., Dervan, P.B. and Rees, D.C. (1998a) Structural basis for GïC recognition in the DNA minor groove. *Nature Structural Biology* 5, 104–109.

Kielkopf, C.L., White, S., Szewczyk, J.W., Turner, J.M., Baird, E.E., Dervan, P.B. and Rees, D.C. (1998b) Structural basis for recognition of A-T and T-A base pairs in the minor groove of B-DNA. *Science* 282, 111–115.

Kim, J.L., Nikolov, D.B. and Burley, S.K. (1993a) Co-crystal structure of TBP recognizing the minor groove of a TATA element. *Nature (London)* 365, 520–527.

Kim, Y., Geiger, J.H., Hahn, S. and Sigler, P.B. (1993b) Crystal structure of a yeast TBP/TATA-box complex. *Nature (London)* 365, 512–520.

Kostrewa, D., Granzin, J., Stock, D., Choe, H.W., Labahn, J. and Saenger, W. (1992) Crystal structure of the factor for inversion stimulation FIS at 2.0 A resolution. *Journal of Molecular Biology* 226, 209–226.

Kozlov, I.A., Nielsen, P.E. and Orgel, L.E. (1998) A method for the 32P labeling of peptides or peptide nucleic acid oligomers. *Bioconjugate Chemistry* 9, 415–417.

Kurakin, A., Larsen, H.J. and Nielsen, P.E. (1998) Cooperative strand displacement by peptide nucleic acid (PNA). *Chemistry & Biology* 5, 81–89.

Lane, M.J., Dabrowiak, J.C. and Vournakis, J.W. (1983) Sequence specificity of actinomycin D and Netropsin binding to pBR322 DNA analyzed by protection from DNase I. *Proceedings of the National Academy of Science, USA* 80, 3260–3264.

Leijon, M., Sehlstedt, U., Nielsen, P.E. and Graslund, A. (1997) Unique base-pair breathing dynamics in PNA-DNA hybrids. *Journal of Molecular Biology* 271, 438–455.

Lerman, L.S. (1961). *Journal of Molecular Biology* 3, 18–30.

Li, L.H., Kelly, R.C., Warpehoski, M.A., McGovren, J.P., Gebhard, I., DeKoning, T.F. (1991) Adozelesin, a selected lead among cyclopropylpyrroloindole analogs of the DNA-binding antibiotic, CC-1065. *Investigational New Drugs* 9, 137–148.

Li, L.H., DeKoning, T.F., Kelly, R.C., Krueger, W.C., McGovren, J.P., Padbury, G.E., Petzold, G.L., Wallace, T.L., Ouding, R.J., Prairie, M.D., Gebhard, I. (1992) Cytotoxicity and antitumor activity of carzelesin, a prodrug cyclopropylpyrroloindole analogue. *Cancer Research* 52, 4904–4913.

Low, C.M.L., Drew, H.R. and Waring, M.J. (1984) Sequence-specific binding of echinomycin to DNA: evidence for conformational changes affecting flanking sequences. *Nucleic Acids Research* 12, 4865–4879.

Marchand, C., Bailly, C., McLean, M.J., Moroney, S.E. and Waring, M.J. (1992) The 2-amino group of guanine is absolutely required for specific binding of the anti-cancer antibiotic echinomycin to DNA. *Nucleic Acids Research* 20, 5601–5606.

Mitchell, M.A., Kelly, R.C., Wicnienski, N.A., Hatzenbuhler, N.T., Williams, M.G., Petzole, G.L., Slightom, J.L. and Siemieniak, D.R. (1991) *Journal of the American Chemical Society* 113, 8994–8995.

Mollegaard, N.E., Bailly, C., Waring, M.J. and Nielsen, P.E. (1997) Effects of diaminopurine and inosine substitutions on A-tract induced DNA curvature. Importance of the 3'-A-tract junction. *Nucleic Acids Research* 25, 3497–3502.

Mrksich, M., Wade, W.S., Dwyer, T.J., Geierstanger, B.H., Wemmer, D.E. and Dervan, P.B. (1992) Antiparallel side-by-side dimeric motif for sequence-specific recognition in the minor groove of DNA by the designed peptide 1-methylimiidazole-2-carboxamide netropsin. *Proceedings of the National Academy of Science, USA* 89, 7586–7590.

Nielsen, P.E. (1997) Peptide nucleic acid (PNA) from DNA recognition to antisense and DNA. *Biophysical Chemistry* 68, 103–108.

Nielsen, P.E., Egholm, M., Berg, R.H. and Buchardt, O. (1991) Sequence selective recognition of DNA by strand displacement. *Science* 254, 1497–1500.

Nielsen, P.E., Egholm, M. and Buchardt, O. (1994) Peptide nucleic acid (PNA). A DNA mimic with a peptide backbone. *Bioconjugate Chemistry* 5, 3–7.

Pabo, C.O. and Sauer, R.T. (1992) Transcriptional factors: Structural families and principles of DNA recognition. *Annual Review of Biochemistry* 61, 1053–1095.

Pavletich, N.P. and Pabo, C.O. (1991) Zinc finger-DNA recognition: Crystal structure of a Zif268-DNA complex at 2.1 _. *Science* 252, 809–817.

Pelton, J.G. and Wemmer, D.E. (1989) Structural characterization of a 2:1 distamycin A.d(CGCAAATTGGC) complex by two-dimensional NMR. *Proceedings of the National Academy of Science, USA* 86, 5723–5727.

Pelton, J.G. and Wemmer, D.E. (1990) *Journal of the American Chemical Society* 112, 1393–1399.

Rich, A., Nordheim, A. and Wang, A.H.-J. (1984) The chemistry and biology of left-handed Z-DNA. *Annual Review of Biochemistry* 53, 791–846.

Seaman, F.C. and Hurley, L.H. (1993) Interstrand cross-linking by bizelesin produces a Watson-Crick to Hoogsteen base-pairing transition region in d(CGTAATTACG)2. *Biochemistry* 32, 12577–12585.

Seaman, F.C. and Hurley, L.H. (1996) Manipulative interplay of the interstrand cross-linker bizelesin with d(TAATTA)2 to achieve sequence recognition of DNA. *Journal of the American Chemical Society* 118, 10052–10064.

Sun, D. and Hurley, L.H. (1993) *Journal of the American Chemical Society* 115, 5925–5933.

Thompson, A.S. and Hurley, L.H. (1995) Solution conformation of a bizelesin A-tract duplex adduct: DNA-DNA cross-linking of an A-tract straightens out bent DNA. *Journal of Molecular Biology* 252, 86–101.

Trauger, K.W., Baird, E.E. and Dervan, P.B. (1996) Recognition of DNA by designed ligands at subnanomolar concentrations. *Nature (London)* 382, 559–561.

Van Dyke, M.W., Hertzberg, R.P. and Dervan, P.B. (1982) Map of distamycin, netropsin, and actinomycin binding sites on heterogeneous DNA: DNA cleavage-inhibition patterns with methidiumpropyl-EDTA.Fe(II). *Proceedings of the National Academy of Science, USA* 79, 5470–5474.

Van Dyke, M.W. and Dervan, P.B. (1984) Echinomycin binding sites on DNA. *Science* 225, 1122–1127.

Wade, W.S., Mrksich, M. and Dervan, P.B. (1992) *Journal of the American Chemical Society* 114, 8783–8794.

Wade, W.S., Mrksich, M. and Dervan, P.B. (1993) Binding affinities of synthetic peptides, pyridine-2-carboxamidonetropsin and 1-methylimidazole-2-carboxamidonetropsin, that form 2:1 complexes in the minor groove of double-helical DNA. *Biochemistry* 32, 11385–11389.

Wakelin, L.P.G. and Waring, M.J. (1990) in *Comprehensive Medicinal Chemistry*, Vol. 2 (ed. P.G. Sammes), Pergamon Press, Oxford, pp. 703–724.

Waring, M.J. (1970) Variation of the supercoils in closed circular DNA by binding of antibiotics and drugs: evidence for molecular models involving intercalation. *Journal of Molecular Biology* 54, 247–279.

Waring, M.J. (1981) DNA modification and cancer. *Annual Review of Biochemistry* 50, 159–192.

Waring, M.J. and Bailly, C. (1994) The purine 2-amino group as a critical recognition element for binding of small molecules to DNA. *Gene* 149, 69–79.

Waring, M.J. and Bailly, C. (1997) The influence of the exocyclic amino group characteristic of GC base pairs on molecular recognition of specific nucleotide sequences in DNA by berenil and DAPI. *Journal of Molecular Recognition* 10, 121–127.

Waring, M.J. and Wakelin, L.P.G. (1974) Echinomycin: a bifunctional intercalating antibiotic. *Nature (London)* 252, 653–657.

Watson, J.D. and Crick, F.H.C. (1953) *Nature (London)* 171, 737.

White, S., Baird, E.E. and Dervan, P.B. (1997) On the pairing rules for recognition in the minor groove of DNA by pyrrole-imidazole polyamides. *Chemistry & Biology* 4, 569–578.

White, S., Szewczyk, J.W., Turner, J.M., Baird, E.E. and Dervan, P.B. (1998) Recognition of the four Watson-Crick base pairs in the DNA minor groove by. *Nature (London)* 391, 468–471.

Wong, J.M. and Bateman, E. (1994) TBP-DNA interactions in the minor groove discriminate between A:T and T:A base pairs. *Nucleic Acids Research* 22, 1890–1896.

Yuan, H.S., Finkel, S.E., Feng, J.A., Kaczor-Grzeskowiak, M., Johnson, R.C. and Dickerson, R.E. (1991) The molecular structure of wild-type and a mutant Fis protein: relationship between mutational changes and recombinational enhancer function or DNA binding. *Proceedings of the National Academy of Science, USA* 88, 9558–9562.

Part II

Immobilisation of biomolecules

4 Immobilisation chemistry of biological recognition molecules

Andrew G. Mayes

Introduction

For application in chemical sensing, the most important groups of biomolecules that need to be considered are enzymes and antibodies, although DNA and RNA are rapidly emerging into the sphere of chemical sensing, and will be an area of great future activity in relation to healthcare, defence and other applications. Since the primary focus of this book is immunosensing, this chapter will concentrate mainly on the immobilisation of antibodies, although most of the techniques discussed are equally applicable to other proteins such as enzymes. The majority of the surface activation and coupling chemistry is also applicable to small molecules, such as those used as haptens in competition-based immunoassays. In this case however, there is often no need for aqueous coupling chemistry so the range of options is much broader.

The literature relating to the immobilisation of proteins and ligands is extensive. Much of it comes from the fields of protein and enzyme technology, immunochemistry and affinity chromatography, where many of the issues relating to maintaining activity and optimising accessibility/orientation were addressed and developed between the 1960s and the 1980s (Jakoby and Wilchek 1974; Lowe and Dean 1974; Mosbach 1976; Scouten 1981; Dean *et al.* 1985; Wong 1991; Hermanson *et al.* 1992; Hermanson 1995). The same issues obtain when considering the immobilisation of biomolecules on transducers to make sensors, so this older literature provides many useful methods and insights. The 1990s have also provided some alternative techniques, such as the widespread adoption of self-assembled thiolate monolayers on gold surfaces, which will be discussed in a later section. Many other, newer (and often more convenient) adaptations of older methods/ideas also continue to emerge, and this chapter will attempt to summarise the main techniques available in a way that will hopefully guide the reader towards the most appropriate methods for a particular application.

Creation of the molecular recognition interface on the transducer

Creating a molecular recognition interface involves localisation or attachment of a molecular receptor near to or onto the transducer surface. Many different approaches are possible, and the best one for a particular application depends on a variety of factors, including the transducer type, the nature of the biological component and the nature of the sample. The exact approach chosen also depends on the way the device will be used, and the surface chemistry of the device itself. Immobilisation can be achieved in several different ways:

- adsorption of the receptor directly to the transducer surface;
- physical entrapment near the transducer surface (e.g. in a polymer layer or by use of a membrane);
- covalent coupling of the receptor directly to the transducer surface;
- covalent coupling to a polymer layer on the transducer surface;
- use of a chemical/biochemical 'capture system'.

These methods are discussed on pages 60–78, where particular advantages and disadvantages of each approach are addressed. Methods are also available that enable the spatial geometry of deposition to be controlled so that arrays of molecules can be produced for multi-analyte sensing applications. This is a research area that is likely to develop rapidly during the coming years, and many elegant new methods and chemistries will undoubtedly emerge. Pages 78–83 present reviews of some of the current methods developed for making arrays.

An overview of the advantages and disadvantages of the various methods available for immobilisation is presented in Table 4.1. This table is intended to guide the reader towards the technique that is most likely to be appropriate for a specific application. Each of the categories is discussed in greater detail in one of the main sections of this chapter.

The method chosen for immobilisation depends on the type of sensor being produced and the required stability. For instance, will it be regenerated and reused or just used once and then thrown away? What is the nature of the receptor; is it hydrophilic or hydrophobic, large or small? Is it sensitive to handling? Whole organisms must be kept alive during immobilisation, enzymes must retain their activity after immobilisation and antibodies must retain their binding ability. What is the nature of the transducer surface, and what is the mechanism of signal generation? It is pointless to have a thick layer of receptor if the transducer only 'sees' a small fraction of it. Is the development intended as a proof of principle or a research tool, or is it intended to be translated to commercial mass-production and sale? Commercial production might define an acceptable cost per device, and could rule out some chemistries for safety reasons. Arriving at the correct compromise for any particular situation is a matter of knowledge, experience and careful consideration of the known parameters in relation to the types of issues outlined above. There are no hard and fast rules or rigid decision trees that can be applied. This chapter is intended as a guide to some options and general principles that need to be considered, and it is hoped that it will prove helpful in arriving at a final solution.

The transducer surface

Desirable properties of surfaces for covalent immobilisation

If a protein is to be attached to a surface, it is vital to consider the nature and physico-chemical properties of the surface. In general, proteins adsorb strongly to hydrophobic surfaces (this has often been used as an immobilisation procedure in

Table 4.1 Advantages and disadvantages of different immobilisation strategies

Method	Advantages	Disadvantages
Adsorption	Simple, inexpensive Good for single-use applications	Relatively unstable Proteins denature on hydrophobic surfaces Adsorption highly pH, temperature, solvent, surface and bio-molecule dependent – may need extensive optimisation
Entrapment: behind 　　　　membrane	Simple universal approach for macromolecules Very mild conditions Large excess of protein can be trapped Long working life	Difficult to mass-produce Diffusion barrier slows response time
in polymer gel	Mass production potential	Protein denaturation by free radicals
Covalent coupling	Stable coupling Intimate contact with transducer Low diffusion barrier – rapid response	Complexity and cost of derivatisation steps Limited sites for attachment leads to shorter lifetime
Covalent coupling to surface immobilised polymer	Larger number of coupling sites Increased signal size May give lower steric hindrance to binding	More complex preparation More complex kinetics and diffusion
Use of 'capture system'	Generic surfaces where specificity can be switched Many options for regeneration Opportunities for antibody orientation	Expensive and complex multi-step derivatisation procedures Multi-layer structure may reduce signal Non-specific binding to components of capture system

immunological techniques such as RIA and ELISA). Adsorption is often followed by a slow thermodynamically-driven unfolding of the protein structure, in order to increase the amount of hydrophobic polypeptide chain (usually from the protein interior) in contact with the surface (Wahlgren and Arnebrant 1991). This phenomenon leads to substantial denaturation and deactivation in many cases. Antibodies are rather resistant to this type of unfolding since they have a very rigid tertiary structure, which has evolved to stand the rigours of life in the circulatory system, so this is not an issue of such great concern for many immunosensor devices.

The second implication of a hydrophobic surface is, however, of universal concern. Strong adsorption means that *any* protein present in solution will tend to bind to the surface. For many types of transducer, this 'non-specific' binding (NSB) cannot be differentiated from the intended specific binding, and thus leads to high background signals. Even using multi-channel devices with a reference, this means that the result is interpreted by subtracting a large figure from an even larger one, which is intrinsically error prone.

It is not only hydrophobic surfaces that cause problems with NSB. Charged surfaces also bind proteins due to ion-pair interactions between immobilised charges and ionised surface groups on the protein. This phenomenon occurs on many types of surfaces, even those not obviously charged. For instance, glass has cation exchange capacity above about pH 4 due to ionisation of surface silanol groups. Ion-pair formation is not always a bad thing – if used carefully and deliberately it can be of great value, for instance in binding DNA to positively charged membranes, or for creating exclusion barriers to prevent undesirable analytes reaching a transducer surface (particularly with electrochemical devices).

The most favourable type of surface to prevent NSB and denaturation seems to be one most closely mimicking an aqueous solution, i.e. a flexible uncharged environment rich in hydroxy groups. These do not ionise under any physiological conditions and are also highly hydrated, thus minimising the possibility of hydrophobic effects.

Unfortunately, the hydroxy group is relatively difficult to activate for subsequent covalent attachment of proteins under aqueous conditions. In general, alternative functional groups are usually introduced that have more aqueous-compatible coupling chemistry. Although hydroxylated surfaces are quite effective in reducing NSB, increasing the mobility of the tethered groups reduces NSB even further and short chains of oligoethyleneglycol are even more effective in this respect (Yang and Yu 1997).

The best type of surface for covalent immobilisation of a protein is probably one containing a mixture of OH and oligo(ethylene glycol) and just enough of another type of functional group to achieve the required degree of surface derivatisation. This functional group must be displayed at the surface in a configuration that is accessible both to activating chemicals and the final coupling partner such as an antibody.

Both the physical and chemical nature of the transducer surface is clearly critical when considering the attachment of biological or chemical material with nm dimensions. Since a vast array of different types of transducers is used, manufactured by many different processes, it is impossible to consider every type. Some of the surface properties of typical materials used in transducers for immunosensors are discussed below, pages 53–57. The actual process of chemical attachment of antibodies or ligands is detailed later, on pages 64–73.

Modifying the surface properties of typical transducer materials

A very wide range of different materials is used to fabricate transducers used for biosensor devices. This makes it very difficult to discuss potential methods for attachment of biomolecules. On pages 53–58, some of the key characteristics of the most common materials likely to be encountered are summarised, together with methods to alter the surface chemistry of these materials to produce functional groups suitable for covalent immobilisation strategies.

Gold

Gold is the most popular noble metal for use in biosensors. It is used both as an electrode material in electrochemical and acoustic sensors, and as an integral part of most SPR sensor devices. Its popularity arises from its inert nature (it does not oxidise in air and hence is easy to handle) and its easy compatibility with many semiconductor manufacturing processes. It can be coated onto surfaces by evaporation or sputtering, although a thin chromium or titanium interlayer is required to get good adhesion to many types of surface (e.g. glass). Thin layers of gold also retain a degree of transparency. Freshly evaporated or cleaned gold is quite hydrophilic in nature, but it rapidly becomes hydrophobic on exposure to the atmosphere due to adsorption of various organic molecules.

One of the most important attributes of gold is its suitability as a substrate for forming self-assembled monolayers (SAMs). The phenomenon of self-assembly of thiols on gold was first reported in the early 1980s (Nuzzo and Allara 1983), and has subsequently developed into a major research area. SAMs can be formed on a number of different types of metals, including silver, platinum and copper. Gold is favoured, however, since it is relatively inert and thus does not suffer from surface oxidation. Silver, on the other hand, oxidises rapidly in ambient atmospheres. The most commonly used gold surfaces for SAMs are those obtained from low pressure thermal evaporation of gold, which produces a polycrystalline material with terraces of Au(111), although other materials such as sputtered gold surfaces or polished gold rods can also be used. The highest degree of SAM order is obtained on the Au(111) crystal orientation.

Thiols (R—SH), sulphides (R—S—R) and disulphides (R—S—S—R) all self assemble on gold. They are adsorbed directly from solution, usually in an alcohol or water (if the thiol has a very hydrophilic ω-functionality). The solution is typically a few mM, and self-assembly takes less than 1 hour (much less at higher concentrations) to form a dense layer. The gold surface is simply immersed in the solution and left for a period of time before being removed and thoroughly washed.

Thiols with long alkyl chains ($n > 10$) assemble into dense monolayers with a 2D crystalline order. The SAM undergoes maturation and re-organisation to produce this perfect crystalline monolayer, and this process sometimes takes several hours after the initial SAM deposition. Extensive characterisation has shown that the chains pack with a tilt angle of about 30° to the surface normal. Pinholes have been detected in the structure by both electrochemical measurements and AFM/STM, but coverage is usually around 99 per cent over quite large areas (Finklea 1996).

Thiols deprotonate upon adsorption to create strong gold thiolate bonds:

$$R—SH + Au \rightarrow R—S—Au + e^- + H^+ \tag{4.1}$$

These bonds are very strong and show little tendency to dissociate under normal conditions. The reactive coupling to the surface has some implications, since adsorbed molecules tend to remain in place and do not migrate around the surface. This means that mixed monolayers can be formed from chains of different lengths, or with different ω-functionality. Adsorption is random and statistical, hence the composition of the monolayer reflects the concentration of the different species in solution (Bain *et al.* 1989), although this may be influenced by the relative solubilities of the different thiols. There is little tendency to phase separate to produce

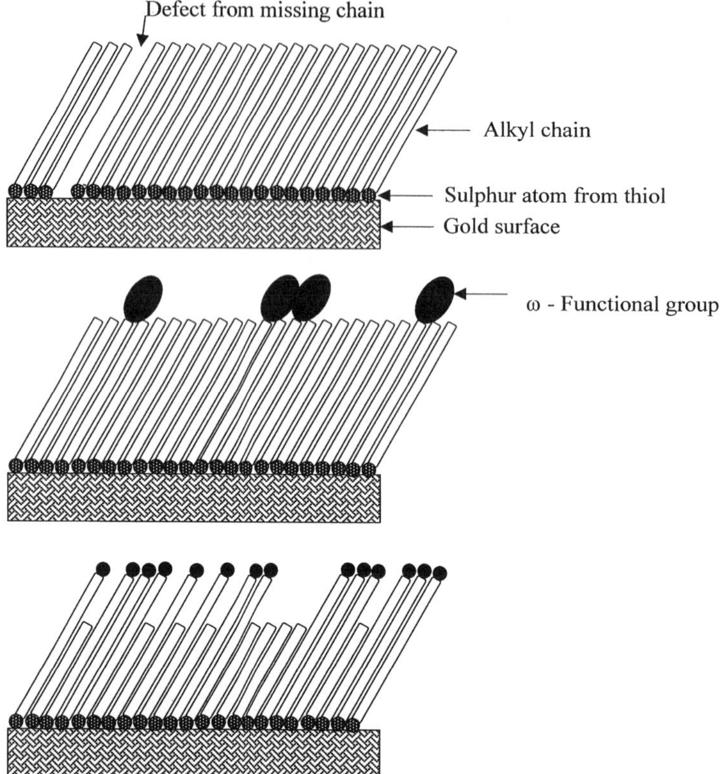

Figure 4.1 Self-assembled monolayers of thiols on gold surfaces. (Top) a typical close-packed alkanethiol monolayer with the chains tilted at about 30°. A defect caused by a missing chain is also indicated. (Middle) a mixed monolayer of alkanethiol and a thiol with a bulky ω-functional group (e.g. COOH). (Bottom) a mixed monolayer of a shorter chain alkanethiol and a longer chain thiol with a small ω-functional group (e.g. OH)

'islands' enriched in one species (a major problem with Langmuir Blodget films, where the greater dynamic mobility of molecules tends to favour phase separation). This behaviour is particularly helpful in producing 'mixed' monolayers where thiols with ω-functional groups are spaced out on the surface so that they can react without undue steric hindrance. The structure can be likened to a 'lawn' of one type of thiol with 'daisies' of another functional group standing out from the surface.

Amino and hydroxy head-groups are relatively small and do not significantly affect packing, but carboxylic acid groups are a little bulky and hence are probably best used as mixed monolayers with a slightly shorter spacer. Carboxylic acids also tend to self-associate by hydrogen bonding, and this may affect their incorporation into monolayers.

Many different thiols with different chain lengths and functionality are available. Due to widespread interest in SAM technology, the range is constantly increasing. There are also many reports of the synthesis of specialised thiol terminated compounds for specific applications, so it should be possible to produce a SAM with almost any terminal functionality required for an immobilisation protocol.

Glass, silica and metal oxides

Many glass, silica and metal oxides are characterised by the same general types of surface bonds. They usually contain a mixture of oxygen-bridged metal atoms and hydroxy groups. The latter make such surfaces intrinsically quite hydrophilic, and also provide an opportunity to modify the surface using organosilane reagents.

Although SAMs of thiols on gold are undoubtedly simpler to prepare and have greater uniformity than silane monolayers, the presence of gold is not always appropriate. Many sensor applications rely on absorbance or fluorescence measurements, and in this case transparency of the substrate is important. For such applications, glass or silica is often used, in the form of planar waveguides or fibres. Silicon dioxide is used for micro-fabricated structures such as field-effect transistors. Tin oxide is also quite a popular material since it is transparent, electrically conducting and compatible with semiconductor manufacturing processes.

Many different reaction conditions can be used to couple chloro- or alkoxysilanes to the hydroxy groups on metal oxide surfaces. Silane derivatives are available with different terminal functional groups such as amine, thiol, epoxy and chlorobenzyl. These terminal functional groups can subsequently be used to couple proteins or other receptors to the surface (see pages 65–66).

There is some disagreement in the literature as to which procedure produces the best silane monolayers. The most common procedures involve exposing the cleaned surface to the silane either in the vapour phase or in an inert solvent such as toluene (often under reflux), although aqueous methods are also possible with water-soluble silanes such as 3-aminopropyl-triethoysilane (Weetall 1976; Haller 1978; Kurth and Bein 1993). Silanisation involves a series of (trans)esterification and condensation reactions, leading to an additional layer on the surface carrying the functional side chain.

Artificial and natural polymers

The range of possible polymer materials for use in sensor devices is vast. It is beyond the scope of this chapter to consider individual materials and their properties. Polymers are likely to become increasingly important in sensors, however, due to their low cost, optical properties and the range of processing options such as

Reaction scheme 4.1 Reaction of a silane reagent with silica to produce a modified surface

injection moulding, extrusion and thermal micro-moulding. Using such techniques, complex shapes and surface structures (such as diffraction gratings) can be mass-produced to provide integrated devices with, for example, combined liquid handling and optical components. Developments in this field are moving towards micro-scale analytical systems, which may or may not contain sensing devices.

The properties of polymer surfaces range from very hydrophobic to very hydrophilic, and some contain ion-exchange groups. The surface behaviour may also be altered by additives such as plasticisers. The intrinsic properties of some polymers have been used with great effect. For instance, the strong adsorption of proteins to hydrophobic polystyrene surfaces was used in the development of radio-immunoassay and enzyme-linked immunoassay. Many other polymers also adsorb strongly, such as poly(L-lysine). The large size and additive interactions of such molecules leads to almost irreversible binding under relatively mild aqueous conditions. Adsorption thus provides a convenient route to introduce functional groups such as amines onto polystryene surfaces for subsequent covalent coupling. Similar approaches can also be used with many other polymers.

Other ways in which polymer surfaces can be altered include a range of oxidation methods. This might be achieved by wet chemical oxidation, flaming or treatment with a plasma, and generally makes polymer surfaces much more hydrophilic. New types of polymers can also be grafted chemically onto the surface of a different type of polymer, which allows the structural, processing or optical properties of one polymer to be combined with the surface properties of another (Bamforth and Al-Lamee 1994). This technology is of major importance for producing bio-compatible surfaces for medical implants, and could also have a widespread application in sensor development.

The most important natural polymers are the carbohydrates, such as cellulose, agarose and dextran. These materials are very hydrophilic, due to the dominance of the hydroxy groups, and generally provide a 'protein friendly' non-denaturing surface. This means that there is very little adsorption, which is good for minimising non-specific binding but also means that proteins need to be covalently coupled. As discussed on page 72, hydroxy groups are quite difficult to use for direct coupling, but a number of activation strategies have been developed for affinity chromatography. Popular approaches are the use of epichlorohydrin to introduce epoxides (which can be opened with ammonia to generate amines) and trichloro-s-triazine, which activates the surface ready for direct coupling of an amine-containing molecule. Several other methods are also available. Details of all these methods will be found in the affinity chromatography literature (Lowe and Dean 1974; Mosbach 1976; Scouten 1981; Dean *et al.* 1985).

Carbon

Carbon is a common electrode material for electrochemical sensors. It is used in a variety of forms, including rods, fibres, pastes and colloidal particles, which are used in many ink formulations. The latter are used in screen-printing processes to lay down conducting tracks for low-cost disposable electrochemical sensor devices. Very thin carbon fibres are often used to make micro-electrodes.

Carbon surfaces are generally very hydrophobic, and readily adsorb many species. This can be quite useful (see page 60) but it may also lead to many unwanted components binding on the surface leading to non-specific signals. A certain number of oxygen-containing functional groups, such as carboxylic acids, are also present. The surface is quite difficult to modify chemically, although limited surface oxidation is possible. It is generally better to adsorb an interlayer of some description to the surface. This could be a protein or a polymer containing a functional group of interest, or even a hydrophobic small molecule (this is often used to attach electrochemical mediators to carbon electrode surfaces).

Switching surface functionality

Although many different ω-functionalised thiols and silanes are available, not all functional groups are available with every type of surface derivatisation chemistry. For instance, silanes with ω-carboxy groups are not available. In order to change one functional group to another, a wide range of reagents is available. These react with one type of surface functionality and leave another. A selection of such reagents and the switches achieved by them is summarised in Table 4.2. This concentrates on surface functional groups easily achieved with silane chemistry, although the principles are clearly broadly applicable. The list is by no means exhaustive and many other alternatives with, for instance, different spacer lengths, are commercially available. A large number of derivatisation reagents have also been developed for protein chemistry, designed to target specific functional groups. Many of these are also suitable for derivatising one type of surface, and presenting a new reactive group that can be used for subsequent coupling. The chemistry used in these linkers is discussed on pages 64–73.

Surface-tethered polymers as immobilisation matrices: extending the surface into a third dimension

Surface-tethered polymers have been used for some time in 'tentacle' ion-exchange media for protein purification (Janzen *et al.* 1990). A polymer grafted to a surface extends out into the solution adjacent to the interface, effectively increasing the concentration of ion exchange groups near to the surface. For biosensors, the approach has been popularised by the highly successful carboxymethyl dextran surface introduced for the BIAcore SPR machine. This surface is created in four steps (Lofas and Johnsson 1990):

i   a self assembled monolayer of 16-mercaptohexadecanol is formed on the gold SPR surface;
ii   the hydroxylated surface is activated with epichlorohydrin to produce a surface containing reactive epoxide groups;
iii   dextran is coupled to the epoxide groups by nucleophilic attack under strongly basic conditions;
iv   the coupled dextran is carboxymethylated using bromoacetic acid, under strongly basic conditions yielding a substitution of about one carboxy group per glucose ring.

The carboxymethyl dextran surface produced has some very attractive properties. Dextran is known to have very low non-specific binding to proteins, hence the

Table 4.2 Summary of some simple reagents that can be used to transform one type of surface functional group into a different one

Surface functionality	Reagent	New surface functionality
R—NH$_2$		R'—COOH
R—NH$_2$		R'—CHO
R—(epoxide)	NH$_3$	R'—NH$_2$
R—(epoxide)	H$_2$N—CH$_2$—COOH	R'—COOH
R—SH	I—CH$_2$—COOH	R'—COOH
R—OH	Br—CH$_2$—COOH	R'—COOH
R—OH	Cl—(epoxide)	R'—(epoxide)
R—OH	(trichlorotriazine)	R'—(dichlorotriazine)

environment favours the specific interactions being studied. The carboxy groups serve three valuable functions. First, they provide a convenient group for activation and covalent coupling of proteins or ligands (see pages 64–69). Second, the negative charge of the unactivated carboxy groups causes electrostatic concentration of proteins (which are usually positively charged at a suitable coupling pH – generally between 3.5 and 7) in the vicinity of the dextran layer via ion exchange mechanisms. This places the protein in the correct region for covalent coupling and thus makes the process more efficient (important for scarce and valuable proteins that have to be used at low concentrations). Finally, the residual negative charge tends to keep the dextran matrix extended at the interface by charge repulsion mechanisms, thus maximising diffusion and protein–protein interactions.

The tethered dextran layer extends about 100–150 nm above the surface and thus provides a three-dimensional space in which protein/ligand interactions can occur. This has the effect of both increasing the number of immobilised molecules, thus increasing the size of the sensor signal (up to 50 times compared with the same component immobilised on a two-dimensional surface) and also changing the geometry of the system. The geometric effect may make it possible for molecules to interact in ways that would be sterically constrained on a two-dimensional surface. On the other hand, the presence of the charged dextran gel also introduces some complications and uncertainties into the system, which can make data interpretation more complex.

Although it is by far the best known example, CM-dextran is not the only surface-tethered polymer that has been described. A number of other systems have been developed and tested for similar applications, mostly with the aim of simplifying the preparation procedure or the subsequent covalent immobilisation step for the bio-molecule of interest.

A hydrophilic poly(vinylpyrollidone) copolymer containing NHS ester functionalities has been prepared and attached to cysteamine SAMs on gold surfaces through amide bond formation (Millot *et al.* 1995). The remaining NHS ester groups can then be used for protein immobilisation without the need for prior activation, as is required for CM-dextran coatings. It was demonstrated that antibodies could be covalently bonded in this way and used for SPR immunoassay. The polymer layer was calculated to be about 20 nm thick, and was shown to have low non-specific protein binding after deactivation of residual NHS groups with ethanolamine.

Dextran sulphate has been used as an alternative to CM-dextran in SPR studies (Toyama and Ikariyama 1997). In this case the dextran was oxidised with periodate to generate aldehydes, which were used to attach the polymer to a cysteamine SAM and subsequently to attach a protein of interest, both through Schiff base formation with amine groups. The Schiff bases were then stabilised by borohydride reduction before the devices were tested. This immobilisation method worked well, but has the disadvantage that the protein to be coupled has to be available at the time of surface preparation, since aldehydes are relatively labile in storage.

A terminally thiolated poly(vinyl alcohol) carrying poly(acrylic acid) grafts has also been prepared and tested as an immobilisation matrix for SPR-immunosensing (Disley *et al.* 1998). The terminal thiol group enabled facile direct polymer coupling to a silver surface. The carboxylic acids of the acrylic acid grafts were then activated with NHS/EDC chemistry for IgG immobilisation. The polymer was hydrophilic, had a high binding capacity for covalent coupling and showed low non-specific binding. This type of direct polymer immobilisation through polymer thiolation is much simpler than the multi-step procedure used for CM-dextran coating, and could be an attractive approach for commercial device fabrication.

Poly(L-lysine) has also been covalently bonded to a SAM of 11-mercaptoundecanoic acid on gold (Frey and Corn 1996). The poly-lysine was subsequently derivatised with an amino-reactive maleimide reagent to produce groups suitable for thiol coupling, although it could also be used for direct attachment via carbodiimide or dialdehyde chemistry. Poly-lysine has also been used extensively for adsorption to polystyrene in immunoassay methods, to provide a surface amenable to derivatisation (see page 56).

Adsorption to the transducer surface

Many molecules adsorb to a variety of surfaces, due to hydrophobic, ionic and Van der Waals interactions. Thus many proteins bind well to certain types of plastic, or to carbon electrodes. Many other molecules (particularly aromatic ones) also show this tendency. A number of problems associated with surface adsorption were highlighted on pages 50–52, but several useful and beneficial applications of adsorption were described on page 56, allowing surface properties to be changed. Surface adsorption is thus both a blessing and a curse in biosensor development, and the most important issue is to understand and control undesirable adsorption while utilising beneficial interactions to aid sensor design and construction.

Physical adsorption is the simplest approach to immobilisation of receptors, but it tends not to be very stable, and activity of enzymes is often lost in a time-dependent denaturation process, as the protein unfolds on the surface. For this reason, adsorption has to be carefully controlled, and the technique is generally only suitable for disposable, one-shot devices. This is the case in much antibody-based assay technology, such as RIA and ELISA, where extensive use of spontaneous protein adsorption to plastic (usually polystyrene) surfaces has been made. Depending on the assay configuration, either an antibody or a hapten-protein conjugate is adsorbed on the surface. Such surfaces are usually post-treated with a protein solution such as albumin or casein to block any remaining adsorption sites. This reduces background signals in subsequent assay steps.

Many of the approaches first developed for immunoassay have been extended to develop immunosensor devices. A typical recent example is the adsorption of a 2,4-dichlorophenoxyacetic acid-protein conjugate onto a screen-printed carbon paste electrode to produce a disposable, single-use, competitive immunosensor for herbicide measurement (Kroeger *et al.* 1998). A second recent example, representing a very different type of sensor, was based on diffraction from a micro-contact printed antibody grating produced by direct adsorption of antibodies on a gold surface (St John *et al.* 1998). Clearly, if used with thought and care, adsorption is a simple, inexpensive and highly effective method of protein immobilisation. Careful reading of the vast immunoassay literature provides many insights into the use and control of surface adsorption for immobilisation purposes.

Entrapment methods

Physical entrapment behind membranes

Physical entrapment is a method whereby a semi-permeable membrane, such as cellulose dialysis tubing, is used to retain a solution containing the receptor molecules in a compartment adjacent to the transducer. Small molecules can diffuse freely in and out of this compartment, but the high molecular weight component is held inside. This approach was used in the first demonstration of a biosensor device based on an oxygen electrode (Clark and Lyons 1962), and thus has an important place in the history of biosensor development. The technique is still quite widely used, and is particularly suitable for cells and organelles although it can also be used for enzymes and antibodies.

For large bio-molecules such as enzymes and antibodies, dialysis tubing is often used as the semi-permeable membrane. It typically has a molecular weight cut-off around 5000–10,000 Da for globular molecules and thus completely retains typical proteins while allowing relatively unhindered diffusion for small molecules. Many other types of membranes are also available, made from other materials and with different porosities, which might be more suitable for some applications: typical examples are polycarbonate and nylon.

Physical entrapment has some very attractive features. It provides a very simple and broadly applicable system for a wide range of macromolecules, and the conditions used are very mild, requiring no chemicals or procedures liable to cause denaturation or loss of viability. The biological components can be maintained hydrated in a suitable buffer system throughout. The amount of volume occupied is defined physically rather than chemically, so a large excess of biological component can be included in the device, if required, to maintain activity over a long working life. If the volume is too large, it will increase response time due to increased diffusion distances, and it may also affect sensitivity due to dilution effects, so a sensible compromise must be sought. The approach is, of course, limited to certain types of transducers. For instance, it would not be appropriate for systems using surface effects such as SPR, but it has been applied in a wide range of electrochemical and optical sensor designs. A key limitation is the difficulty of transferring this approach to mass production, due to problems of sealing wet membranes over devices; thus while it remains a useful approach for laboratory demonstrations, it is unlikely to find its way into many mass-produced products.

Entrapment in hydrogels

Enzymes and antibodies can be physically entrapped within the volume of a polymer hydrogel, rather than behind a membrane, while retaining substantial activity. This can be achieved in a variety of ways (O'Driscol 1976). The simplest approach is to use polymers that can be dissolved in warm or hot water but gel on cooling due to hydrogen-bond formation. One widely-used polymer is agarose, a polysaccharide obtained from marine algae. Low melting-point versions are available that gel around 40 °C, so the biological component does not need to be exposed to a high temperature. Agarose has quite high porosity so it is most suitable for whole micro-organisms or organelles, rather than isolated proteins, since these would leach out of the gel structure. An alternative is poly(vinyl alcohol). This synthetic, hydrophilic polymer comes in many molecular weights and degrees of hydrolysis (it is made from poly(vinyl acetate)). High molecular weight fully hydrolysed polymer has the useful property that it dissolves in boiling water, and once dissolved it stays in solution at room temperature, but it is insoluble in cold water. A solution can be made and cooled. Proteins can then be mixed with the polymer solution and deposited on surfaces, by printing, droplet deposition, etc. The film is then dried, entrapping the enzyme. When subsequently rehydrated, the film swells rapidly and the protein becomes active, but cannot diffuse out of the polymer. Other types of polymers have been used in a similar way.

The hydrogel can also be made *in situ* by free-radical polymerisation of suitable monomers and cross-linkers in the presence of the protein. The enzyme gets caught in a network of polymer chains from which it cannot escape. Typical monomer systems include acrylamide/methylenebisacrylamide and hydroxyethyacrylate/ ethylene glycol dimethacrylate. Some activity is usually lost during the polymerisation procedure, probably due to the free radicals used to initiate polymerisation, but the remaining activity is often much more stable in the gel than it would be in free solution. The polymerisation may be directly at the transducer surface, or blocks of hydrogel may be formed and subsequently physically placed on the transducer. When used for biological molecules, polymerisation is usually initiated photochemically rather than thermally, since this tends to cause less denaturation.

Covalent incorporation into hydrogels

In order to reduce the amount of protein that can leach out of an entrapped hydrogel, it may be desirable to covalently link the protein to the growing polymer chains. This probably happens naturally to a certain extent due to radical abstraction and chain transfer processes during polymerisation. In order to ensure covalent incorporation, however, it is necessary to derivatise the protein so that it contains polymerisable functional groups that can be covalently incorporated into the growing polymer chains. The most common way to do this is to form acrylamide groups on a protein surface by reacting the protein with the N-hydroxysuccinimide (NHS) or nitrophenyl ester of acrylic acid.

For enzymes, covalent attachment allows much less cross-linked polymers to be used, when compared with direct entrapment within a polymerised hydrogel without prior derivatisation. This has some advantages in terms of diffusion properties and accessibility to the immobilised species, especially if it involves recognition of macromolecules.

This approach can also be used for low-MW receptors or ligands (e.g. biotin for an avidin/biotin capture system), which could not be physically entrapped in a hydrogel.

An example of this is the use of methacrylate derivatives of crown ethers, which were covalently incorporated into cross-linked hydroxyethylmethacrylate films for use in the development of ion-selective holographic sensors (Mayes *et al.* 2000). A

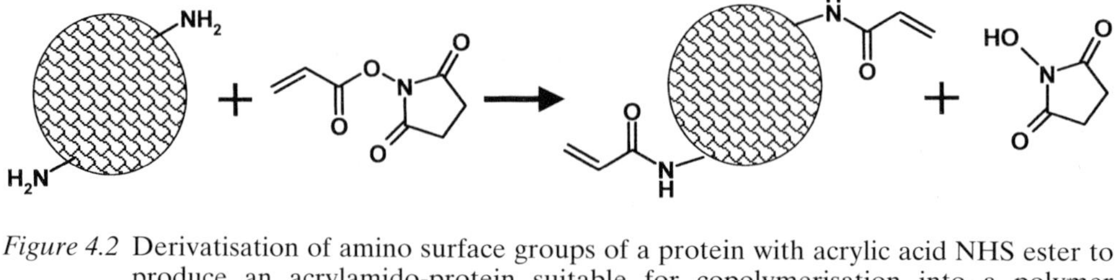

Figure 4.2 Derivatisation of amino surface groups of a protein with acrylic acid NHS ester to produce an acrylamido-protein suitable for copolymerisation into a polymer hydrogel

similar approach is taken with many other receptors for producing ISFET devices and ion-selective electrodes with long operational life (Reinhoudt *et al.* 1994; Heng and Hall 2000). Receptors are gradually lost from plasticised PVC membranes, leading to drift in the response and a limited life. Covalent coupling of the receptor to a self-plasticising acrylate membrane gives much greater long-term stability and a longer working life.

Entrapment within conducting polymers

For some types of electrochemical sensors, an electrically conducting polymer is used in the entrapment procedure. The conducting properties of the polymer may then be used as part of the transduction mechanism, relaying charge to the electrode surface. The most widely used approach is with pyrrole, which can be electropoly-merised under relatively mild aqueous conditions at a potential that does not damage biomolecules. This method has been used with a variety of enzymes and also with antibodies (Barisci *et al.* 1996). Applying a similar philosophy to that used to covalently bind proteins to hydrogels, proteins have also been derivatised with 3-carboxymethyl pyrrole so that they become covalently connected to the poly(pyrrole) structure during electrochemical polymerisation (Yon-Hin *et al.* 1993). Electropolymerisation is an attractive method for mass production, since it is compatible with microelectronic production methods. The immobilisation procedure can also be directed to specific areas to produce arrays (see pages 82–83).

Entrapment within silica sol-gel glasses

The sol-gel method creates porous silicate networks under mild aqueous-compatible conditions. They are made by acid or base hydrolysis of an alkoxide precursor such as tetramethylorthosilicate. A series of hydrolysis and condensation reactions proceed, leading eventually to formation of a silicate network. Once the hydrolysis/condensation reaction is underway, the solution can be buffered at moderate pH values allowing proteins to be introduced without denaturation. Proteins present in the solution during the condensation phase become entrapped in the final network (Dave *et al.* 1994).

The reagents start off as a homogeneous solution, which progressively becomes more viscous as colloidal sol particles form and finally gel together to form a continuous network. In the viscous state it can be spin-coated or dip-coated. The latter is less controllable but is useful for curved surfaces such as optical fibres. Thin films can thus be formed on a variety of different substrates. The best adhesion is achieved with glass, silica and metal oxides to which it covalently bonds. The final gel undergoes 'maturation', where further solvent loss and network condensation occurs.

Sol-gel glasses are optically transparent and hence are particularly useful for optical sensor devices. Biological molecules entrapped in sol-gels seem to be highly stable and retain their activity over long time periods. The disadvantage for immunosensing is that the network has rather small pores. These allow relatively free diffusion of small molecules, but not of higher molecular weight antigens. There are thus few reports of antibodies entrapped in sol-gel materials, although it has been successfully applied to a competitive fluorimetric assay of a low molecular weight hapten (Jordan *et al.* 1996).

Covalent coupling chemistry

In this section, a variety of methods will be outlined for covalently attaching one functional group to another. If one of the functional groups is attached to a surface (see previous section, page 58, for methods to achieve this), then the soluble protein or small molecule will be immobilised. Some of the procedures are two-step methods, where the surface is first activated and then reacted with the soluble partner: others are completed in a single step. Some of the procedures can be achieved by either approach, and there may be specific advantages to splitting the action into two stages. This is discussed in the sections relating to particular methods.

Almost all of the chemistry presented below has been developed from the fields of protein chemistry and affinity chromatography, and is dominated by the need to carry out the coupling reactions under mild conditions in water or buffer solutions. This rules out a great many reactions used commonly in organic synthesis, even though these are often simple, inexpensive and high-yielding. The reader should be aware of the origin and reason for the selected chemistry presented, however, since the immobilisation strategy can often be simplified if small molecules are being immobilised. If these are soluble in organic solvents there is no reason to use aqueous conditions, and a much wider range of options becomes available, as do much cheaper reagents.

In Table 4.3, a wide variety of possible coupling reactions is summarised, covering the majority of functional groups that are likely to be encountered. From this the reader can select a reaction type suitable for their needs and then refer to the appropriate section for a more detailed discussion of the reaction. In most cases, the 'R and R' groups are interchangeable, i.e. it does not matter which one is on the surface and which one is in solution. Cases where it makes a difference are discussed in more detail in the text.

Coupling to carboxylic acids

Carboxylic acids on surfaces can be readily produced via SAMs, by oxidation, through surface hydrolysis of polymers and via polymer grafting; hence these are some of the most frequently used coupling procedures. In most cases the soluble molecule has an available amine group (preferably primary since these are the most nucleophilic), but other nucleophilic species can also be coupled such as phenols and thiols. If more than one functional group is present, then a variety of coupling reactions will occur, with the relative proportions of each dictated by the relative reactivity, accessibility and abundance of the available nucleophiles. If a typical protein such as an antibody is used, the majority of linkages will be made to surface amine groups (mostly ϵ-amino groups of lysine), even if a few surface tyrosine, cysteine or histidine residues are present since the latter are usually much less abundant. For small molecules, a wide range of nucleophiles can be used for displacement, including phenols, anilines, thiols and even alcohols (under non-aqueous conditions), although the stability of the resulting bonds is variable (e.g. phenyl esters are relatively base-labile).

Table 4.3 Summary of different functional groups that can be coupled together, and the structure of the resulting bonds. Each of the reactions in the table is discussed in the text in the appropriate section

Functional group 1	Functional group 2	Bond formed
carboxylic acid $R-C(=O)OH$	H_2N-R'	$R-C(=O)-NH-R'$ amide
	$HS-R'$	$R-C(=O)-S-R'$ thioester
	$HO-R'$	$R-C(=O)-O-R'$ acyl ester
	$HO-\text{C}_6\text{H}_4-R'$	$R-C(=O)-O-\text{C}_6\text{H}_4-R'$ aryl ester
aldehyde $R-C(=O)H$	H_2N-R'	$R-CH=N-R'$ imine
		$R-CH_2-NH-R'$ secondary amine (after reduction)
	$H_2N-NH-R'$	$R-CH=N-NH-R'$ hydrazone
	$HS-R'$	$R-S-S-R'$ disulphide

continued

Table 4.3 continued

Functional group 1	Functional group 2	Bond formed
thiol **R—SH**	*(maleimide structure)*	*(thioether adduct)* thioether
	H₂N—R' (via tosylate)	*(secondary amine structure)* secondary amine
	(via triflate)	*(amido sulphate ester structure)* amido sulphate ester
alcohol **R—OH**	(with epichlorohydrin activation)	*(secondary amine structure)* secondary amine
	(with triazine activation)	*(substituted triazine structure)* substituted triazine
amine **R—NH₂**	*(isothiocyanate structure)* S=C=N—R'	*(thiourea structure)* thiourea
	(sulphonyl chloride structure)	*(sulphonamide structure)* sulphonamide

Use of carbodiimide reagents

The most used immobilisation reaction involving carboxylic acids is probably amide bond formation using carbodiimide reagents. This reaction has been studied widely due to its importance in organic chemistry in general and in solid-phase peptide synthesis in particular (Sheehan and Hess 1955). The reaction involves two stages; activation by the carbodiimide reagent to form an O-acyl isourea intermediate and nucleophilic displacement of the intermediate by the amine to form the final amide bond. In organic solvents the carboxylic acid is simply mixed with the amine in the presence of a mole equivalent amount of a suitable carbodiimide. If dicyclohexyl carbodiimide is used the resulting dicyclohexyl urea precipitates out, which is helpful for work-up of solution reactions. If surfaces are being used, then it is better to use diisoproyl carbodiimide to avoid surface fouling with precipitate.

Under aqueous conditions, things are a little more complex. Water soluble carbodiimides must be used, the most common of which is 1-ethyl-3-(3-dimethlyamino-propyl)carbodiimide (EDC). The mechanism of the activation of the carboxylic acid by the carbodiimide is acid catalysed, and hence proceeds most rapidly at low pH. The subsequent displacement of the complex by the amine nucleophile is favoured at high pH, however, where the amine is deprotonated. If the reaction is done in a single step, a compromise must be reached and a pH in the slightly acidic range (5–6.5) generally gives the best overall rates. At higher pH the hydrolysis of the carbodiimide complex by hydroxide ion to regenerate the acid becomes a significant competitor and the efficiency of coupling drops off. The hydrolysis reaction is always present, however, so even under mildly acidic conditions a large excess of carbodiimide reagent should be used.

One way to increase the efficiency of the reaction is to use a 'sacrificial nucleophile'. This can be present at high concentration, and hence reacts with the carbodiimide complex as soon as it forms, minimising the competing hydrolysis reaction. A similar approach is often applied in peptide synthesis, although in this case the primary objective is to reduce racemisation. The two most common reagents for this are N-hydroxy succinimide (NHS) and nitrophenol. These form 'active esters' *in situ* which have much longer solution half-lives (hours at neutral pH) than the o-acyl isourea formed from the carbodiimide. The 'active esters' can then be displaced by the amine nucleophile (see below, page 68). NHS is preferred since it is less toxic, but nitrophenol has the advantage that when it is subsequently displaced it generates a bright yellow species (the nitrophenolate anion) that is readily visualised, and provides an indicator of coupling if it is used in a two-stage procedure (see below, page 68). The EDC/NHS procedure has been popularised and extensively optimised by Biacore, since it is the most common method of coupling proteins to carboxymethyl-dextran SPR chips. Typical conditions would use about 0.1 M EDC and NHS in a mild buffer at around pH 5–6. The nature of the buffer is important; it must not contain any carboxylic acid or nucleophile groups that might interfere with the reaction.

Reaction scheme 4.2 Formation of an amide bond by activation of a carboxylic acid with a carbodiimide reagent followed by nucleophilic displacement with the amine

Displacement of active esters

Esters can be synthesised using a wide range of different methods, but only a few are important for immobilisation chemistry. The use of *in situ* generation of active esters has already been discussed, but the same active esters can be used in a two-step approach. This has some advantages. First, compared with the one-pot carbodiimide method, protein cross-linking is eliminated. If all the reagents are present together, a degree of cross-linking and aggregation is unavoidable since proteins contain both amine and carboxylic acid functional groups on their surfaces and hence condense together. Secondly, the conditions can be optimised for each of the two steps independently. Since the ester formation is separated from the subsequent coupling step, non-aqueous conditions can be adopted for the former, giving a wider choice of methods. For instance, the ester formation step can be done using a carbodiimide reagent (see pages 66–67) or by going via the acid chloride. In this case, the carboxylic acid might be converted to the acid chloride using thionyl chloride, phosphorus pentachloride or oxalyl chloride in an aprotic solvent, and then reacted with a solution of NHS or nitrophenol in a suitable organic solvent to form the active ester. The active ester can be displaced by an amine to form the final amide.

For low molecular weight compounds, this is a convenient route to immobilising many types of molecules containing amines, phenols, anilines, thiols and alcohols. Note, however, that DMSO cannot be used as a solvent for ester formation via acid chlorides, since it leads to oxidation of the alcohol to an aldehyde or ketone and reduction of DMSO rather than the desired esterification reaction.

Once formed the active ester species is relatively stable and can be washed, dried and stored if required, prior to coupling of the protein in an aqueous non-nucleophilic buffer at a pH of about 8 (bicarbonate or borate buffers are often used although TRIS can also be used in this case since it is sterically hindered and hence non-reactive).

In some circumstances, formation of the active ester under aqueous conditions is required, as in the case of the CM-dextran used in Biacore SPR chips. In this case the aqueous conditions are required to swell and solvate the dextran polymer, making the carboxyl groups accessible to the derivatisation reagents. In general, not

Reaction scheme 4.3 Two-step amide bond formation. A carboxylic acid is first converted to an 'active ester' via an acid chloride, then reacted with an amine to form the amide

all of the carboxy groups are activated. This leaves a degree of negative charge on the polymer, which helps to keep the polymer chains extended by charge repulsion and also ionically attracts protein into the region where it is needed for reaction. In order to make use of this effect the protein must be net positively charged and the CM-dextran net negatively charged at the pH used for the coupling step, hence a lower pH is often used than might be thought optimal for nucleophilic displacement by an amine (often around pH 5–6).

Mixed anhydride formation

A final method for generating amides is via mixed anhydrides, most often with carbonic acids. This is another procedure that has been widely adopted in peptide synthesis, but non-aqueous conditions are required for the first step, which involves reaction of the carboxylic acid with a reagent such as ethyl chloroformate to generate the mixed anhydride. Once formed this is relatively unstable and must be used immediately for displacement by an amine to form an amide. The fact that one of the carbonyl groups in the mixed anhydride is flanked by two oxygen atoms diminishes its reactivity, directing nucleophilic attack to the original carboxy group.

Reaction scheme 4.4 Amide bond formation via a mixed anhydride intermediate, formed by reaction of the carboxylic acid with ethyl chloroformate

Reactions involving aldehydes

Aldehydes are, in general, not especially stable, but this group of coupling reactions is important because of the widespread use of dialdehydes such as glutaraldehyde as coupling agents, and because aldehydes can be generated easily by oxidative cleavage of vicinal diols by periodate:

This reaction is often used to produce aldehydes for coupling from the carbohydrate moities of glycoproteins. Due to the positioning of the carbohydrate chains on antibodies, this provides a valuable route to achieve both immobilisation and a degree of orientation (see page 74).

Aldehydes react with amines to form Schiff bases (imines). This is an equilibrium reaction and has an equilibrium constant quite close to 1 under typical reaction conditions (aqueous buffer, pH about 5–6), hence the reaction cannot proceed to completion.

Due to this equilibrium the imines are also inherently unstable, so they are often reduced to secondary amines to overcome this problem. Reduction can be done after

$$R\text{--}C(OH)(H)\text{--}C(OH)(H)\text{--}R' + HIO_4 \longrightarrow R\text{--}CHO + OHC\text{--}R' + HIO_3 + H_2O$$

Reaction scheme 4.5 Reaction of a vicinal diol (e.g. in a sugar) with periodic acid to generate aldehyde groups

imine formation using sodium borohydride, but is possibly best done *in situ* using sodium cyanoborohydride. This is a very useful reducing reagent, since it will selectively reduce imines in the presence of aldehydes, and hence can be used both to stabilise the coupling bonds and to force the equilibrium towards completion at the same time.

$$R\text{--}CHO + H_2N\text{--}R' \rightleftharpoons R\text{--}CH\text{=}N\text{--}R'$$

Reaction scheme 4.6 Reaction of an aldehyde with an amine to form a Schiff base (imine)

$$R\text{--}CHO + H_2N\text{--}R' \rightleftharpoons R\text{--}CH\text{=}N\text{--}R' \xrightarrow{NaBH_3CN} R\text{--}CH_2\text{--}NH\text{--}R'$$

Reaction scheme 4.7 Formation of an imine followed by *in situ* reduction to a secondary amine using cyanoborohydride

As well as amines, hydrazides can also be coupled to aldehydes in a similar way. In general, both alkyl and acyl hydrazides are somewhat more reactive than amines and the coupled products are more stable than imines. Like imines, however, they can also be reduced after formation to further increase stability.

Reactions involving thiols

Like aldehydes, free thiol groups are not very stable, being easily oxidised to a variety of species including disulphides, sulphones and sulphonates. Oxidations are catalysed by some transition metal ions, so it is generally best to carry out thiol coupling reactions under inert atmospheres using buffers containing chelators such as EDTA where this is practical. Thiol groups are important in antibody coupling, due to the possibility of producing Fab′ fragments with free thiols (see page 75). Thiol groups can also be introduced into other proteins, or at other sites via reaction of surface amines with Trauts reagent (2-iminothiolane). In general, however, this is unnecessary for immobilisation purposes unless direct attachment to gold or reversible covalent immobilisation is required (see below, page 71), since the amine groups themselves can be equally well utilised.

Thiols are generally coupled either by disulphide formation or by creating thioethers. Disulphides are relatively easily formed and have the unique property that, while stable under normal conditions, they can be readily cleaved and reformed via thiol–disulphide exchange. This offers the possibility of replacing the biological component on a transducer surface by mild chemical stripping followed by reactivation and recoupling.

The first stage of immobilisation via disulphide bond formation involves synthesis of a disulphide at the surface, by reacting a surface thiol group with a symmetrical disulphide such as 2,2′-dithiodinitrobezene or 2,2′-dithiodipyridine under mildly basic aqueous conditions. Thiol disulphide exchange takes place to form a surface-bound disulphide. The driving force is the stability of the pyridine thiolate or nitrobenzenethiolate anion formed during the exchange.

Once formed, the surface-bound asymmetric disulphide can undergo a second exchange reaction with a thiol group on the protein or ligand to be immobilised. The driving force is again the stability of the displaced anion. A useful feature of this reaction is that the anions formed have quite strong absorbance and can be used to quantitate the amount of thiol ligand coupled.

To remove the protein, the surface is treated with a thiol containing reducing agent such as dithiothreitol (DTT). This removes the protein, leaving free thiol

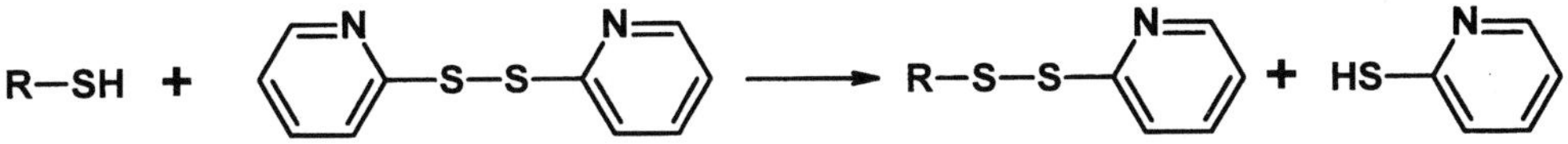

Reaction scheme 4.8 Thiol-disulphide exchange between a surface-bound thiol group and 2,2′-dithodipyridine

Reaction scheme 4.9 Thiol-disulphine exchange between a surface disulphide and a ligand with a free thiol groups, leading to ligand immobilisation

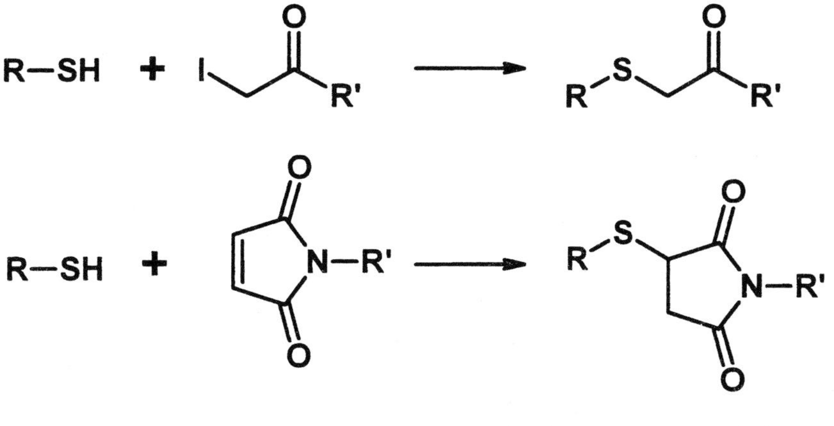

Reaction scheme 4.10 Coupling to thiols via thioether formation with (top) and acyl iodide and (bottom) a maleimide

groups on the surface again, so that the cycle can be repeated. DTT is the reagent of choice for this step since it 'excises' itself from the surface by an internal thiol-disulphide exchange leading to a cyclic disulphide and a free surface-bound thiol.

Thioether formation is usually achieved either by Michael addition to the reactive double bond of a maleimide or by displacement of an alkyl or (more often) acyl iodide. Both reactions proceed quite rapidly at room temperature under mildly alkaline aqueous conditions.

Reactions involving hydroxy groups

Alcohol groups are common functional groups in typical carbohydrates, and as such they often appear in the form of membranes (e.g. cellulose) or hydrogels (e.g. dextran, agarose) in biosensor devices. Due to the chemical similarity between alcohols and water, alcohols cannot generally be activated and coupled under mild aqueous conditions. If it is possible to use a two-step procedure, however, alcohols may be activated and coupled with nucleophile-containing ligands and proteins in a procedure first introduced by Nilsson and Mosbach for activation of carbohydrate-based chromatography matrices.

The first step of alcohol activation involves treatment with tosyl chloride or tresyl chloride in dry pyridine.

These activated products can then be reacted with amines or thiols under mildly basic aqueous conditions. With the tosylated surface, elimination of toluenesulphonic acid occurs yielding secondary amines and thioethers respectively.

With the tresylated surface, however, the situation appears to be more complex under these conditions and the reaction proceeds via a different mechanism. The

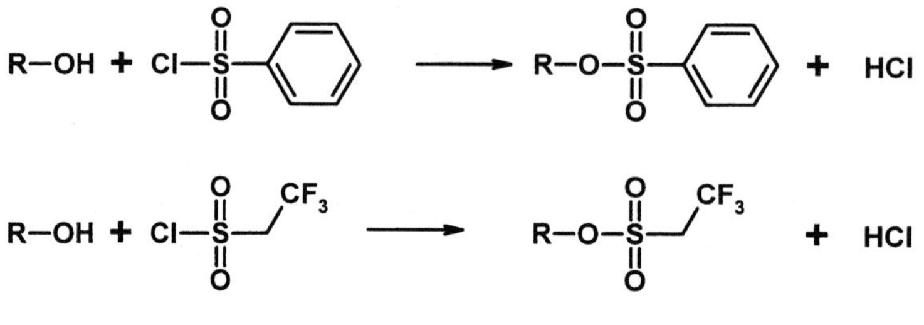

Reaction scheme 4.11 Formation of (top) a tosylate and (bottom) a tresylate by reaction of the sulphonyl chloride with an alcohol

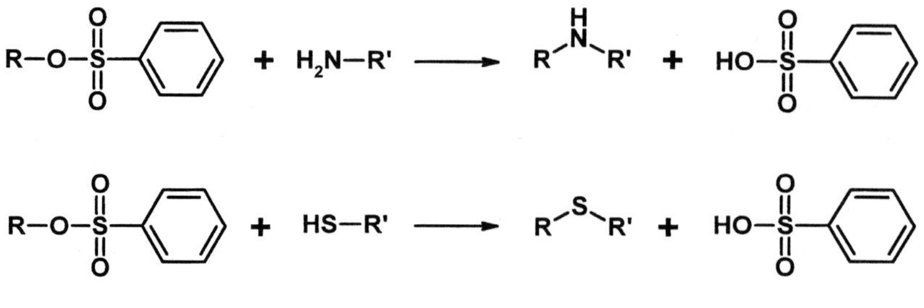

Reaction scheme 4.12 Displacement of a tosylate with (top) a primary amine to generate a secondary amine and (bottom) a thiol to generate a thioether

Reaction scheme 4.13 Reaction of tresylates with (top) amines and (bottom) thiols to produce stable sulphuric esters

final coupled products appear to be N-alkylamidosulphuric and S-alkylthiosulphuric esters respectively (Demiroglou *et al.* 1994), but these are also stable in use so in practical terms it is of little consequence unless it has a deleterious effect on non-specific binding.

Many other procedures have been developed for activating carbohydrate surfaces and introducing new functional groups that can be used in alternative coupling procedures. Some of these were discussed above in relation to control and modification of the transducer surface (page 58). Others will be found in the affinity chromatography literature.

Reactions involving amines

Coupling of proteins and ligands to amine surfaces has already been covered extensively in the sections on carboxylic acids and aldehydes. The reactive groups are reversed in this case, but the same chemistry and conditions can be used. For coupling of proteins it is generally better to have a carboxylic acid surface, since the activation and coupling can then be separated to avoid aggregation. With an amine surface this cannot be achieved. For some small ligands, however, coupling to amine surfaces is very convenient. For instance, the carboxylic acid of biotin can be directly coupled using NHS/EDC chemistry, avoiding the need for expensive biotin derivatives.

Two other reactions that are available for coupling to amines are the reactions with isothiocyanates and sulphonyl chlorides. Isothiocyanates rearrange to form thioureas and sulphonyl chlorides form stable sulphonamides.

These two reactions are used extensively for quantitating amines by formation of fluorescent derivatives, and they can be used in a similar way on surfaces to confirm that amine-containing surfaces have been produced. Isothiocyanates were widely used for fluorescently labelling antibodies historically, but they have been largely superseded by other reagents such as NHS-esters of fluorophores. Sulphonly chlorides such as dansyl chloride are still used to quantitate amines in various applications. As far as protein coupling is concerned, however, they have little use.

Reaction scheme 4.14 Reaction of an amine with (top) an isothiocyanate to form a thiourea and (bottom) a sulphonyl chloride to form a sulphonamide

Techniques giving some control over the orientation of the immobilised antibody

Covalent coupling through attached glycosides

Natural antibodies produced in eukaryotic systems are almost always glycosylated to a greater or lesser extent (typically 3–4 per cent for IgG). The exact nature and degree of glycosylation depends on the species and class of antibody. The sugar units are attached to the Fc part of the antibody structure, well away from the antigen binding sites, and provide a simple opportunity to target this region by using a well-established two-step covalent coupling procedure. This approach leads to antibodies coupled such that their antigen binding sites are less sterically hindered than when using random coupling procedures, giving a binding capacity closer to the theoretical maximum for the amount of antibody coupled.

In the first step, a controlled oxidation is carried out using sodium periodate. This cleaves vicinal diols in the carbohydrate structure and generates aldehydes. The side chains of terminal sialic acid residues are particularly vulnerable to attack. The aldehydes can then be used to couple the antibody to an amine derivatised surface via Schiff base (imine) formation. The Schiff bases are relatively labile, but rearrange under mildly acidic conditions to form more stable ketoamines via Amadori rearrangement. A more rapid and better-defined method for stabilisation is to reduce the Schiff bases to secondary amines using a post-treatment with sodium borohydride. Alternatively, sodium cyanoborohydride can be used *in situ* to reduce the Schiff bases as they form, without reducing the aldehydes. This procedure is both mild and efficient, since it drives the equilibrium towards completion and leads to a stable coupled product in a single step (see pages 69–70).

Generation of specifically-located thiol groups

The heavy chains of IgG molecules are linked together in the hinge region via one or more disulphide bonds. These can be reduced using simple thiols such as 2-mercaptoethanol or dithiothreitol to generate antibody fragments containing specifically located thiol groups (Johnston and Thorpe 1987a). These thiols can subsequently be used to couple the fragments to a suitably activated surface. In this way the coupling is directed away from the antigen-combining site and orientation is achieved. Coupling is usually achieved through maleimide or thiol-disulphide chemistry, or by direct adsorption to gold surfaces.

Another common procedure is first to digest the antibody with pepsin. The antibody is cleaved just below the hinge region to generate a F(ab)$_2$ fragment. The F(ab)$_2$ fragment can be isolated and subsequently reduced as above to produce Fab' fragments with thiol groups available for coupling (Johnston and Thorpe 1987b). This approach removes the Fc part of the molecule, thus removing both unnecessary bulk and opportunities for non-specific binding in the final device. The maximum surface density of binding sites is thus increased, as is the accessibility (Spitznagel and Clark 1993). The drawback is the relatively demanding and time-consuming protein chemistry required. Different antibody classes vary in the ease with which they can be fragmented and reduced, and yields are very variable, hence this approach is probably only viable for very demanding and high-value applications.

A more recent approach to introducing thiol groups is the use of enzymatic C-terminal cysteinylation using carboxypeptidase-Y-catalysed transpeptidation (You *et al.* 1995). This is a mild procedure, and has the advantage that it selectively introduces thiol groups at the exposed C-terminus of the heavy chain, near the end of the Fc region of the antibody. This is an ideal position to achieve oriented coupling, and it has been shown by STM studies that the resulting immobilised antibodies do indeed stand up from the surface with their antigen-combining sites pointing out into solution.

Using genetic engineering approaches, proteins can be produced with amino-acid alterations at specific sites on the protein. Thiol groups can thus be introduced at appropriate sites on the protein surface for subsequent use in coupling by changing codons to express a cysteine residue at the required position. This approach was successfully used to introduce tethered enzyme cofactors onto glucose dehydrogenase (Persson *et al.* 1991). In principle, it should also be possible for any antibody fragment produced by protein engineering. Care must be taken, however, to ensure that the introduced thiol does not disrupt correct protein folding pathways and appropriate disulphide bond formation. It should also be noted that thiol groups introduced onto protein surfaces in this way are often very susceptible to oxidation, and the proteins must be handled with appropriate care.

Use of antibody-binding proteins

Perhaps the most obvious class of proteins capable of binding to antibodies are antibodies themselves. Species and class selective antibodies, such as goat anti-human IgG, could be used to capture the antibody of interest. This type of approach is used extensively in ELISA technology. Monoclonal capture antibodies could be selected that bind to a region remote from the antibody combining site. In practice, this method is rarely used in immunosensor development, probably because the capture antibodies are bulky and have themselves to be immobilised, hence it is better simply to immobilise the antibody of interest directly.

There are a number of proteins, isolated from bacteria, that have the ability to bind to the Fc region of immunoglobulins. The best known examples are Protein A from *Staphylococcus aureus* and Protein G from *Streptococcus* strain G148, although there are some others. Protein A and Protein G are complementary because they have different species and sub-class specificities (Breitling and Duebel 1999). These proteins have been used extensively for antibody purification, and engineered versions have been produced with modified properties to improve selectivity, specificity and column capacity.

When immobilised, Protein A and Protein G can be used as capture systems. Since they bind the Fc region of antibodies, the captured antibodies have the correct orientation to maximise access to the binding sites. Unfortunately, if the capture proteins are immobilised randomly, they may not all be able to bind antibody, hence, although the captured antibody will function efficiently, it may be present at a lower density than if it had been immobilised directly, thus negating some of the advantage.

The most important property of using Protein A or Protein G capture systems is not so much the orientation effect, but rather the ability to remove the active antibody layer and replace it with a different one under relatively mild conditions. These surfaces thus act as generic immunosensor platforms, the selectivity of which can be changed at will by substituting different antibodies. The sensor surface can also be regenerated by desorbing the antibody from the binding protein rather than trying to break the Ab/Ag interaction. This is often much easier to achieve (usually by treatment with mild acidic buffers such as 20 mM citrate pH 3.5), and leads to more reproducible surfaces, since the antibody is replaced each time without the use of aggressive conditions that could cause surface or bio-molecule degradation.

Avidin/streptavidin capture systems

The use of the interaction between biotin and avidin or streptavidin is widespread in assay and labelling methodology, because of the extremely high affinity of the avidin/biotin interaction, which is essentially non-reversible under normal assay conditions. The various uses of avidin/biotin technology have been reviewed extensively in the literature (Wilchek and Bayer 1988) and will not be discussed at length here. Streptavidin and avidin are generally interchangeable, though streptavidin is usually preferred (despite its higher cost) due to its lower non-specific binding characteristics.

A variety of methods for immobilising avidin have been used including general covalent strategies (see pages 64–73) and the capture of the avidin using covalently-immobilised biotin. Perhaps more interesting, however, is the ability of avidin to form two-dimensional crystalline arrays on suitable surfaces (Mueller *et al.* 1993; Spinke *et al.* 1993). Avidin is ideal for packing in this way, due to its 'block-shaped' protein structure, with four biotin-binding sites. The binding sites are distributed with two on each of the near-parallel faces of the molecule. If an appropriate surface is prepared with immobilised biotin molecules, avidin can bind to this surface to form a tightly-anchored array, with a high site density for biotin binding (Morgan and Taylor 1992). This should be an ideal surface for capture of biotinylated antibodies, using mild conditions that cause little denaturation.

The use of an avidin/biotin capture system relies, of course, on the biotinylation of antibodies. This can be done in a variety of different ways. Simple use of EDC/NHS chemistry or direct use of biotin-NHS ester will couple biotin molecules at random onto the antibody, mostly onto ϵ-amino groups of surface lysine residues. All the same arguments about orientation that have been discussed in the direct covalent coupling section will thus apply. A better alternative would be to oxidise the carbohydrate moities and couple biotin at these sites through hydrazide chemistry or reductive amination. This will lead to more selectively located biotin molecules and will lead to capture by avidin in a more controlled orientation.

Use of tags with engineered antibody fragments

As the technology for producing engineered antibodies and antibody-fragments in bacteria, plants and other organisms improves, the opportunity to alter some parts of the antibody structure to aid in purification or subsequent use arises. While such modifications are detrimental for clinical use of the antibodies, they are quite acceptable for *in vitro* diagnostic purposes and provide a number of useful handles that can be exploited for immobilisation purposes. Most of these were originally developed to aid recombinant protein purification by affinity techniques, but the same approaches can be used for antibody immobilisation.

The use of a heterologous fusion component to produce a bifunctional molecule containing the antibody functionality and a binding-protein functionality has been used successfully for protein purification. Binding proteins such as the maltose-binding protein (Bregegere *et al.* 1994), which binds to amylose and starch, the cellulose-binding protein (Shpigel *et al.* 2000) and the functional core of streptavidin (Duebel *et al.* 1995) have been used for this purpose. The latter system is very attractive since it allows purification of the construct under mild conditions using relatively low-affinity imminobiotin, and subsequent high-affinity immobilisation on any biotinylated surface. The cellulose-binding protein is useful to attach the conjugate protein to cellulose fibres or filters directly without the need for any pre-activation or coupling chemistry. A Protein A/cellulose-binding protein conjugate has been demonstrated as a useful tool for capturing antibodies and linking them to cellulose filters for immunological applications (Ramirez *et al.* 1993). A similar approach should be possible with engineered antibody fragments, and might be appropriate for high-volume production of low-cost disposable systems.

The main drawback of the fusion-protein approach is the bulk of the second protein domain in the final structure. This can be overcome by the use of short peptide tags instead of whole functional protein domains. By far the most widely used system is the oligo(histidine) tag (his-tag), although many others have also been developed. The his-tag is a sequence of 3–6 histidine residues attached to one terminus of the protein. For antibody Fv fragments it is attached to the C-terminus, since this is furthest from the combining site. The his-tag sequence has the ability to bind to transition metals such as Ni, Cu and Zn, and can bind to partially chelated metal ions in the process of immobilised metal chelate chromatography (IMAC) (Sulkowski 1985). This has become the method of choice for purifying engineered scFv fragments (Molloy *et al.* 1995). In IMAC, a metal chelator such as iminodiacetic acid or nitrilotriacetic acid is immobilised on a support resin and loaded with a suitable transition metal. This complex then interacts with the his-tag, binding the protein. The protein can be eluted with a competing metal chelator (e.g. EDTA) or ligand (e.g. free imidazole). Due to the stability of the interaction, quite a high concentration of competing compound is required. The complexes are also stable at high ionic strength and in the presence of denaturants.

Using the same methodology, immobilised chelators can be attached to activated surfaces and used to capture his-tagged proteins for use in sensor applications (Sigal *et al.* 1996). The advantage of the metal–chelate interaction as an immobilisation method is that the affinity can be regulated. For instance, the number of his residues in the tag can be increased to take advantage of avidity effects caused by multiple binding interactions. The transition metal in the immobilised chelator can also be switched to increase or decrease the affinity. The captured proteins are bound with good stability under normal assay conditions, but (as in the case of the Protein A system) can be removed under quite mild conditions when required in order to regenerate the surface for reuse.

Spatial control of surface immobilisation

There is currently a great deal of interest in constructing sensor arrays, driven by the desire to analyse samples for a variety of different components simultaneously. Application areas include medicine (e.g. checking hormone levels), environmental monitoring (e.g. monitoring pesticides in ground-water), customs and excise (e.g. detecting drugs of abuse), food science (e.g. measuring vitamins or banned growth promoters) and military uses (e.g. detecting biological warfare agents). In some cases, arrays are constructed from a number of separate components, which are subsequently combined together. In other cases, it is more appropriate to make a single device with different elements of the array defined by their geometric position. Such systems would, for instance, be appropriate for subsequent analysis by fluorescence or SPR microscopy techniques. In order to create such an array, spatial control of antibody immobilisation is essential. A number of different approaches are available, including spatial control of the antibody deposition itself and a number of photochemical methods, including direct photo-immobilisation and the use of photo-deprotection strategies. The substrate can also be patterned in order to define the areas of deposition. These techniques are discussed in the following sections, pages 78–83.

Patterning the substrate to introduce appropriate surface functionality

The use of positive and negative photoresist procedures has been developed into a highly reproducible and high-resolution (sub-micrometre) technique by the semiconductor industry (Reiser 1989). Such approaches can also be used to protect certain areas of a substrate while others are derivatised by an appropriate surface-modification procedure. This method can work well if only a few different types of surface functionality are required for subsequent attachment of biological molecules. Hydrophilic/hydrophobic patterns have been produced in this way for directing the growth of neuronal cell cultures at surfaces (Clark *et al.* 1993).

The problem with photoresist methodology is that the current types of photoresist available are not compatible with aqueous conditions. The organic solvents, high temperatures and high radiation intensities used would denature proteins if they were attached to the surface, hence the patterning must be completed before the bio-molecules are introduced. Sequential procedures, where one bio-molecule is attached at one position and then protected with resist while a different area is exposed and treated, are not possible. This means that the same chemistry for surface derivatisation cannot be used for different areas, unless it is combined with spatial deposition (see below). Lithography is a useful technique, however, when simple patterns such as circles or squares of activated surface surrounded by borders of inactive surface are required. Lithography is also compatible with a wide range of surfaces found on transducers, such as glass, quartz, silicon, metals and metal oxides. Its main limitation is the need for flat surfaces that can be spin-coated with resist.

An alternative technique to conventional lithography has been developed during the last few years called 'micro-contact printing' (Mrksich and Whitesides 1995). This is a very simple and rapid technique, particularly suitable for patterning gold surfaces although it can also be combined with other types of chemistry. The basic procedure comprises making a flexible silicone rubber stamp containing the required pattern. The stamp is then coated with a suitable thiol compound and used to contact print the thiol onto the required areas of the gold substrate, where it spontaneously self-assembles. Untreated areas can subsequently be coated with another thiol by immersion in a solution of the thiol.

The micro-contact printing method has been used to define the surface chemistry of substrates used, for example, to control the growth of cell cultures in particular patterns (Singhvi *et al.* 1994). It can also be used in many ways to provide spatially defined coupling sites for bio-molecules in biosensor arrays (Gooding and Hibbert 1999).

Apart from its simplicity (good results can be achieved in the laboratory without the need for complex equipment and clean-room facilities), micro-contact printing has another advantage over conventional lithography: it can be used with curved surfaces and other three-dimensional surfaces, providing there is access for an appropriately shaped rubber stamp. This considerably increases the scope of the technique.

Control of deposition by physical placement

A variety of methods exists for spatial delivery of liquid samples, and the most appropriate one depends on the scale of the array being produced and the through-put required. The simplest is manual deposition using a suitable pipette or syringe, either directly onto the surface or using a suitable mask with wells formed into it. Manual deposition is really only suitable for relatively large spot sizes. Deposition is made easier if the boundaries of the spots have been defined, either by printing a mask pattern onto the surface, or by using hydrophobic (e.g. hydrocarbon or fluoro-carbon) surface chemistry to ring-fence the active deposition areas. This maintains an aqueous droplet within the boundary due to its reluctance to wet the hydropho-bic surface.

Syringe-based deposition can be automated to produce small batches of arrays suitable for research or pre-production testing. By the use of computer controlled X–Y translation stages and syringe pumps it is possible to make quite large and complex arrays, or larger numbers of small arrays. A system built from commercial components capable of depositing ninety-six spots per slide on thirty-two micro-scope slides in about 200 minutes has been described in detail (Graves *et al.* 1998). Commercial instrumentation is also available.

Inkjet printing (Badach *et al.* 1999) is both rapid and precise for larger-scale applications, and is compatible with both organic and aqueous conditions. Very small droplet sizes are possible (down to nl), and the deposition pattern can be varied at will using computer control. Changing from one bio-molecule to another tends to be rather wasteful and time-consuming with some printing systems, however, so this approach may not be very suitable for research-scale activity. The equipment is also rather expensive, although with appropriate knowledge it is probably possible to adapt low-cost commercial ink-jet printers.

Light directed immobilisation and patterning

Pattern generation using classical lithography with photoresists to define exposed and protected areas has been discussed above (page 78). In this section, the use of more direct photochemical immobilisation methods will be outlined. There are two basic strategies that can be adopted, direct photocoupling reactions and photo-deprotection reactions, and these will be discussed separately.

Direct photocoupling

A number of chemical functional groups, when activated by photons of light, transform into highly reactive intermediates. The three most commonly used groups for photoimmobilisation are aryl azides, aryl diazirines and benzophenones.

Simple aryl azides were the first photo-crosslinkers to be used in bioconjugate chemistry (Fleet and Porter 1972), and a number of different substituted aryl azides have subsequently been introduced. Aryl azides absorb a photon, usually in the UV range around 330–370 nm (convenient for mercury discharge lamps, which have a strong peak at 360 nm), to generate a nitrene, with loss of nitrogen. The nitrene is highly reactive and will abstract a proton from a nucleophile or a C–H bond, producing an insertion product.

The high reactivity and short half-life means that the nitrene reacts instantly with any group in the vicinity including solvent molecules, hence the efficiency of insertion is not always very high. More substituted aryl rings (e.g. tetrafluorophenyl azides) are claimed to give higher C–H bond insertion rates, since they form true nitrenes rather than rearranging into (still very reactive) dehydroazepines (Keana and Cai 1990). Another useful substitution is to add one or two nitro groups to the

Reaction scheme 4.15a Photo-decomposition of an aryl azide to generate a highly reactive nitrene, which inserts via proton abstraction

aryl ring. This increases the wavelength at which the molecule absorbs towards the visible, giving more flexibility of light sources for photo-activation.

Aryl diazirines, usually in the form of trifluoromethyl derivatives, are relatively unstable compounds and undergo UV induced photo-decomposition to generate highly reactive carbene species. The carbenes can insert into a wide range of bonds, particularly acidic species such as R—NH and R—OH bonds. The carbenes thus react favourably with proteins and many other types of polymers, but also with water, which reduces the coupling efficiency somewhat.

Reaction scheme 4.15b Photo-decomposition of an aryl diazirine to form a carbene, which inserts into a wide variety of (preferably slightly acidic) bonds

Reaction scheme 4.15c Photo-activation of a benzophenone derivative to form a biradical, which couples via proton abstraction

Benzophenone derivatives also absorb photons in the UV range around 360 nm to generate a biradical at the ketyl centre. The 'O'-radical abstracts a proton, preferentially from a C—H group α to an electron withdrawing group. These are abundant in proteins but, importantly, absent in typical solvents, particularly water. The result is that the radical does not react with solvent; it either couples or returns to the ground state, from where it can be reactivated by a new photon. Overall, this makes benzophenone based photo-coupling rather more efficient than aryl azide-based systems. They are also much more stable in storage and under ambient light conditions.

The general procedure for using any of the above reagents for photo-chemical coupling would be to immobilise the reagent on the surface via any suitable coupling method. A variety of amino and thio reactive derivatives with aryl azide or benzophenone groups are commercially available, as are simple compounds such as benzoylbenzoic acid that can be easily coupled to an amine surface by carbodiimide coupling. The surface is covered with a solution of the bio-molecule to be coupled (preferably reasonably concentrated), and with a mask which defines the areas that will be irradiated. The structure is then exposed to UV light for a few minutes (typically 10–30 minutes at an intensity of about $10\,mWcm^{-2}$) to achieve coupling. Using this type of approach a wide range of small and large molecules can be attached to a surface, including biotin (which can subsequently be used as a capture system – see pages 76–77) (Pritchard *et al.* 1995), polymers (Barie *et al.* 1998) and antibodies (Rozsnyai *et al.* 1992; Gao *et al.* 1994; Sundarababu *et al.* 1995; Delamarche *et al.* 1996). By repeating the coupling cycle with different biomolecules and masks, an array of different immobilised molecules can be produced. The procedure might also be very good for chemically coupling arrays of molecules produced physically in other ways, e.g. by ink-jet deposition.

Photo-deprotection

The alternative to direct photo-activation and coupling is to use a photo-deprotection strategy to 'unmask' a functional group and make it available for a conventional covalent coupling reaction. This approach is used extensively in light-directed combinatorial synthesis to make arrays of molecules such as peptides at a surface (Jacobs and Fodor 1994).

Methods for photo-deprotecting amine groups are thus particularly well developed, and the amines can subsequently be used in a wide range of covalent coupling methods (see pages 65–66). The same strategy was used in the development of a photo-activatable polyacrylamide gel containing protected amino groups that could be photo-exposed and used for coupling of antibodies and antigens (Sanford *et al.* 1998).

A related approach to photo-deprotection is the use of 'caged' biotin. This molecule can be immobilised using conventional methodology, but it is unable to bind avidin due to a bulky substituent (methyl nitropiprionyloxycarbonyl) that sterically precludes binding. Upon irradiation with UV light, however, the cage is removed, liberating free biotin that can then bind avidin. Using this approach together with the avidin/biotin capture system described on pages 76–77 allows arrays of antibodies or other biomolecules to be constructed (Blawas *et al.* 1998).

Electro-chemical control of deposition

Using modern micro-fabrication or printing methods it is possible to produce two-dimensional arrays of electrodes that can be individually addressed. Using such structures, it is possible to produce arrays of immobilised bio-molecules, including antibodies, by use of electrochemical polymerisation of pyrrole or other suitable monomers (see page 63). The entire device surface can be bathed in the pyrrole/biomolecule solution, since polymerisation will only occur at electrodes where the circuit is complete (i.e. those that are 'switched on' at the time).

An alternative way to achieve the same effect is to use a single continuous electrode surface poised at a potential, and to deposit liquid droplets that are physically separated from one another. Deposition is controlled by placing a counter electrode wire into the droplet to be polymerised in order to complete the circuit. This wire could, for instance, be fixed into a pipette tip or other similar device. If the monomer is drawn into the tip and the tip placed in contact with the surface, a micro-electrochemical cell is produced and material will be deposited in a spot defined by the internal diameter of the end of the pipette tip. Using an X–Y translator, this type of methodology could easily be automated (see page 79).

Conclusion

In the preceding pages, many different techniques and approaches to protein and ligand immobilisation have been outlined. This might at first seem bewildering, but by focusing on the surface properties of the device and the available functional groups on the material to be immobilised the choices will be greatly reduced. Consideration of the stability required for the device and the mechanism of transduction being employed will further limit the options, probably only leaving a few to consider. At this point the best way forward is to try out a few procedures to see which one gives the best compromise between speed, cost and performance for the particular system under development. There are usually too many subtle variables to allow accurate predictions to be made and a pragmatic approach is the only option. With immobilisation chemistry, the proof of the pudding is truly in the eating.

References

Badach, W., Abel, A.P., Bruno, A.E. and Beuschafer, D. (1999) 'Planar waveguides as high-performance sensing platforms for fluorescence-based multiplexed oligonucleotide hybridisation assays.' *Anal. Chem.* 71, 3347–3355.

Bain, C.D., Evall, J. and Whitesides, G.M. (1989) 'Formation of monolayers by the coadsorption of thiols on gold: variation in the head group, tail group and solvent.' *J. Am. Chem. Soc.* 111, 7155–7164.

Bamforth, C.H. and Al-Lamee, K.G. (1994) 'Studies in polymer surface functionalization and grafting for biomedical and other applications.' *Polymer* 35, 2844–2852.

Barie, N., Rapp, M., Sigrist, H. and Ache, H.J. (1998) 'Covalent photolinker-mediated immobilisation of an intermediate dextran layer to polymer-coated surfaces for biosensing applications.' *Biosensors and Bioelectronics* 13, 855–860.

Barisci, J.N., Conn, C. and Wallace, G.G. (1996) 'Conducting polymer sensors.' *Trends Pol. Sci.* 4, 307–311.

Blawas, A.S., Oliver, T.F., Pirrung, M.C. and Reichert, W.M. (1998) 'Step-and-repeat photopatterning of protein features using caged/biotin-BSA: characterisation and resolution.' *Langmuir* 14, 4243–4250.

Bregegere, F., Schwartz, J. and Bedouelle, H. (1994) 'Bifunctional hybrids between the variable domains of an immunoglobulin and the maltose-binding protein for *Escherichia coli*: production, purification and antigen binding.' *Protein Eng.* 7, 271–280.

Breitling, F. and Duebel, S. (1999) *Recombinant Antibodies.* John Wiley and Sons Inc., New York, p. 136.

Clark, L.C. and Lyons, C. (1962) 'Electrode system for continuous monitoring in cardiovascular surgery.' *Ann. N.Y. Acad. Sci.* 102, 29–45.

Clark, P., Britland, S.T. and Connolly, P. (1993) 'Growth cone guidance and neuron morphology on micropatterned laminin surfaces.' *J. Cell Sci.* 105, 203–212.

Dave, B.C., Dunn, B., Valentine, J.S. and Zink, J.I. (1994) 'Sol-gel encapsulation methods for biosensors.' *Anal. Chem.* 66, 1120A–1127A.

Dean, P.D.G., Johnson, W.S. and Middle, F.A. (1985) *Affinity Chromatography: A Practical Approach.* IRL Press, Oxford.

Delamarche, E., Sundarababu, G., Biebuyck, H., Michel, B., Gerber, C., Sigrist, H., Wolf, H., Ringsdorf, H., Xanthopoulos, N. and Mathieu, H.J. (1996) 'Immobilisation of antibodies on a photoactive self-assembled monolayer on gold.' *Langmuir* 12, 1997–2006.

Demiroglou, A., Bandel-Schlesselmann, C. and Jennissen, H.P. (1994) 'A novel reaction sequence for the coupling of nucleophiles to agarose with 2,2,2-trifluoroethanesulphonyl chloride.' *Angew. Chem. Int. Ed. Engl.* 33, 120–123.

Disley, D.M., Blyth, J., Cullen, D.C., You, H.X., Eapen, S. and Lowe, C.R. (1998) 'Covalent coupling of immunoglobulin G to a poly(vinyl alcohol) poly(acrylic acid) graft polymer as a method for fabricating the interfacial-recognition layer of a surface–plasmon resonance immunosensor.' *Biosensors and Bioelectronics* 13, 383–396.

Duebel, S., Breitling, F., Kontermann, R., Schmidt, T., Skerra, A. and Little, M. (1995) 'Bifunctional and multimeric complexes of streptavidin fused to single-chain antibodies (scFv).' *J. Immunol. Methods* 178, 201–209.

Finklea, H.O. (1996) In *Electroanalytical Chemistry: A Series of Advances*, Vol. 19, Bard, A.J. and Rubinstein, I. (eds), Marcel Dekker, New York, pp. 110–317.

Fleet and Porter (1972) 'The antibody binding site–labelling of a specific antibody against the photo-precursor of an arly nitrene.' *Biochem. J.* 128, 499.

Frey, B.L. and Corn, R.M. (1996) 'Covalent attachment and derivatisation of poly-L-lysine monolayers on gold surfaces as characterised by polarisation–modulation FTIR spectroscopy.' *Anal. Chem.* 68, 3187–3193.

Gao, H., Kislig, E., Oranth, N. and Sigrist, H. (1994) 'Photolinker-polymer-mediated immobilisation of monoclonal-antibodies, f(ab')2 and f(ab') fragments.' *Biotechnology and Applied Biochemistry* 20, 251–263.

Gooding, J.J. and Hibbert, D.B. (1999) 'The application of alkanethiol self-assembled monolayers to enzyme electrodes.' *Trends Anal. Chem.* 18, 525–533.

Graves, D.J., Su, H.-J., McKenzie, S.E., Surrey, S. and Fortina, P. (1998) 'System for preparing microhybridization arrays on glass slides.' *Anal. Chem.* 70, 5085–5092.

Haller, I. (1978) 'Covalently attached organic monolayers on semiconductor surfaces.' *J. Am. Chem. Soc.* 100, 8050–8055.

Heng, L.Y. and Hall, E.A.H. (2000) 'Producing "self-plasticising" ion-selective membranes.' *Anal. Chem.* 72, 42–51.

Hermanson, G.T. (1995) *Bioconjugate Techniques.* Academic Press, New York.

Hermanson, G.T., Mallia, A.K. and Smith, P.K. (1992) 'Immobilised affinity ligand techniques.' Academic Press Inc., New York.

Jacobs, J.W. and Fodor, S.P.A. (1994) 'Combinatorial chemistry – applications of light-directed chemical synthesis.' *Tibtech* 12, 19–26.

Jakoby, W.B. and Wilchek, M. (1974) 'Methods in Enzymology.' *Affinity Techniques – Enzyme Purification: Part B.* Academic Press, New York.

Janzen, R., Unger, K.K., Muller, W. and Hearn, M.T.W. (1990) Adsorption of proteins on porous and nonporous poly(ethyleneimine) and tentacle-type anion exchangers *J. Chromatogr.* 522, 77–93.

Johnston, A. and Thorpe, R. (1987a) *Immunochemistry in Practice.* Blackwell Scientific Publications, Oxford, pp. 55–59.

Johnston, A. and Thorpe, R. (1987b) *Immunochemistry in Practice.* Blackwell Scientific Publications, Oxford, pp. 59–65.

Jordan, J.D., Dunbar, R.A. and Bright, F.V. (1996) 'Aerosol-generated sol-gel-derived thin films as biosensing platforms.' *Anal. Chim. Acta.* 332, 83–91.

Keana, J.F.W. and Cai, S.X. (1990) 'New reagents for photoaffinity-labeling – synthesis and photolysis of functionalized perfluorophenyl azides'. *J. Org. Chem.* 55, 3640–3647.

Kroeger, S., Setford, S.J. and Turner, A.P.F. (1998) 'Immunosensor for 2,4-dichlorophenoxyacetic acid in aqueous/organic solvent soil extracts.' *Anal. Chem.* 70, 5047–5053.

Kurth, D.G. and Bein, T. (1993) 'Surface reactions on thin layers of silane coulping agents.' *Langmuir* 9, 2965–2973.

Lofas, S. and Johnsson, B. (1990) 'A novel hydrogel matrix on gold surfaces in surface-plasmon resonance sensors for fast and efficient covalent immobilisation of ligands.' *J. Chem. Soc. Chem. Commun.* 21, 1526–1528.

Lowe, C.R. and Dean, P.D.G. (1974) *Affinity Chromatography.* John Wiley and Sons, London.

Mayes, A.G., Blyth, J., Millington, R.B. and Lowe, C.R. (2000) 'Holographic ion sensors based on crown-ether containing hydrogels.' *Anal. Chem.* submitted.

Millot, M.C., Martin, F., Bousquet, D., Sebille, B. and Levy, Y. (1995) 'A reactive macromolecular matrix for protein immobilisation on a gold surface. Application in surface plasmon resonance.' *Sensors and Actuators B* 29, 268–273.

Molloy, P.E., Graham, B.M., Cupit, P.M., Grant, S.D., Porter, A.J. and Cunningham, C. (1995) 'Expression and purification strategies for the production of single-chain antibody and T-cell receptor fragments in *E. coli.*' *Mol. Biotechnol.* 4, 239–245.

Morgan, M. and Taylor, D.M. (1992) 'A surface plasmon resonance immunosensor based on the streptavidin-biotin complex.' *Biosensors and Bioelectronics* 7, 405–410.

Mosbach, K. (1976) 'Methods in Enzymology XLIV.' *Immobilised Enzymes.* Academic Press, New York.

Mrksich, M. and Whitesides, G.M. (1995) 'Patterning self-assembled monolayers using microcontact printing: a new technology for biosensors.' *Tibtech* 13, 228–235.

Mueller, W., Ringsdorf, H., Rump, E., Widburg, G., Zhang, X., Angermaier, L., Knoll, W., Liley, M. and Spinke, J. (1993) 'Attempts to mimic docking processes of the immune system: recognition-induced formation of protein multilayers.' *Science* 262, 1706–1708.

Nuzzo, R.G. and Allara, D.L. (1983) 'Adsorption of bifunctional organic disulfides on gold surfaces.' *J. Am. Chem. Soc.* 105, 4481–4483.

O'Driscol, K.F. (1976) 'Techniques of enzyme entrapment in gels.' In *Immobilised Enzymes*, Vol. XXXIV, Mosbach, K. (ed.), Academic Press, New York, pp. 169–243.

Persson, M., Mansson, M.O., Bulow, L. and Mosbach, K. (1991) 'Continuous regeneration of NAD(H) covalently bound to a cysteine genetically engineered into glucose dehydrogenase.' *Bio/technology* 9, 280–284.

Pritchard, D.J., Morgan, H. and Cooper, J.M. (1995) 'Micron-scale patterning of biological molecules.' *Angew. Chem. Int. Ed. Engl.* 34, 91–92.

Ramirez, C., Fung, J., Jr., R.C.M., Antony, R., Warren, J. and Kilburn, D.G. (1993) 'A bifunctional affinity linker to couple antibodies to cellulose.' *Bio/technology* 11, 1570–1573.

Reinhoudt, D.N., Engbersen, J.F.J., Brzozka, Z., Vlekkert, H.H.v.d., Honig, G.W.N., Holterman, H.A.J. and Verkerk, U.H. (1994) 'Development of durable K+-selective chemically modified field effect transistors with functionalised polysiloxane membranes.' *Anal. Chem.* 66, 3618–3623.

Reiser, A. (1989) *Photoreactive Polymers: the Science and Technology of Resists.* John Wiley and Sons, New York.

Rozsnyai, L.F., Benson, D.R., Fodor, S.P.A. and Schutz, P.G. (1992) 'Photolithographic immobilisation of biopolymers on solid supports.' *Angew. Chem. Int. Ed. Engl.* 31, 759–761.

Sanford, M.S., Charles, P.T., Commisso, S.M., Roberts, J.C. and Conrad, D.W. (1998) 'Photoactivatable cross-linked polyacrylamide for the site-selective immobilisation of antigens and antibodies.' *Chem. Mater.* 10, 1510–1520.

Scouten, W.H. (1981) *Affinity Chromatography: Bioselective Adsorption on Inert Matrices.* John Wiley and Sons, New York.

Sheehan, J.C. and Hess, G.P. (1955) 'A new method of forming peptide bonds.' *J. Am. Chem. Soc.* 77, 1067–1068.

Shpigel, E., Glodlust, A., Eshel, A., Ber, I.K., Efroni, G., Singer, Y., Levy, I., Dekel, M. and Shoseyov, O. (2000) 'Expression, purification and applications of staphylococcal Protein A fused to cellulose-binding domain.' *Biotechnology and Applied Biochemistry* 31, 197–203.

Sigal, G.B., Bamdad, C., Barberis, A., Strominger, J. and Whitesides, G.M. (1996) 'A self-assembled monolayer for the binding and study of histidine-tagged proteins by surface plasmon resonance.' *Anal. Chem.* 68, 490–496.

Singhvi, R., Kumar, A., Lopez, G.P., Stephanopoulos, G.N., Wang, D.I.C., Whitesides, G.M. and Ingber, D.E. (1994) Engineering cell shape and function.' *Science* 264, 696–698.

Spinke, J., Liley, M., Guder, H.-J., Angermaier, L. and Knoll, W. (1993) 'Molecular recognition at self-assembled monolayers: the construction of multicomponent multilayers.' *Langmuir* 9, 1821–1825.

Spitznagel, T.M. and Clark, D.S. (1993) 'Surface-density and orientation effects on immobilised antibodies and antibody fragments.' *Bio/technology* 11, 825–829.

St John, P.M., Davies, R., Cady, N., Czajka, J., Batt, C.A. and Craighead, H.G. (1998) 'Diffraction-based detection using a microcontact printed antibody grating.' *Anal. Chem.* 70, 1108–1111.

Sulkowski, E. (1985) 'Immobilised metal affinity chromatography.' *Trends Biotechnol.* 3, 1–7.

Sundarababu, G., Gao, H. and Sigrist, H. (1995) 'Photochemical linkage of antibodies to silicon chips.' *Photochemistry and Photobiology* 61, 540–544.

Toyama, S. and Ikariyama, Y. (1997) 'Design and characterisation of polymer matrix anchored to gold thin layer for the effective usage of evanescent field in SPR-based immunosensor.' *Chem. Lett.* 1083–1084.

Wahlgren, M. and Arnebrant, T. (1991) 'Protein adsorption to solid surfaces.' *Trends in Biotechnology* 9, 201–208.

Weetall, H. (1976) 'Covalent coupling methods for inorganic support materials.' In *Methods in Enzymology*, Vol. XLIV, Mosbach, K. (ed.), Academic Press, New York.

Wilchek, M. and Bayer, A.E. (1988) 'The avidin–biotin complex in bioanalytical applications.' *Anal. Chem.* 171, 1–32.

Wong, S.S. (1991) *Chemistry of Protein Conjugation and Cross-Linking.* CRC Press, Boca Raton.

Yang, Z. and Yu, H. (1997) 'Preserving the globular protein shape on glass slides: a self-assembled monolayer approach.' *Adv. Mater.* 9, 427–429.

Yon-Hin, B.F.Y., Smolander, M., Crompton, T. and Lowe, C.R. (1993) 'Covalent electropolymerisation for glucose oxidase in polypyrrole. Evaluation of methods of pyrrole attachment to glucose oxidase on the performance of electropolymerised glucose sensors.' *Anal. Chem.* 65, 2067–2071.

You, H.X., Lin, S.M. and Lowe, C.R. (1995) 'A scanning tunneling microscope study of site-specifically immobilised immunoglobulin G on gold.' *Micron.* 26, 311–315.

5 Binding isotherms and kinetics of immobilized biological systems

Claus Duschl

Introduction

In order to characterize the binding between a ligand and a receptor, three properties are of particular interest: the equilibrium binding constant K describes the ratio between bound and unbound ligands for a given concentration of receptor under steady-state conditions. The two reaction rate constants k_a and k_d determine the association rate of complex formation and the dissociation rate of its disintegration respectively. Knowing two of the three parameters K, k_a and k_d allows the determination of the third one via $K = k_a/k_d$. The knowledge of K gives a very global range of information on the binding behaviour at equilibrium conditions, the knowledge of k_a and k_d can reveal some details about molecular mechanism (see for example Cantor and Schimmel 1980).

In biological systems binding events occur both in solution and at interfaces, in particular on cell membranes. Examples of the former are immunological antibody antigen reactions and enzymatic reactions in the blood or cytoplasm. At interfaces, immunological reactions such as antigen presentation by major histocompatibility complexes (MHC) or signal transduction events are well investigated examples. In practice, it is equally possible to measure the binding parameters in solution and at interfaces. Ideally, for the characterization of a given ligand-receptor pair, one would choose its native environment and geometry. But for many reasons this is not always the most straightforward experimental approach. The measurements of the binding parameters in solution profit from well-defined and thermodynamically equilibrated boundary conditions such as an isotropic distribution of the binding partners, well-characterized translational and rotational diffusion constants and homogeneous micro-environments of the binding sites. They are straightforward to control and, therefore, easy to reproduce in different laboratories. A major problem of the measurement of the binding parameters in solution concerns the difficulty to separate 'unbound' from 'bound' binding partners and, therefore, tedious washing and separation procedures may be needed. In addition, *in situ* monitoring of reaction kinetics is often not possible.

The measurements of binding constants and rate constants at interfaces, in particular on functionalized sensor surfaces where the ligand or the receptor is immobilized avoids these limitations (here, we will assign the surface-bound entity always as being the receptor and the solubilized molecules always as being the ligand what may sometimes contradict the biological terminology). The sensor surface serves two purposes which, in combination, allow the acquisition of signals only from the interaction between ligand and receptor: first, the surface enables the spatial separation of bound compounds from the free compounds in solution and, second, it allows the employment of surface sensitive techniques such as optical evanescent wave techniques, acoustic mass balances or electro-chemical impedance spectroscopy. These techniques have the intrinsic ability to discriminate between 'bound' and 'unbound' since they probe molecules bound to the surface with high sensitivity and are rather insensitive to the unbound species in solution. As a consequence, *in situ* monitoring comes as a natural feature of this approach. The intrinsic separation of bound and unbound analyte consider-ably shortens the measurement time and this is why this approach became very popular in recent years. In a typical experiment, the measurement of the relative amount of bound ligand to the surface allows the determination of the binding para-meters, provided that the concentration of free ligand $[L]$ in the sample cell is known. The depletion of free ligands in solution through binding to the surface may be avoided through the use of sufficiently high concentrations of free ligand, or alternatively through the application of a constant flow of ligand containing buffer through the sample cell. This experimental approach is very simple and straight-forward.

However, the introduction of an interface makes the interpretation of the meas-ured signals rather complicated and, in order to obtain the binding parameters from them, careful analysis is necessary. Using an interface for the immobilization of one species of the binding partners results in its high local concentration. This local enrichment on the surface may lead to considerable concentration gradients of the other binding partner in solution since the formation of bound complexes causes a depletion of the freely diffusing partner in the vicinity of the interface. Secondly, this approach leads to situations in which a binding event cannot be considered to be independent from its environment and from the states of binding of the neigh-bouring molecules. The other important consequence of utilizing a surface for the measurements of the binding parameters is that an interface simply breaks the sym-metry of the sample, therefore, orientation, alignment and accessibility of binding sites of the immobilized species become important factors which may affect the binding parameters. Depending upon the various methods of attachment, the degrees of freedom of movement of the immobilized species are reduced. Another, rather practical problem is to perfect the engineering of functionalized surfaces in the sense that they are laterally isotropic and homogeneous, so that each binding site is equivalent. An important aspect is the need to suppress nonspecific binding of the ligand to the sensor surface.

The immobilization of receptors on a substrate may resemble the situation at the membrane of a biological cell. This holds in particular if surface immobilization techniques are employed which are based on the formation of supported lipid bilay-ers (Kalb *et al.* 1992; Terrettaz *et al.* 1993; Duschl *et al.* 1994; Sackmann 1996; Cornell *et al.* 1997; Heyse *et al.* 1998; Bieri *et al.* 1999). There, the two-dimensional mobility of receptor molecules or sub-units of receptors at the interface may lead to a very effective 2D mass transport. It may also accelerate the transport of molecules between different anchor sites at lipid membranes (reduction-of-dimensionality (RD) (Axelrod and Wong 1994)). Such scenarios are addressed in numerous publications (Adam and Delbrück 1968; Thompson *et al.* 1981; Axelrod and Wong 1994) but will not be the subject of this chapter.

This chapter is organized as follows. First, a short description of measurements of binding parameters on surfaces are given where binding sites on the surface are independent and the transport of ligands to the surface is not affecting the reaction rates. This situation is often referred to as the 'rapid mixing model'. As a next step, the more realistic situation is discussed where the density and accessibility of the binding sites of surface immobilized receptors may considerably influence the outcome of a binding experiment. In this context, the role of surface engineering is addressed. Next, a finite mass transport rate of the freely diffusing ligand will be added to the most simple kinetic model and it will be discussed how it affects kinetic measurements for the determination of the association and dissociation rate constants. Criteria will be established which allow a semi-qualitative evaluation of the extent of mass transport effects in an experiment. As a result, some experimental strategies will be discussed which minimize the influence of mass transport. Recent work using computer simulations and fitting procedures is presented and it is shown how it contributes to a more detailed understanding of this topic. Finally, experimental approaches will be described which are based on competitive reactions and which either circumvent the problem of mass transport or, alternatively, make use of it.

The determination of binding constants and of kinetic rate constants

The simple Langmuir model and the rapid mixing model

Modern transducer techniques such as total internal reflection fluorescence (TIRF) (Thompson *et al.* 1997)), optical waveguides (Pawlak *et al.* 1998), surface plasmon resonance (Fägerstam *et al.* 1992) or acoustic wave devices (Gizeli *et al.* 1997; Keller and Kasemo 1998) provide an on-line signal which is proportional to the number of adsorbed molecules at the sensor surface. If the adsorption is purely caused by the formation of ligand–receptor complexes, then the signal of the transducer technique is directly proportional to the number of ligand–receptor complexes formed. The calibration of the transducer signal in respect to mass or number of molecules per unit area can be carried out experimentally by immobilizing known amounts of molecules and measuring the response, or theoretically by applying well-established theories (such as Fresnel theory for optical techniques based on the measurements of the refractive index). The sensitivity of these techniques is sufficient to monitor the formation of submonolayers (to which extent depends on the techniques used; fluorescence techniques are able to detect as little as 100 fluorescent molecules/cm^2, techniques based on the measurements of the refractive index allow the measurement of $\sim 10^{-9}$ g/cm^2, given a molecular weight MW = 50 kD, this corresponds to $\sim 10^{10}$ molecules/cm^2).

An idealized experiment for measuring binding constants and kinetic rate constants on surfaces consists of a sensor surface on which the binding sites can be considered as independent of each other and are freely accessible for the binding partner in solution (see, for example, Cantor and Schimmel 1980); the surface anchored molecules are completely 'decoupled' from nonspecific surface interactions and from their neighbouring molecules. Finally, the concentration of the molecules on the surface should be low enough so that, at any time interval, the binding reaction does not lead to any significant depletion of molecules in any space segment of the solution and therefore to any concentration gradients. (Such low concentrations of immobilized molecules reduce, of course, the sensitivity of the assays considered here.) An infinitely fast diffusion constant of the solubilized binding partner is equivalent to the last condition. This case is very often referred to as the 'rapid mixing model' (Karlsson *et al.* 1991), this is also the term used in this chapter.

The reaction between receptors of concentration $[R]$ and ligands of concentration $[L]$ can be expressed by

$$[R] + [L] \underset{k_d}{\overset{k_a}{\rightleftharpoons}} [RL] \tag{5.1}$$

with k_a and k_d being the kinetic reaction rate constants of association and dissociation. For such a reaction, the law of mass action relates the relative concentrations of the constituents $[R]$ and $[L]$ and of the product $[RL]$ to the binding constant K at equilibrium conditions ($d[RL]/dt = 0$):

$$K = \frac{[RL]}{[R][L]} \tag{5.2}$$

and

$$K = \frac{k_a}{k_d} \tag{5.3}$$

The binding constant (often the dissociation constant $K_D = 1/K$ is used) is the ratio of the two rate constants k_a and k_d.

When the reaction takes place on a surface, it is convenient to introduce $[R_{tot}] = [R] + [RL]$, the total number of immobilized receptors on the surface. In this case, (5.2) can be transformed into the classical Langmuir equation:

$$\Gamma = \frac{[RL]}{[R_{\text{tot}}]} = \frac{K[L]}{1 + K[L]} \tag{5.4}$$

with Γ being the relative surface coverage.

Γ is typically plotted as a function of the logarithm of the ligand concentration. Very often a Scatchard plot is drawn (see Figure 5.1) which is a linearization of equation (5.4):

$$\frac{\Gamma}{[L]} = K(1 - \Gamma) \tag{5.5}$$

It gives a straight line, with the binding constant K being the intercept of the ordinate and with the maximum coverage Γ_{max} being the intercept of the abscissa. In this representation deviation from the ideal binding behaviour causes a bending of the curve which can easily be identified. Later in this chapter, reasons are presented which lead to curved Scatchard plots.

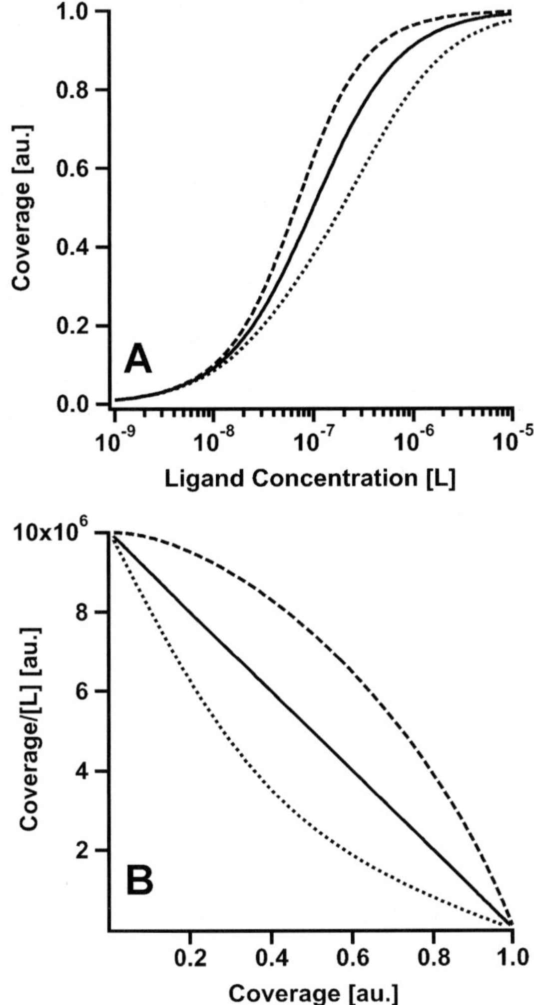

Figure 5.1 Adsorption isotherms (A) and the corresponding Scatchard plots (B). The full line (—) is calculated on the base of the simple Langmuir model. The dashed line (- - -) represents the Langmuir model including cooperative binding and the dotted line (⋯) represents the Langmuir model with anti-cooperative binding

The time course of this reaction can be described by:

$$\frac{d[RL]}{dt} = k_a[R][L] - k_d[RL]$$

(5.6)

Again, introducing $[R_{tot}]$, (5.6) can be rewritten:

$$\frac{d[RL]}{dt} = k_a([R_{tot}] - [RL])[L] - k_d[RL]$$

(5.7)

$$= k_a([R_{tot}][L] - (k_a[L] + k_d\,[RL]$$

We can easily integrate this equation and obtain:

$$[RL](t) = \frac{k_a[R_{tot}][L]}{k_a[L] + k_d}\,(1 - e^{-(k_a[L]+k_d)(t-t_0)}) \qquad (5.8)$$

Dividing the prefactor in (5.8) by k_d, we immediately see that (5.4) represents the response for infinite times:

$$[RL](t)_{t \to \infty} = \frac{K[R_{tot}][L]}{1 + K[L]} \qquad (5.9)$$

The equations (5.3–5.9) now offer recipes for the design of two simple experiments which allow the determination of the binding constant of a receptor ligand pair. (i) First, the amount of bound ligand $[RL]$ or $\Gamma = [RL]/[R_{tot}]$ is measured as a function of the concentration of free ligand in solution ($[L]$) at times when a steady state has been reached ($t \to \infty$). $[R_{tot}]$ and K are then obtained from simply fitting a plot of Γ versus $[L]$ using a routine based on the Langmuir equation (equation (5.4)). The value which is obtained for $[R_{tot}]$ has to be considered with some care. Only when the density of receptors is low enough, so that the size of the complexes does not determine the maximum coverage, does $[R_{tot}]$ reflect the actual number of receptors on the surface.

The most accurate results are obtained when the range of concentrations of ligands covers a few orders of magnitude around $1/K$. High ligand concentrations allow the accurate determination of $[R_{tot}]$. The time needed for such a measurement is considerable. This makes this approach rather unsuited for fast screening of large numbers of different ligands. In addition, due to long measuring times, the consumption of material might be quite high when a flow-through system is used.

(ii) A faster way of determining the binding constant is the measurement of the kinetic reaction rates (see equation (5.3)) (Karlsson *et al.* 1991; Altschuh *et al.* 1992; Fägerstam *et al.* 1992; Panayotou *et al.* 1993). This strategy has the advantage of obtaining the whole set of binding parameters: k_a, k_d and K. The time course of the binding reaction during the association phase is monitored *in situ* for several ligand concentrations. The dissociation phase is measured with no ligand in solution. From the association part, both rate constants can be determined. Plotting the rate of binding $d[RL]/dt$ versus the amount of binding $[RL]$ for the various ligand concentrations $[L]$ gives straight lines whose slopes are equal to $-(k_a[L] + k_d)$ (see equation (5.7)). These slope values ($(d[RL]/dt)/[RL]$ are then plotted versus the concentration of ligands in solution $[L]$. Again, a straight line is obtained. Its slope yields directly k_a whilst k_d is given by the intercept on the ordinate. k_d can also be determined separately from the dissociation phase. After the adsorption of ligands, pure buffer is added to the measuring cell. In this case, equation (5.6) is reduced to

$$\frac{d[RL]}{dt} = -k_d[RL] \qquad (5.10)$$

and the dissociation time course can easily be fitted by a single exponential to obtain k_d.

Binding isotherms and surface engineering

The environment of a binding site of a molecule anchored to a surface is highly anisotropic and this may substantially influence its binding properties such that the 'local' binding constant differs from the 'intrinsic' binding constant measured in an isotropic environment. Therefore, the measurement of the binding constant of a ligand–receptor pair on a surface necessitates the careful engineering of the interface on a molecular level.

Three properties of the interfacial design are crucial for a successful determination of the binding constant: (i) the density of the binding sites on the surface, (ii) the accessibility of the binding sites which may be affected by the density of binding sites and (iii) the lateral homogeneity of the surface in the sense that the environment of each binding site is equivalent on a molecular level. A fourth aspect which is not intimately related to the properties of the binding sites themselves, but which affects the evaluation of binding data, is the control of nonspecific binding to the surface. In the following, the four aspects will be discussed in terms of how they have an impact on the measurement of a binding constant in a steady-state experiment.

(i) and (ii) The density of binding sites on the surface and their accessibility is closely related and will be treated together. In order to increase the sensitivity of a device which is able to monitor surface reaction, an obvious strategy is to increase the density of binding sites on the sensor surface. However, a physical binding site at a surface is, in general, a three-dimensional object and its close packing may lead to the obstruction of the whole site or parts of it. This holds, in particular, when the receptor is a relatively small molecule such as a peptide and the 'docking sites' for binding are situated on the exterior of the molecule. A binding site in this context should not be mistaken for a surface area segment as it is often used in theoretical adsorption models. In many cases, the optimal density of binding sites depends on the size of the ligand and on the shape of its docking site. If the ligand is large with respect to the binding site of the surface, a low concentration of receptor molecules at the surface is important such that the access of the ligand is not hindered by neighbouring receptor molecules. In addition, the close packing of small receptor molecules may affect their structure which may be vital for the recognition process. Peptides are likely candidates for such a behaviour. In solution peptides may form random coils whereas, when immobilized on a surface at a certain density, their structure may change to a more ordered peptide motif. Close packing of large receptor molecules such as proteins, where the docking site lies in the interior and where the size of the receptor leads to a intrinsic dilution of surface binding sites, may not affect the binding properties to the same extent. In the following, two instructive examples are discussed, which demonstrate the influence of accessibility of the receptor. In Figure 5.2, a mixed self-assembled monolayers are shown containing an antigenic peptide (the repetitive amino acid sequence $(NANP)_6$ is part of a coat protein which surrounds the human malaria sporozoite and is believed to be targeted in an immune response (Godson 1985; Nussenzweig and Nussenzweig 1989)). Binding of a monoclonal antibody to such a layer was monitored as a function of the peptide content of the monolayer using surface plasmon resonance (Duschl *et al.* 1996). The graphs in Figure 5.2b show clear maxima of the amount of bound antibody for three different antibody concentrations. The maximum is most prominent for the lowest antibody concentration. These results were interpreted according to the following lines: for low peptide concentrations, the amount of anti-body binding is limited by the number of binding sites. For high antigen concentrations, the binding constant between peptide and antibody is reduced through self-obstruction between neighbouring peptides; in other words, the epitopes are more difficult to access by the antibodies because of steric hindrance. This interpretation was supported by the fact that only the association rate constant decreased with increasing peptide density of the SAM, whereas the dissociation rate constant remained constant. The decrease of antibody binding efficiency starts at a molecular area per peptide of approximately $1400\,\text{Å}^2$ which is comparable to the cross-sectional area of a Fab fragment. This consideration suggests that, in order to have optimal binding, the antigen to be recognized should be isolated enough so that no neighbouring peptide can hinder the approach of the antibody.

The optimal presentation or exposure of binding sites at the surface requires additional important features of the surface architecture. This includes the optimal orientation of the molecule; if the binding site points towards the substrate, no binding will take place. In many cases a spacer is needed which separates the binding site from the substrate. The chemical link must not interfere with the binding unit and must make sure that it is optimally oriented.

The role of a spacer unit which separates the binding site from the substrate was discussed in detail by Spinke *et al.* 1993). They studied the biotin streptavidin system which is a well-established ligand–receptor pair. It is widely used for anchoring molecules to substrates because of its high binding constant. Biotinylated thiols, each comprising a spacer unit of variable length between an alkyl chain and the biotin site, were immobilized in mixed SAMs on gold surfaces. The binding of streptavidin was monitored as a function of the biotin density. Again, a peak in streptavidin binding was found at a certain biotin density. However, this behaviour was only observed when biotinylated thiols were used bearing a hydrophilic spacer unit of sufficient length. In this case, it is expected that the biotin binding sites of the mixed SAMs are distant from the densely packed alkyl chains and well-exposed to the solution containing the receptor protein. This finding was explained by looking at the structure of streptavidin: the binding sites of streptavidin are situated in fairly deep pockets ($9\,\text{Å}$ deep) which have to be entered by the biotin when proper binding should occur. Hence, the ligand must protrude out of the SAM to fully access the binding pocket. In Figure 5.3, a simple structural representation of the ligand–protein complex at a SAM demonstrates the importance of exposing the biotin group in order to achieve optimal binding.

In some cases, the binding constant depends on the number of ligand–receptor complexes already formed at the surface. For example, the tendency of a ligand to aggregate at high concentration may favour the formation of a complex in the presence of high ligand coverages on the surface. Such behaviour is analysed through the representation of the binding data in a Scatchard plot; the coverage divided by the ligand concentration $\Gamma/[L]$ is plotted as a function of the ligand concentration $[L]$ (Figure 5.1b). Any deviation from the ideal behaviour leads to a positive or negative bending of the otherwise straight Scatchard plot. This may easily be understood if we look at the free energy difference ΔG^0 between the bound and unbound state of a ligand–receptor pair:

$$\Delta G^0 = \Delta G_0^0 + RT\phi(\Gamma) \text{ where } \Delta G_0^0 = RT\ln K_0 \tag{5.11}$$

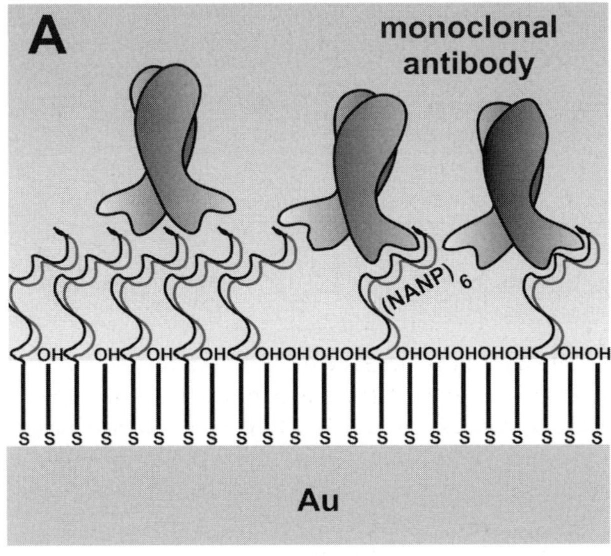

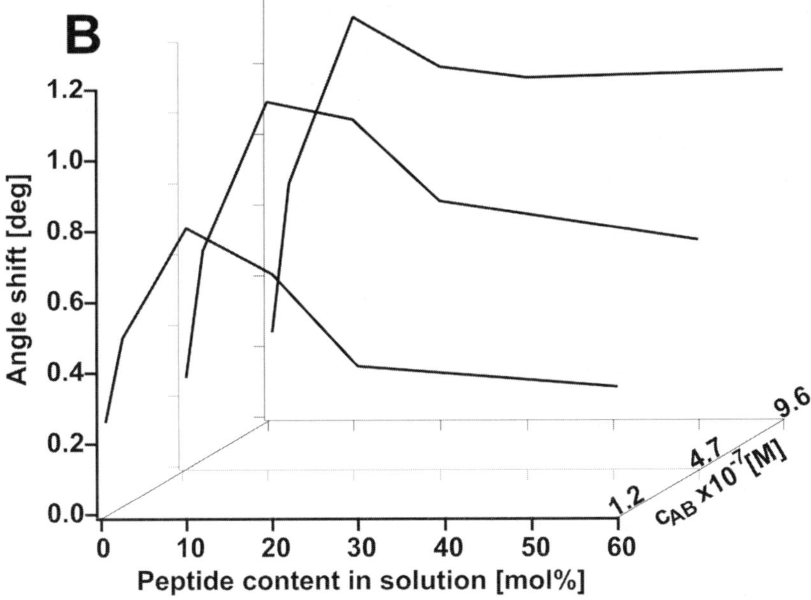

Figure 5.2 A) Schematic representation of a self-assembled monolayer containing 11-mercaptoundecanol and $HS(CH_2)_{11}$-$(NANP)_6Y$. The lower peptide content (right) should show improved accessibility of the binding sites of the antibodies. B) Three-dimensional representation of the specific binding of the monoclonal antibody E9 to peptide-containing SAMs. The angle shift due to the binding of antibodies versus the molar composition of $HS(CH_2)_{11}$-$(NANP)_6Y$ and HS-$(CH_2)_{11}$-OH in solution is plotted for different antibody concentrations in solution (the concentrations of peptide in solution do not reflect the content of peptide in the SAM there, the relative concentration is approximately a factor 10 smaller; both concentrations do not scale linearly). Taken from Duschl (1996)

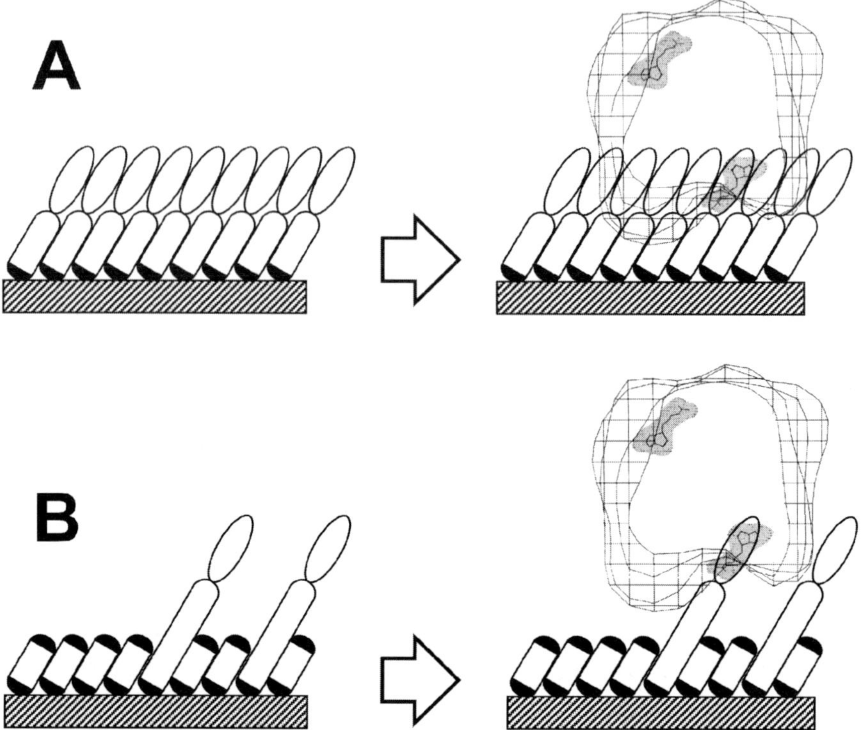

Figure 5.3 A schematic representation of a self-assembled monolayer of thiols and the binding of streptavidin to them. The streptavidin data are taken from Darst (1991) with only two of the four binding sites shown. The mixed monolayer contains biotinylated thiols with no spacer unit between the alkyl chain and the biotin. Binding of streptavidin is sterically hindered (A). The mixed monolayer contains biotinylated thiols with a spacer unit between the alkyl chain and the biotin. The addition of the spacer allows binding of the streptavidin with no steric hindrance (B). Taken from Spinke (1993)

with $\phi(\Gamma)$ describing the interaction potential as a function of the actual coverage Γ. The binding constant K can be represented in the following way:

$$K(\Gamma) = K_0 \exp(\phi(\Gamma)) \tag{5.12}$$

where $\phi(\Gamma)$ is zero at $\Gamma = 0$. In the case of cooperative binding, ϕ is positive which leads to hump-shaped Scatchard plot ($\cap$), when anti-cooperativity is present, ϕ is negative, the curve has a positive curvature ($\cup$, see also Cantor and Schimmel 1980: 860). Here, the problem is to find a reasonable model for the interaction of ligands.

It should be pointed out that besides the issues discussed above, it is very important to make sure that the binding sites are equivalent to each other on a molecular scale. A homogeneous distribution of binding sites on the surface is essential to obtain high quality data. Although this sounds trivial, it is not always easy to engineer a molecular assembly on a surface with sufficient control. Already the surface roughness of most standard substrates (oxide surfaces or metal layers) may produce variable micro-environments of the immobilized binding sites. As a consequence, a distribution of different 'effective' binding constants makes a rigorous characterization of the thermodynamics of a particular binding pair rather difficult. In the case of nonequivalent binding sites, the Scatchard plot also shows a positive curvature ($\cup$); the binding sites with the higher binding constant are preferentially saturated, leaving the low-affinity sites to be bound at high ligand concentrations. In general, this means that the presence of nonequivalent binding sites is indistinguishable from anti-cooperative binding if no further information is available. This again emphasizes the importance of a careful engineering of the surface properties. A related topic is described in a series of papers by Stankowski (Stankowski 1983), who addressed the question of what kind of binding curve, in particular, of what kind of Scatchard plot, is obtained when the binding of a ligand deactivates other binding sites nearby. This may occur if the size of the ligand is large enough to cover several binding sites on the surface (large ligand effect). This effect has been taken account of in the treatment of one-dimensional systems such as biopolymers, but it is rather ignored in biosensor research. The reason for this may be that for the effect to be noticed, the absolute number of binding sites on the surface must be known which is, however, rarely the case. Up to now, there is no stringent test of this theoretical concept when applied to the binding of ligands on a surface.

If the receptor to be immobilized is a protein, a uniform orientation of its binding site towards the ligand containing solution is, in general, not straightforward to achieve. Proteins have a variety of surface domains with hydrophobic, charged or polar character. Each domain may interact with the substrate, leading to a distribution of orientations of the protein on the surface. In order to orient proteins on surfaces, ideally, one would like to introduce surface anchors opposite to their binding sites and to suppress the nonspecific interactions. Genetically engineered proteins bearing histidine extensions as anchor groups which can form stable complexes with metal chelating groups are well-suited for such a strategy. Its usefulness was demonstrated through the immobilization of Fab fragments on a chelator containing self-assembled monolayer (Keller 1995; Kröger 1999). The problem of orientation is very much relaxed when one binding partner is immobilized on a surface via a water soluble polymer network which is accessible by the other binding partner in solution. Of course, the mesh size of the network must allow for free diffusion of the analyte and its free access to the receptor sites.

Biomolecules interact with interfaces via a variety of nonspecific forces, such as van-der-Waals, electrostatic, hydrophobic, steric or entropic forces. These forces can cause nonspecific binding or may lead to an enrichment of the free binding partner in the vicinity of the interface (Andrade and Hlady 1986). Therefore, in order to obtain good binding data, an important aspect in the engineering of sensor surfaces is the minimization of nonspecific attractive interactions (van-der-Waals, electrostatic, hydrophobic) and since it is not possible to completely suppress them, they must to be counterbalanced by introducing defined repulsive interactions (electrostatic, steric or entropic) (Andrade and Hlady 1986; Jeon and Andrade 1991; Jeon *et al.* 1991).

In summary, when one partner is immobilized at an interface, a reasonable thermodynamic characterization of a binding reaction should be preceded by a careful design of the surface architecture. This stresses the importance of surface chemistry for the development of assays which allow the characterization of molecular recognition events.

Measuring rate constants when mass transport limitations are present

Up to here, when kinetic experiments were discussed, it was assumed that the mass transport rate constant k_+ is much faster than the association rate constant. Now, we will explore the implications when this condition is not fulfilled and the transport of ligands from the solution to the surface has to be included (Berg and Hippel 1985) to the reaction scheme according to

$$[R] + [L_{sol}] \underset{k_+}{\overset{k_+}{\rightleftharpoons}} [R] + [L_{surf}] \underset{k_d}{\overset{k_a}{\rightleftharpoons}} [RL] \tag{5.13}$$

with $[L_{sol}]$ being the concentration of ligands far away from the sensor surface and $[L_{surf}]$ being the ligand concentration in the immediate vicinity of the surface (Karlsson 1994; Karlsson *et al.* 1994; O'Shannessy *et al.* 1994; Myszka *et al.* 1996; Nieba *et al.* 1996).

Before we proceed, a short intermediate remark should be added. As we shall see later, in a first order approximation, mass transport affects both rate constants, k_a and k_d, equally. This, we should bear in mind since much of the discussion to follow will evolve around the effect mass transport has on determination of the dissociation rate constant k_d. This is simply because there, the principles are more intuitive and easier to describe. As a consequence of the equal impact of mass transport on the rate constants, mass transport effects cancel out when the binding constant is to be determined.

In recent years, the measurement of the kinetic rate constants between surface immobilized receptors and solubilized ligands became more and more popular. The reason for this is twofold: first, the determination of the association and the dissociation rate constants gives more detailed information, and second, instead of waiting for an equilibrium situation to be established, a short measurement period is in principle sufficient to characterize both rate constant (as a review, see for example Szabo *et al.* 1995). The reliable measurement of reaction rate constants necessitates the use of sophisticated cell designs and fluid handling systems for a precise control over the ligand supply to the surface. Improved fluid cells also result in measurement procedures which reduce the material consumption with respect to steady-state measurements. All these factors were known to play a role when the first commercial instrument which used surface plasmon resonance as a detection principle was launched on the market (Fägerstam *et al.* 1992).

When the first results were published using this instrument it was argued that a permanent flow of ligands towards the sensor surface is sufficient to overcome the problems which arise from a limiting mass transport in solution (Karlsson *et al.* 1991; Altschuh 1992; Fägerstam 1992; Panayotou *et al.* 1993). However, as will be shown in the following, these arguments were not justified and only later, it turned out that the successful and reliable characterization of the kinetic parameters of a ligand receptor pair needs a very careful design of the experimental conditions in order to avoid mass transport effects (Karlsson 1994; Karlsson *et al.* 1994; O'Shannessy 1994;

Myszka *et al.* 1996; Nieba *et al.* 1996). In this part, some simple measurements are discussed which exemplify the problems which one is confronted with when designing a kinetic experiment. This is followed by a short introduction on diffusion and of the consequences it has on reactions at interfaces. Afterwards, some tests are discussed which allow the assessment of kinetic measurements. Computer simulations, adapted to the most used sample cell designs, are presented which give criteria for the evaluation of measurements. These simulations also offer strategies for the design of experiments where mass transport effects are to be avoided or to be carefully controlled.

Let's look at a simple experiment which demonstrates how mass transport interferes with the measurement of reaction rate constants. The system we chose is the same as already discussed in the pervious section (Duschl *et al.* 1996). It consists of the $(NANP)_6$ peptide which serves as antigen and which is immobilized on a gold substrate and of a monoclonal antibody which binds specifically to the peptide. However, in this example, the functionalized gold substrate is a so-called sensor chip which is used in a BIAcore surface plasmon resonance instrument. This instrument is equipped with a state-of-the-art flow through a sample cell and a sophisticated fluid handling system.

The density of the immobilized NANP-peptides which represent the receptors, is 4.8×10^{12} peptides/cm^2 or 8×10^{-12} mol/cm^2. The surface concentration corresponds to four times the value which was recommended by BIAcore in the early times (Karlsson *et al.* 1991) (for the given diffusion constant of antibodies of 5×10^{-7} cm^{2}s^{-1}) in order to allow a reliable determination of the kinetic rate constants even for low flow rates (note that, in this case, the surface architecture is different from the standard surface of a BIAcore chip on which an approximately 100 nm thick water soluble dextran network contains the receptor sites. Therefore, no additional mass transport effects due to the polymer mesh, as discussed by Schuck, are present (Schuck 1996)). We concentrate on the desorption process of the monoclonal antibodies since the effect of mass transport limitation are more straightforward to observe here than during the association phase.

In Figure 5.4 the adsorption of the monoclonal antibody on the antigen coated surface is shown followed by desorption processes initiated by rinsing at various flow rates using buffers containing different concentrations of the soluble antigen $(NANP)_6$. The rinsing of the antibody layer with plain buffer removes only a minor part of the bound antibody even when applying the maximal flow rate of $100 \, \mu$lmin^{-1}. An analysis of this time course shows that the desorption process is much slower ($k_{des} = 10^{-3}$ s^{-1}) than one would expect from the rate constants. Faster desorption is observed when free $(NANP)_6$ antigen is added up to a concentration of 10^{-4} M and a flow of $10 \, \mu$lmin^{-1} where saturation seems to occur ($k_{des} = 7.5 \times 10^{-3}$ s^{-1}). The experiment shows clearly that the rinsing with plain buffer is by far not sufficient to lead to a desorption of antibodies with the expected rate constants. The observed curves can be explained by the competition between the free antigens in solution and the antigens immobilized on the sensor surface for free binding sites of the antibody. The free peptides reduce the probability of rebinding of the antibodies to the surface since the binding sites of the antibody become occupied once they unbind from the antigens on the surface (Goldstein *et al.* 1989). These experiments demonstrate that even for receptor densities at the sensor

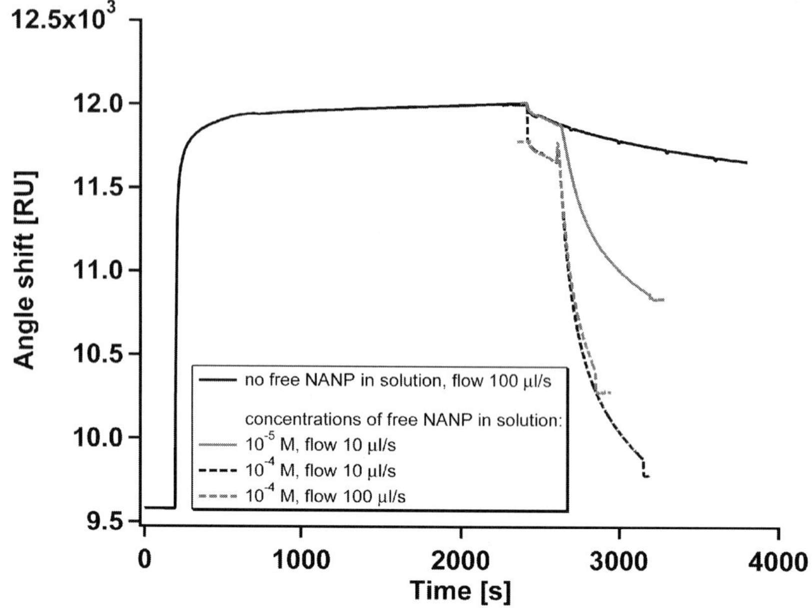

Figure 5.4 Adsorption and desorption of a monoclonal antibody on mixed SAMs containing 4.8×10^{12} antigenic NANP-peptides/cm^2 (see Figure 5.1). Desorption is initiated either by a flow of pure buffer or by a flow of buffer containing various concentrations of soluble peptide. Assuming that the desorption curves follow a simple single exponential decrease, the measured rate constants give $k_{des} = 10^{-3}\,\text{s}^{-1}$ for the desorption initiated by a flow of pure buffer and $k_{des} = 7.5 \times 10^{-3}\,\text{s}^{-1}$ for the fastest desorption process using soluble peptide

surface which are at the lower limit of detectability, the rebinding process causes a considerable retardation of the desorption process.

The bivalency of the antibody should not play a role in these experiments. Irrespective of whether mono- or bivalent binding of the antibodies to the surface is predominant, the peptides in solution can bind to the antibody binding sites only when the antibodies have already dissociated. It is not expected that the free peptides affect the molecular dissociation process on the surface itself.

Binding reactions and diffusion: some simple considerations

Let's consider a ligand which just dissociates from a receptor immobilized on a surface with a given density of free receptors. What is the further destiny of such a ligand or being more precise, what are the probabilities that the ligand is found at a certain position at the surface or in solution after a certain time period has elapsed? This problem is nicely discussed in the book of Berg (Berg 1993) and here, a few important facts and results from this book will be used. As a starting point, a look at a computer simulation of a two-dimensional random walk gives some insight on the spatial distribution of the trajectory of a freely diffusing particle (Figure 5.5). Berg describes the result as follows:

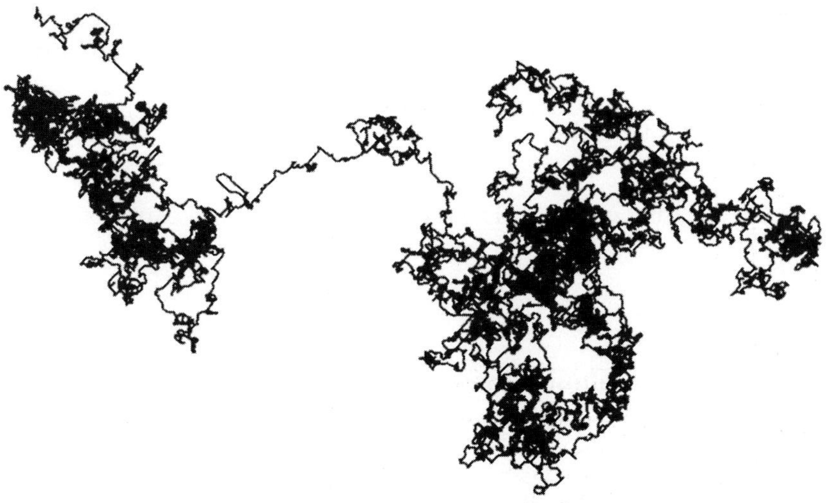

Figure 5.5　An x, y plot of a two-dimensional random walk of $n = 18{,}050$ steps. The computer pen started at the upper left corner of the track and worked its way to the upper right edge of the track. It repeatedly traversed regions that are completely black. It moved, as the crow flies, 196 step lengths. The expected root-mean-square displacement is $(2n)^{1/2} = 190$ step lengths. Taken from Berg (1993)

Since explorations over short distances can be made in much shorter time than explorations over long distances, the particle tends to explore a given region of space rather thoroughly. It tends to return to the same point many times before finally wandering away.

(Berg 1993)

Translating this description to the situation just described above, where the dissociated ligand is found in the immediate vicinity of the sensor surface suggests that this ligand will remain in the interfacial region for some time until it will by chance diffuse away. This, of course, makes it probable that it will bind again on one of the free receptors and at some time later occasionally dissociates and the same cycle may occur again. This simple picture already highlights the experimental parameters which are essential for describing the probability that rebinding occurs. The higher the density of free receptors on the surface, the more likely it is that the ligand will bind to a receptor when coming into contact with the surface; the probability that a complex will form is proportional to the association rate constant; and third, the faster the diffusion, the higher is the chance for a ligand to escape the interfacial region and to avoid rebinding.

Now, a more quantitative but still very simple analysis will be given. We again follow the ideas of Berg's book (Berg 1993) but we concentrate on planar geometry.

Let's take a surface in contact with a solution of ligand of concentration $[L]$. The surface acts as a perfect adsorber of ligands. This implies that in the immediate vicinity of the surface, the concentration of ligands is zero $[L_{surf}] = 0$. At a finite distance b away from the surface, a concentration $[L_0]$ is maintained. Being only interested in the time-independent behaviour – the steady-state case – we use Fick's first equation in the one-dimensional version (z corresponds to the surface normal) to calculate the flux of ligands J from a plane at a distance b and with a ligand concentration $[L_0]$

$$J = -D\frac{\partial[L]}{\partial z} = -D\frac{[L_0]}{b} \tag{5.14}$$

with D being the diffusion constant. The rate of ligands hitting a certain surface area A is given by:

$$I_{surf} = JA = -D\frac{[L_0]A}{b} \tag{5.15}$$

I_{surf} can be considered as a diffusion current which describes the rate of ligands being absorbed by the surface of an area A. From it, a diffusion rate constant k_+ can be calculated:

$$K_+ = \frac{I_{surf}}{[L_0]} = \frac{DA}{b} \tag{5.16}$$

Next, the current (in steady-state) into a small planar disc with radius 'a' in contact with the ligand solution will be given. The disc again acts as a perfect absorber. The calculation is a bit complicated (see Crank 1975: 42) hence only the result (Berg and Purcell 1977; Shoup and Szabo 1982) is given:

$$I_{disc} = 4Da[L_0] \tag{5.17}$$

(This result is obtained using cylindrical coordinates, therefore $[L_0] = [L](\infty)$.)

As a model for a surface with a given receptor density $[R_{tot}] = N/A$ (number of receptors per area), we chose a planar nonadsorbing surface on which a number N of perfectly absorbing discs are evenly distributed. The ligand current to such a model surface can be calculated using a very intuitive approach which is presented in Berg (1993): the time-independent second Fick's law

$$D\frac{\partial^2 C}{\partial z^2} = 0 \tag{5.18}$$

is analogous to the Laplace equation for the electrostatic potential in the charge-free space. If now the electrical current is interpreted as the ligand current and the electrical potential difference as the ligand concentration difference, we may assume that the diffusive resistance is analogous to the electrical resistance and is given by:

$$R_{disc}^{diff} = \frac{1}{4Da} \quad \text{and} \quad R_{surf}^{diff} = \frac{b}{DA} \tag{5.19}$$

It is easily seen that the problem described in this manner has the same mathematical structure than the electrical equivalent circuit shown in Figure 5.6 and, hence, is straightforward to solve (Berg uses spherical geometry because his interest is concerned with biological cells rather than with planar sensor surfaces).

As a result, the modified ligand current is reduced relative to the ligand current towards a perfectly absorbing surface according to:

$$I = \frac{I_{surf}}{1 + \dfrac{A}{Nab}} \tag{5.20}$$

Now, we interpret the expression $k_{disc} = 4Da$ as the association rate constant k_a of a ligand–receptor pair (Shoup and Szabo 1982) and then we can write an effective association rate constant which is affected by the diffusive resistance which accounts for the transport of ligands from solution to the surface. It has been shown (Berg 1978; Shoup and Szabo 1982) that the same can be applied for the dissociation rate constant:

$$k_{a/d}^{\,eff} = \frac{k_{a/d}}{1 + \dfrac{Nk_a}{k_+}}$$

With $k_+ = DA/b$ from equation (5.16) and with $[R] = N/A$, being the density of free receptors on the sensor surface, we obtain:

$$k_{a/d}^{\,eff} = \frac{k_{a/d}}{1 + \dfrac{[R]bk_a}{D}} \tag{5.21}$$

This result allows us to obtain some insight into the binding and unbinding kinetics between ligands and surface-bound receptors, in particular, the desorption process of ligands from surface-immobilized receptors. It is obvious that if

$$\frac{Nbk_a}{DA} = \frac{[R]bk_a}{D} \ll 1 \quad \text{or} \quad [R] \ll \frac{D}{bk_a} \tag{5.22}$$

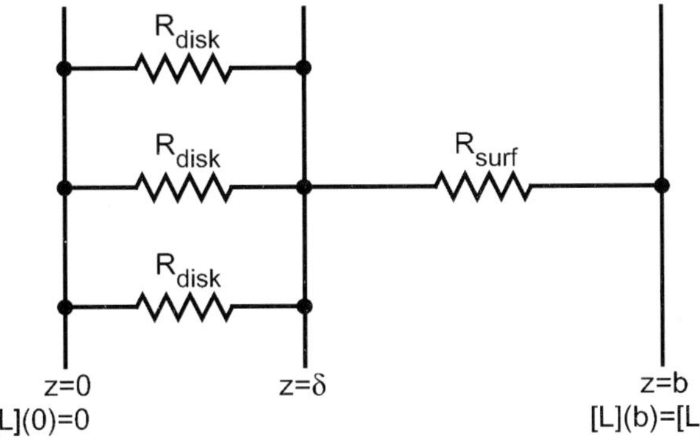

Figure 5.6 An electrical model for the problem of N adsorbers of radius a on a planar sensor surface. $R_{disk} = 1/4\,Da$ is the diffusion resistance of a disk-like adsorber of radius a. $R_{surf} = b/DA$ is the diffusive resistance of the whole surface. The original model for spherical surface geometry from Berg (1993) has been adapted to a planar surface geometry

mass transport limitation should not play a role (note, that $[R]$ is not the total receptor density $[R_{tot}]$, $[R]$ is varying constantly during a binding or a desorption experiment). From this inequality we can estimate the maximal free receptor density on the surface for any set of 'standard' parameters for a kinetic binding experiment in order to avoid mass transport effects: the only ill-defined parameter is b which describes the distance between the surface and a boundary at which the concentration gradient becomes zero, or in other words, where the ligand concentration $[L]$ corresponds to the macroscopic bulk concentration $[L_0]$. However, we can use some simple estimation: in the most simple cell design which is an unstirred sample cell where no continuous flow of solution is present, b corresponds to the thickness of the unstirred layer which is typically several hundreds of micrometer (between 100 and 500 µm). As the 'best' case we use the cell design of the BIAcore instrument which contains a cell of a width of 50 µm (Stenberg *et al.* 1991). We assume b to be half of this width, which is 25 µm. For k_a we use a typical value of $k_a = 10^5 \, \mathrm{M^{-1} s^{-1}}$ and for D, the diffusion constant of the antibody, $D = 5 \times 10^{-7} \, \mathrm{cm^2 s^{-1}}$. These values allow the estimation of an upper limit of the density of total receptor on the surface for two standard cell designs when mass transport effects are to be avoided ($[R_{tot}]bk_a/D \approx 0.1$): for an unstirred cell (thickness of the unstirred layer is 500 µm) a value for $[R_{tot}] \approx 2 \times 10^9$ free receptors/cm² is obtained, and for the flow-through cell from BIAcore 10^{11} receptors/cm² is found. The latter value may be compared to what the instruction manuals of BIAcore recommend: the information given in earlier times (Karlsson *et al.* 1991) of $[R_{tot}] \approx 0.3 - 1.2 \times 10^{12}$ receptors/cm² is surely too high. Only recently, an upper limit of receptor densities of $[R_{tot}] \leq 10^{11}$ receptors/cm² were considered to be reasonable in order to avoid mass transport effect. It is interesting to see that these latter values which are derived from experimental data are in good agreement with the simple evaluation given above. At this point, one could argue that, if the binding experiments preceding the dissociation of ligand lead to a complete saturation of the surface-immobilized receptors at the beginning of the desorption process, $[R]$ would be zero. Therefore, the initial part of the time course of dissociation – let's say 20 per cent of it – can be utilized for the determination of the dissociation rate constant. However, practical experience shows that in many cases the first part of the time course is corrupted by strong variations of the signal due to changes of buffer conditions or of flow velocity. These effects may lead to unacceptable errors. In addition, for the sake of high processing speed and minimal compound consumption, many adsorption experiments are carried out such that saturation of ligand binding will not occur and, therefore, the density of receptors with free binding sites is already finite at the beginning of the dissociation phase.

The simple expressions above were derived from the time-independent differential equation. Their application to processes which vary in time is limited and depends on the experimental conditions. We may ask whether there is a more suitable experimental situation at which the equations (5.21) and (5.22) can be immediately tested. Let's have a look at the following situation: the dissociation process of ligands from the surface is triggered through rinsing with plain buffer. If the receptor density is high, the desorption process does not lead to full separation but will end at a surface coverage which is non-zero (see Figure 5.4). The experiment reaches a quasi-steady regime which now corresponds to the boundary conditions which were used to derive the equations (5.22) and (5.23). Under these conditions, the probability of a ligand which just dissociated from the receptor to escape from the interfacial region goes to zero. Berg has shown that the fraction F_{diss} of ligands which achieve true separation from the surface after rinsing with buffer for long periods of time can be calculated with the formalism used above (Berg 1978; Goldstein and Dembo 1995) and the result has a structure which is very similar to equation (5.21).

$$F_{diss} = \frac{1}{1 + \dfrac{NK_a}{k_+}} = \frac{1}{1 + \dfrac{[R]bk_a}{D}} \tag{5.23}$$

F_{diss} is an easily accessible experimental property and allows us to test a few assumptions and to recover some parameters. As an example, we now evaluate the thickness of an unstirred layer b in a desorption experiment under steady-state conditions. We use again those experimental data from NANP-peptide system obtained using the BIAcore instrument. The association rate constant k_a for this system was measured as $10^5 M^{-1} s^{-1}$ (Duschl *et al.* 1996) and the diffusion constant D of the antibodies is given in literature by $5 \times 10^{-7} cm^2 s^{-1}$. Surface densities of the NANP-peptides of 2×10^{12} receptors/cm² and 4.8×10^{12} receptors/cm² lead to a fraction F_{diss} of antibodies which achieve true separation from the surface of 50 per cent and 10 per cent respectively using the maximal flow rate of $100 \, \mu l min^{-1}$. In the case of the lower surface coverage of NANP peptides, the calculation of the thicknesses of the unstirred layer gives a value of $b \approx 15 \, \mu m$, for the higher coverage a value of $b \approx 50 \, \mu m$ is obtained. These values agree well with the dimension of the flow cell of this instrument (width $= 50 \, \mu m$).

Equation (5.23) puts tougher conditions on the design of an experiment for the accurate measurements of the rate constants than the previous treatment. Let's assume that a certain (small) percentage of the ligands never come off the surface at infinite time, the dissociation rate constant must have already decreased for a considerable time period before the equilibrium was reached. In this sense the two results are not equivalent.

Now, we investigate the effect on a kinetic experiment when a soluble receptor is added to the rinsing solution which competes for ligand binding with the surface immobilized receptors. The presence of sufficiently high concentration of a free soluble receptor in the rinsing buffer may reduce rebinding because the free receptor competes for free binding sites on the ligand. We focus again on the desorption process and use the same assumptions and boundary conditions as in the previous example. Goldstein *et al.* (1989) have described an approach which includes the presence of free receptors in solution. However, it was formulated for spherical geometry using the time-independent version of second Fick's law (in the same way as described above). Two terms were added which account for the interaction of the free ligands with the free receptor in solution. The term

$$k_a[S][L]$$

describes the binding between free ligands and free receptors in solution with $[S]$ being the concentration of soluble receptor and $[L]$ being the concentration of free ligand (we assume a large $[S]$ such that a dissociation term can be neglected). In this configuration, it is no longer possible to maintain a non-zero steady-state by holding the ligand concentration constant at a distance b. Therefore, a second term G was added which is responsible for maintaining a given free ligand concentration uniformly throughout space. The loss of free ligand through binding to free or immobilized receptors is compensated through an injection rate of free receptors G. It must be such that $G/k_a[S][L_0] = 1$, other values would give unphysical spatial ligand distributions. A fixed concentration $[S_0]$ at a macroscopic distance b away from the sensor surface as it was used in the simpler case with no free receptors in solution, would not lead to reasonable solutions.

From the time-independent differential equation

$$\frac{\partial^2[L]}{\partial z^2} - \frac{k_a[S]}{D}[L] + \frac{G}{D} = \frac{\partial[L]}{\partial t} = 0 \tag{5.24}$$

with $[L](0) = 0$ and $[L](b) = [L_0]$, the spatial distribution of the free ligand concentration is given by:

$$[L](z) = \left([L_0] - \frac{G}{k_a[S]}\right)\frac{\sinh(\mu z)}{\sinh(\mu b)} + \frac{G}{k_a[S]}\left(\frac{\sinh(\mu(z-b))}{\sinh(\mu b)} + 1\right) \tag{5.25}$$

$$\text{with }\left(\mu = \sqrt{\frac{k_a[S]}{D}}\right)$$

μ has the dimension of an inverse length and $1/\mu$ gives the distance a ligand migrates during a given reaction time.

Now, the modified diffusion rate constant k_+ at the interface can be calculated:

$$k_+ = \frac{DA\dfrac{\partial[L]}{\partial z}\bigg|_{z=0}}{[L_0]} = \frac{DA\mu}{\sinh(\mu b)}\left(1 + \frac{G}{k_a[S][L_0]}(\cosh(\mu b) - 1)\right) \tag{5.26}$$

With $b \gg \dfrac{1}{\mu} = \sqrt{\dfrac{D}{k_a[S]}}$ and the condition $G/k_a[S][L_0] = 1$, we obtain

$$k_+ \approx AD\sqrt{\frac{k_a[S]}{D}} = A\sqrt{k_a[S]D} \tag{5.27}$$

Now, we can calculate the free receptor concentration necessary to suppress rebinding for a given receptor density at the sensor surface and a given association rate constant k_a. The diffusion rate constant k_+ may be introduced into equations (5.21) or (5.23) and we obtain from

$$\frac{Nk_a}{k_+} \ll 1, \text{ the inequality } [S] > [R]^2\frac{k_a}{D} \tag{5.28}$$

It allows the estimation of a free receptor concentration in solution $[S]$ which is needed to efficiently suppress rebinding.

In order to test this inequality using experimental data, we take the parameters from the already well-known NANP-experiment discussed at the beginning of this section. The surface density of receptors in this experiment was determined to be $[R_{tot}] = 4.8 \times 10^{12}$ receptors/cm^2. In order to obtain the right units for S, $[R_{tot}]$ is expressed as a molarity, $c = 4.8 \times 10^{12}/6 \times 10^{20}$ M, the association rate constant of this system was measured to be $k_a = 10^5$ M^{-1}s^{-1} and the diffusion constant for antibodies to be $D = 5 \times 10^{-7}$ cm^{2}s^{-1}. We calculate that the concentration of free receptor in solution $[S]$ should be greater than $[S] = 1.3 \times 10^{-5}$ M. This is in good agreement with the experimental results: from the desorption time courses (see Figure 5.4), we recognize that $[S] = 10^{-5}$ M is still too small to prevent rebinding effects, but $[S] = 10^{-4}$ M is able to suppress completely mass transport effects. Note that in this approximation the width of the unstirred layer does not occur anymore. This is a consequence of the inequality $b \gg 1/\mu$ and holds only for sufficiently high concentrations of free receptors.

In this part, a fairly basic concept was used in order to demonstrate the effect of mass transport limitations which lead to the rebinding of ligands when they desorb from surface immobilized receptors. The concept gives some guidance and some quantitative criteria for the choice of the experimental parameters on a kinetic measurement using a planar chip-based detection principle when mass transport limitations are to be avoided. First, it sets an upper limit on the density of the receptor on the sensor chip as a function of the association rate constant (of course, an accurate number for association rate constants should be the result of such an experiment; however, in many cases an educated guess of its size should be possible and should serve as a starting point in any practical approach). Second, if the surface density of the receptors has to be increased above this limit (e.g. for sensitivity reasons), the consequences can be evaluated. In particular, for the design of an experiment using a competitive binding regime, concentrations of free receptors needed to suppress rebinding can be readily estimated. Third, the thicknesses of the unstirred layer can be determined, which may be important for the design of the geometry of sample cells.

However, this model reaches its limitation when an accurate evaluation of the rate constants from the time courses of the adsorption and the desorption processes is needed. For such tasks computer simulation programs and fitting programs have recently been developed. In many cases, they have been adapted to the cell designs of the BIAcore system which is most commonly used. These programs may be applied to measurements obtained from this instrument and some of them are included in its evaluation software. In the following section, two examples of such evaluation programs will be discussed. The first paper was published by Glaser (1993) and the second article by Myszka *et al.* (1998) appeared in 1998. Their major features greatly aid a reasonable interpretation of the measured curves.

Evaluations of kinetic experiments using computer simulations and fitting programs

The mathematical formulation of the binding and dissociation reactions of a ligand–receptor pair, which also includes the transport of the ligand to the receptor through diffusion or applied flow as a function of time, leads to a set of differential equations (see equation (5.29) and (5.30)) which cannot be solved analytically. This is no longer a critical limitation as, nowadays, standard computers are sufficiently powerful to perform a numerical analysis of such tasks without big investments of money or time. Two strategies to solve this problem have been employed: first, simulated binding or dissociation curves can be produced by numerically solving the basic differential equation for certain sets of parameters (association and dissociation rate constants, two-dimensional receptor density, concentration and diffusion constant of the ligand, cell dimensions and flow rate of the solutions). The only approximation is the division of the cell volume into small compartments and the time course into short intervals. The size of the volume elements and the width of the time steps determine the accuracy of the solution. The discrete volumina are not necessarily of equal size, nor must the time intervals be equally spaced; wherever the concentration gradients are large, the compartment sizes must be small and, accordingly, whenever the changes in time are fast the time intervals are to be short. The curves produced can then be compared with experimental results. In the following, the usefulness and the limitations of such a strategy are briefly discussed.

Second, perhaps a practically more fruitful approach is based on the development of a fitting program. However, the many compartment model is unsuited to formulate a fitting routine because of its complexity. The task is then to find a sufficiently simple model for the development of a fitting program. The second step is to evaluate such a fitting program and to test how well it reproduces the input parameters of simulated time courses which are produced as described above. If this test is successful, it can be applied to experimental curves. This approach is discussed in the second part of this section.

In the paper of Glaser (1993), the set of differential equations which must be numerically integrated describe the reaction of receptor and ligand (5.29) and the transport of ligand (5.30).

$$\frac{d[RL]}{dt} = k_a[R][L] - k_d[RL] \tag{5.29}$$

$$\frac{\partial[L]}{\partial t} = v_x \frac{\partial[L]}{\partial x} - D\frac{\partial^2[L]}{\partial z^2} \tag{5.30}$$

x defines the direction along the flow which is parallel to the plane of the sensor surface and z is parallel to the surface normal. In case there is no flow applied, the first term on the right-hand side of equation (5.30) disappears. For the details of the boundary conditions, the reader is referred to the original paper.

In order to evaluate simulations derived from the two differential equations, a quasi-steady state is introduced which is realized by a constant concentration of free ligand in the immediate vicinity of the sensor surface. In this case, mass transport and the binding reaction process can be considered to be sequential events. With this assumption the rate of binding or the binding flux j as it is called in this chapter is proportional to

$$j \propto \frac{L_m}{L_r + L_m} \tag{5.31}$$

with

$$L_r = k_a[R_{free}] = k_a([R_{tot}] - [RL])$$ (5.32)

and

$$L_m \approx \sqrt[3]{\frac{D^2 v}{h^2 bl}}$$ (5.33)

being the term which describes the mass transport in a flow through cell with h, b and l being the height, the width and the length of the sample cell; v is the flow rate applied to the solution. The common structure of (5.31) and (5.21) becomes immediately obvious. Mass transport effects do not play a role in a kinetic experiment when the left side of equation $(5.31) \approx 1$, e.g. when $L_m \gg L_r$. However in contrast to what has been discussed above, in (5.31), the term describing the mass transport is derived for a flow through cell (Lok *et al.* 1981). As an example, we refer again to the receptor–ligand system comprising the NANP peptides: a critical surface receptor density may be estimated, below which mass transport is absent, and it can be compared to the values which were derived from the very simple considerations presented on page 104 and before. D and k_a remain the same ($k_a = 10^5\,\mathrm{M}^{-1}\mathrm{s}^{-1}$, $D = 5 \times 10^{-7}\,\mathrm{cm}^2\mathrm{s}^{-1}$). The cell dimensions of the BIAcore instrument are $h = 50\,\mu\mathrm{m}$, $b = 0.3\,\mathrm{mm}$ and $l = 2.4\,\mathrm{mm}$ (see BIAcore instrument manual). A flow rate $v = 10\,\mu\mathrm{lmin}^{-1}$ is intermediate between the extreme values of $1\,\mu\mathrm{lmin}^{-1}$ and $100\,\mu\mathrm{lmin}^{-1}$ of this instrument (note that L_m is proportional to $v^{1/3}$, increasing the flow rate over two orders of magnitude, changes L_m only by a factor of 4.6). For $L_r/L_m \approx 0.1$, a receptor density of 1.6×10^{11} receptor/cm^2 is obtained. It is practically identical to the very simple estimation presented on page 104. However, this is not too surprising since, as already mentioned at the beginning of this paragraph, both mathematical formulations are rather similar. It shows that the rather crude assumption on the thickness of the unstirred layer in the BIAcore flow cell described on page 104 is equivalent to the more complicated calculation of the diffusion rate constant in equation (5.33).

In the paper by Glaser, simulations are carried out in which the two different limiting cases $L_m \ll L_r$ and $L_m \gg L_r$ were permutated with $1/K \gg [L]$ and $1/K \ll [L]$. The evaluation of the data followed the same strategy as described on page 92: the binding rate $d[RL]/dt$ was plotted as a function of $[RL]$ for several ligand concentrations $[L]$. The slopes of these curves were then plotted against the ligand concentration $[L]$. In the case where the set of parameters corresponded to the rapid mixing approximation, this procedure yielded both: the association rate constant was given by the slope of the linear curve and the value at which the curve crossed the abscissa gave the dissociation rate constant. When parameters were introduced which were supposed to cause mass transport determined reactions and which included high ligand concentrations $[L]$, the shape of the curves deviated from straight lines as expected. For low ligand concentrations, the shape of the curves suggested the nonexistence of mass transport but the values recovered did not correspond to the ones used as input. These results may serve as a guideline for the development of a strategy which is able to identify mass transport limitations. However, this approach is not generally applicable and not very satisfying either. In order to judge whether mass transport plays a substantial role in a binding experiment, the author focused on the determination of L_r and L_m. For a given cell geometry, a sufficiently accurate calculation of L_m is possible using equation (5.33). The determination of L_r according to (5.32) is more difficult. By simply measuring the amount of immobilized receptor, $[R_{tot}]$ can be independently determined. However, the value for k_a in equation (5.33) is, of course, unknown at the beginning. A way out of this dilemma is to perform association experiments with decreasing receptor densities. When mass transport effects are present, the apparent association rate constant, k'_a, derived from an association curve, is always smaller than the intrinsic k_a. Therefore, k'_a increases with decreasing receptor density. Reliable values for k_a may then be obtained, if k'_a asymptotically approaches a limiting value at low receptor densities.

Glaser's paper also describes simulations of competitive reactions during the dissociation phase. Two options are discussed: the addition of a second ligand B is accelerating the dissociation through the suppression of rebinding of ligand A since the free sites on the receptor are efficiently occupied by a ligand of type B and therefore, blocked for ligand A. The concentration of ligand B, $[L_B]$, has to be such that the probability of rebinding of ligand A is much lower than the binding of ligand B to a free receptor site. In a simulation where the receptor sites were fully occupied after the association phase and where $k_a = 5 \times 10^7 \, \mathrm{M}^{-1}\mathrm{s}^{-1}$ and $[R_{tot}] = 1.5 \times 10^{12}$ receptors/cm^2 were used, the concentration of ligand B, $[L_B]$, needed to suppress any rebinding was $[L_B] = 10^{-6} \, \mathrm{M}$. The other option to avoid mass transport effects is the addition of soluble receptors to the buffer solution, identical to what has been already described in the previous parts. For this strategy, concentrations of soluble receptor $[S]$ are needed which must be considerably higher than the concentration of competitive ligand B $[L_B]$ in the previous case. The simulations show that $[S] = 10^{-5} \, \mathrm{M}$ is just below the value which is needed to avoid mass transport limitations in the desorption kinetics. This result may be compared to the critical concentration of soluble receptor sufficient to avoid mass transport effects which can be calculated using equation (5.28). Although equation (5.28) was not derived for a flow-through cell, we may use it for this configuration since, for the relevant free receptor concentrations $[S]$, the inequality $b \gg 1/\mu$ holds for any realistic thickness of the unstirred layer b in this particular flow-through system (see how equation (5.27) was derived). Taking the parameters of the simulation ($k_a = 2 \times 10^6 \, \mathrm{M}^{-1}\mathrm{s}^{-1}$, $[R_{tot}] = 1.5 \times 10^{12}$ receptors/cm^2 and $D = 5 \times 10^{-7} \, \mathrm{cm}^2\mathrm{s}^{-1}$), the calculated concentration of free receptors must be above $[S] = 2.5 \times 10^{-5} \, \mathrm{M}$ in order to avoid mass transport effects. Although no simulation was carried out for concentration of $[S]$ being in this range, the data presented by Glaser seem to be in accordance with this critical value of $[S]$ derived with the use of equation (5.28).

Both approaches may produce reliable data on the dissociation rate constant. On first sight, the addition of a competing ligand B seems to be more advantageous since lower concentrations of competing compounds are needed (note that, in the first case, the association rate constant was even a factor twenty-five times larger than in the second case). However, if techniques are used which are sensitive to the refractive index change at the sensor surface (e.g. SPR), the choice for the optimal approach depends very much on the relative molecular weight of the competing compound in respect to the ligand to be investigated. High molecular weight compounds interfere more strongly with the measurements than lighter ones. Since the detection principle of SPR is based on the interaction of an evanescent field with the molecules in the vicinity of the surface, the instrumental response also depends on the proximity of the additional mass of the competitor (ligand or receptor) to the surface. If the added compound is a ligand, it will compete for binding sites immobilized to the surface whereas, in this second case, the competition about the binding sites of the receptor takes place in solution. Another aspect to consider is that the diffusion constant is roughly proportional to the inverse of the cubic root of the molecular mass; large competing compounds may introduce unwanted additional mass transport effects. Finally, the design of a competitive assay depends on the availability and solubility of a potential competitor. If, for example, immobilized receptors are to be investigated which need a complex environment (transmembrane proteins), the second option is hardly feasible. Therefore, it depends very much on the overall experimental conditions which, of the two approaches, using a competitive reaction scheme is more favourable.

Up to now, the role of the various parameters in a kinetic experiment for the determination of the reaction kinetics has been discussed. However, what we are still lacking is a stringent analysis which allows us to decide whether and to what extent the time course of a binding or a dissociation reaction is influenced by mass transport. In order to appropriately evaluate kinetic data, the obvious thing to do is to fit them to a model which accounts for both, the mass transport and the reaction kinetics. Such a model must be sufficiently simple in order to limit its mathematical complexity to a reasonable level and to avoid ambiguity of the fitting results.

In the following, we will discuss the paper of Myszka *et al.* (1998) in which a simple model is used to fit simulated data. They were produced in very much the same way as in Glaser's paper. The fit routine was tested through evaluation of how reliably the input parameter could be recovered under various degrees of mass transport effects. Additionally, realistic levels of random noise were added to the simulations in order to explore the signal-to-noise ratio needed to reliably recover the original parameters. The dimensions of the sample cell and the flow rates of buffer were taken from the BIAcore instrument.

The numerical procedure for the generation of the simulations is based on treating the hydrodynamic transport with the method of finite elements. The measuring cell is divided into a 32×32 and a 64×64 grid (Figure 5.7a). The compartments close to the sensor surface and to the opposite cell wall are subdivided in order to account for the higher concentration gradients. The differential equations which had to be solved correspond to the ones in Glaser's work but also include the diffusion in the direction of flow.

The authors do not only analyse the time courses of the association and dissociation phases of ligands but also relate them to the time-dependent concentration of ligands in the compartment closest to the surface (Figures 5.7b and c). This shows convincingly how the ligand concentration varies during the reaction processes: during the association, the concentration increases similarly to the amount of adsorbed ligand; during the dissociation phase, the ligand concentration is non-zero, in contrast to what is assumed in the rapid mixing model. The authors also show that the concentration of bound ligand varies along the sensor surface. For the parameter set used ($k_a = 8 \times 10^6 \, \mathrm{M^{-1}s^{-1}}$, $k_d = 0.2 \, \mathrm{s^{-1}}$, $[R] = 7.5 \times 10^{11}$ receptors/cm^2, $D = 10^6 \, \mathrm{cm^2 s^{-1}}$ and flow $= 50 \, \mu\mathrm{l min^{-1}}$) which implies that mass transport effects are present, the lateral variation of bound ligand is approximately 15 per cent. Since the laser beam probes $\frac{3}{4}$ of the total flow path, the signal is an average over the flow path. In contrast, if a parameter set is used such that no mass transport interferes with the binding reaction ($k_a = 8 \times 10^4\,\mathrm{M^{-1}s^{-1}}$, all other parameters as above), the concentration of ligand in the compartment closest to the sensor surface corresponds to the rapid mixing model.

In order to analyse time-dependent adsorption and desorption curves, and to obtain reliable binding data from them, a fitting routine has been developed which is based on the two-compartment model (Myszka *et al.* 1996; Myszka *et al.* 1997). In the two-compartment model, the flow cell is divided into two parts where the ligand concentrations are uniform in space (but vary in time): in the outer compartment, the concentration equals the preset ligand concentration $[L]$, the concentration of the inner compartment (next to the sensor surface) $[L_{in}]$ is dependent on the transport between the compartments and on the exchange of ligand with the sensor surface. The ligand transport between the compartments, the binding and unbinding is described by k_+ (in the paper k_M is used) k_a and k_d, respectively. The expression for k_+ is identical with the one for L_m in Glaser's paper (apart from a prefactor which is close to 1). Its dependence on the height of the compartment can be removed when the quasi-steady-state approximation (it means that $d[L_{in}]/dt = 0$; see page 108) is applied. As the authors show for the range of parameter values adapted from the BIAcore instrument, this approximation is justified.

The rate constants k_a and k_d, $\tilde{k}_+$ ($\tilde{k}_+$ is the parameterized version of k_+ in order to obtain the right units) are used as the free fit parameters. They are determined from a nonlinear least-square fitting routine (the software used in this analysis (CLAMP) may be downloaded free of charge from the website: http://www.hci.utah.edu/cores/biacore/docs/clamp.html). Different surface receptor densities are fitted in order to evaluate the relative roles of the reaction rate and the transport rates – a strategy which has already been successfully used for the analysis of binding data and which is often described as global analysis (Morton 1995).

In the paper, the authors use two different surface receptor densities: $[R_{tot}] = 3 \times 10^{10}$ and 10^{11} receptors/cm^2 and varied the association rate constants such that they cover all the scenarios ranging from the rapid mixing model to strongly mass transport determined behaviour. As expected, for pure reaction rate determined conditions, fits which were either based on the rapid mixing model or on the two-compartment model, both recovered the input parameter equally well. However, the value of $\tilde{k}_+$ which was obtained from the latter fit only, was not correct, simply because, under these conditions, the data did not supply the information for such a fit. Only when k_a was increased to $k_a = 5 \times 10^5\,\mathrm{M^{-1}s^{-1}}$, $\tilde{k}_+$ yielded a meaningful value which indicated that mass transport started to interfere with the binding reaction, although the ratio $L_r/L_m = k_a[R]/\tilde{k}_+ \approx 0.03$, was surprisingly small. At this stage, the other parameters were equally well recovered by both models, the simple fit routine and the two-compartment model. Using a $k_a = 10^8\,\mathrm{M^{-1}s^{-1}}$ for the simulations of the time courses, only the two-compartment model was able to reproduce the experimental parameters well, including $\tilde{k}_+$. $\tilde{k}_+$ also allowed an accurate determination of the diffusion constant of the ligand (in this case $D = 10^{-6}\,\mathrm{cm^2s^{-1}}$) which may be an additional quality control of the fit. This clearly implied that mass transport was interfering with the binding kinetics. By adding realistic random noise levels to simulations which corresponded to a very low mass coverage (down to $15\,\mathrm{RU}$ or 3×10^{10} bound ligands), it was demonstrated that the fitting procedure still yields excellent

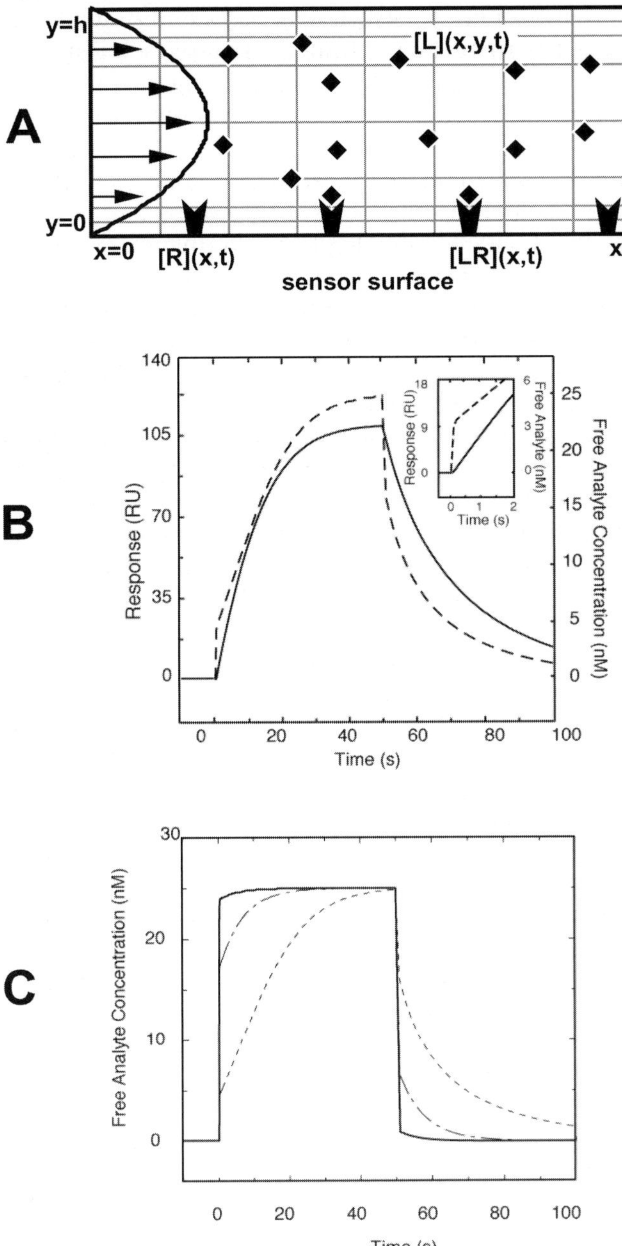

Figure 5.7 Schematic of a BIAcore sample chamber with an 8×8 computational grid in grey (A). Simulated time courses of the average concentration of bound ligand (—) and for the average free ligand concentration just above the sensor surface (-----) during the adsorption and desorption phase. The parameters are such that a mass transport limited kinetics is simulated (B). Simulated time courses of the average free ligand concentration just above the sensor surface for $k_a = 8 \times 10^4 \, \mathrm{M^{-1} s^{-1}}$ (—), $k_a = 8 \times 10^5 \, \mathrm{M^{-1} s^{-1}}$ (−·−) and $k_a = 8 \times 10^6 \, \mathrm{M^{-1} s^{-1}}$ (----) (C). Taken from Myszka (1998)

results under rather unfavourable conditions. Note that the approach does not account for additional mass transport effects which may be caused by the extended dextran network (which reaches approximately 100 nm out into the solution) which is one of the standard matrices for the immobilization of receptors on the sensor surface. The fitting program is now available with the BIAcore software package.

Global analysis

High quality kinetic data can be obtained if a number of experiments are performed with the essential parameters being systematically varied, in particular the receptor surface density and the ligand concentration in solution, and subsequently if all the experimental curves are fitted for one consistent set of values (Morton *et al.* 1995). The requirement for such a global analysis to be successfully carried out is a high degree of control over the experimental system in order to obtain very reproducible raw data. As a good example, the kinetic analysis of the binding between the interleukin-2 receptor (IL-2R) complex (α-, β-, and γ-subunit, the latter is not involved in the binding) with its ligand (IL-2) (Myszka *et al.* 1996) is discussed. On the surface of an activated T-cell, a heterodimer of an IL-2Rα- and an IL-2Rβ-subunit exists which forms a high affinity complex with the ligand. The binding constants of the individual subunits are considerably smaller, albeit there exists a high affinity and a low affinity subunit (see Figure 5.8a). In an attempt to elucidate the role of the ligand-binding extracellular domains in the dimerization process, these domains were immobilized on a dextran-functionalized BIAcore sensor chip via thiol chemistry. The high local concentration of both subdomains on the surface was the precondition for a possible dimerization. First, it was verified that the equilibrium binding constants of the subunits and the complex were close to the ones obtained on T-cells. The kinetic data were recorded from experiments of two different surface densities of each of the two subunits and of the complex for several ligand concentrations. High surface densities of the monomeric receptors had to be applied in order to account for the relatively low molecular weight of the ligand, hence experiments had to be carried out in the mass transport determined regime. The least-square fitting routine was based on the two-compartment model (at that time, this model was not yet tested using simulated data) and was first applied to the data obtained from the individual subunits. From the data of the immobilized high affinity subunit, the mass transport rate of the system could be determined. This was then used to obtain the kinetic rate constants of the even tighter bound heterodimer-ligand complex (again in the mass transport regime) from fits which were based on two models: the 'affinity-conversion' mechanism assumes that the ligand binds first to the high affinity subunit and this complex is then stabilized through the formation of a heterodimer; in the other model which fitted the data much better, the ligand binds directly to the preformed heterodimer of the α- and β-subunit (Figure 5.8b, right). This result which is based on an excellent data set, gave the first indication that the extra-cellular domains are responsible for the association of the full subunits in the membrane of the T-cells, independent of the ligand binding. The paper demonstrates that the global analysis approach is able to determine kinetic rate constants of high quality under conditions where mass transport effects are a consequence of properly reconstituting a biological ligand–receptor system on a sensor surface.

$$\left(\frac{d\Gamma}{dt}\right)_{max} = \frac{D[A]}{\delta} \tag{5.37}$$

The sample cell geometry used was similar to the one of the BIAcore instrument with a cell width of $50\,\mu m$. A laminar flow of $5\,mm/s$ (the flow rate in Myszka (1998). was ten to twenty times faster) gives rise to a diffusion layer thickness of 'a few $10\,\mu m$'. The flow rate was sufficiently high to compensate for antibody consumption at the sensor surface. The major error of this method comes from the additional free antibodies which are produced in the unstirred layer: since the free antibody concentration decreases towards the surface, the equilibrium is disturbed and receptor ligand complexes dissociate. Under the assumption that this process is slow (small dissociation rate constant), the effective dissociation rate in the unstirred layer at distance r from the surface is given by:

$$\frac{d[A]}{dt}(r) = k_d[RA] - k_a[A](r)[R] = k_d[RA]\left(1 - \frac{r}{\delta}\right) \tag{5.38}$$

Integration of this equation over the unstirred layer yields the maximal additional rate of free antibodies produced within the unstirred layer:

$$\left(\frac{d\Gamma}{dt}\right)_{diss} = \frac{k_d[RA]\delta}{2} \tag{5.39}$$

In order to keep the error signal negligible, the authors propose that the dissociation rate of a complex to be measured should not be above $0.02\,s^{-1}$, assuming a binding rate to the surface of 0.1–$0.2\,ng\,cm^2\,s^{-1}$. As can be seen from (5.36) and (5.37), optimal results may be obtained by reducing the thickness of the unstirred layer, hence the design of an optimal flow cell is an essential requirement for performing such experiments.

The authors demonstrated that a linear increase of the binding signal was observed (Figure 5.9a) in agreement with their theoretical treatment of mass transport limited binding of antibodies. From the slope of the signal curve, the concentrations of free antibodies $[A]$ were directly determined. These concentrations were measured for different free receptor (analyte) concentrations $[R_i]$ and the obtained titration curves were fitted to equation (5.34) in order to obtain the binding constant K. The different binding modes of monovalent and bivalent ligands could be clearly distinguished if the proper ligand concentrations were used (Figure 5.9b).

The maximum binding constant which may be detected was estimated to be 10^{10}–$10^{11}\,M^{-1}$. It depends directly on the sensitivity of the instrument: the ligand concentrations $[L_i]$ for producing the optimal results in these experiments are in the range of the inverse binding constant. This condition limits the maximum binding rate towards the surface (see equation 5.35). The dissociation rate is the crucial parameter which determines the rate of production of free ligands in the unstirred layer. Below a binding constant of $10^6\,M^{-1}$, the dissociation rates are so fast that the additionally generated free ligands produce error signals which are comparable to the 'real' signal. Overall, the kinetic detection of equilibrium binding constants in solution is a versatile approach for the fast characterization of ligand–receptor pairs, in particular, when the receptor is a low molecular weight compound and is difficult to immobilize to a sensor surface.

Conclusions

In recent years, a couple of instruments were developed for the on-line measurement of binding reactions at sensor surfaces. These instruments are typically based on optical techniques which allow the detection of changes of the mass coverage with remarkable sensitivity in various media. To optimally exploit this potential for the acquisition of high quality binding data, the experimental conditions must be designed with great care: the surface has to be engineered such that the bindings sites are optimally presented. For kinetic experiments, the transport of ligands to the receptor has to be controlled. In this context three properties are of particular importance, namely the association rate constant k_a, the diffusion constant D of the ligand and the density of unbound receptor on the surface $[R]$. Since only the latter may be freely chosen, interference of mass transport effects is present in most kinetic experiments. In these cases, suitable evaluation software is needed to properly analyse the data. Only when all these requirements are met may useful information on a given receptor–ligand complex be obtained.

Acknowledgement

The measurements of Anne-Françoise Sévin-Landais, the programming of Stephen Heyse and the very helpful discussions with them are gratefully acknowledged. I am grateful to Ralf Glaser for making his simulation program available to us and for his advice on it. I would also like to thank Samuel Terrettaz and Jeremy H. Lakey for critically reading the manuscript.

References

Adam, G. and Delbrück, M., 1968, Reduction of Dimensionality in Biological Diffusion Processes, in Rich, A. and Davidson, N., (eds) *Structural Chemistry and Molecular Biology*, pp. 198–215, San Francisco: W.H. Freeman and Company.

Altschuh, D., Dubs, M.-C., Weiss, E., Zeder-Lutz, G. and Regenmortel, M.H.V.V., 1992, Determination of Kinetic Constants for the Interaction Between a Monoclonal Antibody and Peptides Using Surface Plasmon Resonance, *Biochemistry* 31, 6298–6304.

Andrade, J.D. and Hlady, V., 1986, Protein Adsorption and Materials Biocompatibility: A Tutorial Review and Suggested Hypotheses, in Dusek, K. (ed.) *Advances Polymer Science* 79, pp. 2–63, Berlin: Springer.

Axelrod, D. and Wong, M.D., 1994, Reduction-of-Dimensionality Kinetics at Reaction-Limited Cell Surface Receptors, *Biophysical Journal* 66, 588–600.

Berg, H.C., 1993, *Random Walks in Biology*, Princeton: Princeton University Press.

Berg, H.C. and Purcell, E.M., 1977, Physics of Chemoreception, *Biophysical Journal* 20, 193–219.

Berg, O.G., 1978, On Diffusion-Controlled Dissociation, *Chemical Physics* 31, 47–57.

Berg, O.G. and Hippel, P.H.v., 1985, Diffusion-Controlled Macromolecular Interactions, *Annual Review of Biophysics and Biophysical Chemistry* 14, 131–215.

Bieri, C., Ernst, O.P., Heyse, S., Hofmann, K.P. and Vogel, H., 1999, Micropattern Immobilization of a G-protein-coupled Receptor and Direct Detection of G protein Activation, *Nature Biotechnology* 17, 1105–1108.

Cantor, C.R. and Schimmel, P.R., 1980, *Biophysical Chemistry, Part III: The Behaviour of Biological Macromolecules*, New York: W.H. Freeman and Company.

Cornell, B.A., Braach-Maksvytis, V., King, L.G., Osman, P.D.J., Raguse., B., Wieczorek, L. and Pace, R.J., 1997, A Biosensor that Uses Ion-Channel Switches, *Nature* 387, 580–583.

Crank, J., 1975, *The Mathematics of Diffusion*, 2nd edn, Oxford: Oxford University Press.

Darst, S.A., Ahlers, M., Meller, P.H., Kubalek, E.W., Blankenburg, R., Ribi, H.O., Ringsdorf, H. and Kornberg, R.D., 1991, Two-Dimensional Crystals of Streptavidin on Biotinylated Lipid Layers and their Interactions with Biotinylated Macromolecules, *Biophysical Journal* 59, 387–396.

Duschl, C., Liley, M., Corradin, G. and Vogel, H., 1994, Biologically Addressable Monolayer Structures Formed by Templates of Sulphur-Bearing Molecules, *Biophysical Journal* 67, 1229–1237.

Duschl, C., Sévin-Landais, A.-F. and Vogel, H., 1996, Surface Engineering: Optimization of Antigen Presentation in Self-Assembled Monolayers, *Biophysical Journal* 70, 1985–1995.

Fägerstam, L.G., Frostell-Karlsson, Å., Karlsson, R., Persson, B. and Rönnberg, I., 1992, Biospecific Interaction Analysis Using Surface Plasmon Resonance Detection Applied to Kinetic, Binding Site and Concentration Analysis, *Journal of Chromatography* 597, 397–410.

Gizeli, E., Liley, M., Lowe, C.R. and Vogel, H., 1997, Antibody Binding to a Functionalized Supported Lipid Layer: A Direct Acoustic Immunosensor, *Analytical Chemistry* 69, 4808–4813.

Glaser, R.W., 1993, Antigen-Antibody Binding and Mass Transport by Convection and Diffusion to a Surface: A Two-dimensional Computer Model of Binding and Dissociation Kinetics, *Analytical Biochemistry* 213, 152–161.

Godson, G.N., 1985, Molecular Approaches to Malaria Vaccines, *Scientific American* 5, 32–39.

Goldstein, B. and Dembo, M., 1995, Approximating the Effect of Diffusion on Reversible Reactions at the Cell Surface: Ligand–Receptor Kinetics, *Biophysical Journal* 68, 1222–1230.

Goldstein, B., Posner, R.G., Torney, D.C., Erickson, J., Holowka, D. and Baird, B., 1989, Competition Between Solution and Cell Surface Receptors for Ligand, *Biophysical Journal* 56, 955–966.

Heyse, S., Ernst, O.P., Dienes, Z., Hofmann, K.P. and Vogel, H., 1998, Incorporation of Rhodopsin in Laterally Structured Supported Membranes: Observation of Transducin Activation with Spatially and Time-Resolved Surface Plasmon Resonance, *Biochemistry* 37, 507–522.

Jeon, S.I. and Andrade, J.D., 1991, Protein-Surface Interaction in the Presence of Polyethylene Oxide – II. Effect of Protein Size, *Journal of Colloid and Interface Science* 142, 159–166.

Jeon, S.I., Lee, J.H., Andrade, J.D. and Gennes, P.G.d., 1991, Protein–Surface Interaction in the Presence of Polyethylene Oxide – I. Simplified Theory, *Journal of Colloid and Interface Science* 142, 149–158.

Kalb, E., Frey, S. and Tamm, L.K., 1992, Formation of Supported Planar Bilayers by Fusion of Vesicles to Supported Phospholipid Monolayers, *Biochimica et Biophysica Acta* 1103, 307–316.

Karlsson, R., 1994, Real-Time Competitive Kinetic Analysis of Interactions Between Low-Molecular-Weight Ligands in Solution and Surface-Immobilized Receptors, *Analytical Chemistry* 221, 142–151.

Karlsson, R., Michaelsson, A. and Mattsson, L., 1991, Kinetic Analysis of Monoclonal Antibody–Antigen Interactions with a New Biosensor Based Analytical System, *Journal of Immunological Methods* 145, 229–240.

Karlsson, R., Roos, H., Fägerstam, L. and Persson, B., 1994, Kinetic and Concentration Analysis Using BIA Technology, *Methods: A Companion to Methods in Enzymology* 6, 99–110.

Keller, C.A. and Kasemo, B., 1998, Surface Specific Kinetics of Lipid Vesicle Adsorption Measured with a Quartz-Crystal Microbalance, *Biophysical Journal* 75, 1397–1402.

Keller, T.A., Duschl, C., Kröger, D., Sévin-Landais, A.-F., Cervigni, S.E., Dumy, P. and Vogel, H., 1995, Reversible Oriented Immobilization of Histidine-Tagged Proteins on Gold Surfaces Using a Chelator Thioalkane, *Supramolecular Science* 2, 155–160.

Kröger, D., Liley, M., Schiweck, W., Skerra, A. and Vogel, H., 1999, Immobilization of Histidine-Tagged Proteins on Gold Surfaces Using Chelator Thioalkanes, *Biosensors and Bioelectronics* 14, 155–161.

Lok, B.K., Cheng, Y.-L. and Robertson, C.R., 1981, Protein Adsorption on Crosslinked Polydimethylsiloxane Using Total Internal Reflection Fluorescence, *Journal of Colloid and Interface Science* 91, 104–116.

Morton, T.A., Myszka, D.G. and Chaiken, I.M., 1995, Interpreting Complex Binding Kinetics from Optical Biosensors: A Comparison of Analysis by Linearization, the Integrated Rate Equation, and Numerical Integration, *Analytical Biochemistry* 227, 176–185.

Myszka, D.G., Arulanantham, P.R., Sana, T., Wu, Z., Morton, T.A. and Ciardelli, T.L., 1996, Kinetics Analysis of Ligand Binding to Interleukin-2 Receptor Complexes Created on an Optical Biosensor Surface, *Protein Science* 5, 2468–2478.

Myszka, D.G., He, X., Dembo, M., Morton, T.A. and Goldstein, B., 1998, Extending the Range of Rate Constants Available from BIAcore: Interpreting Mass Transport-Influenced Binding Data, *Biophysical Journal* 75, 583–594.

Myszka, D.G., Morton, T.A., Doyle, M.L. and Chaiken, I.M., 1997, Kinetics Analysis of a Protein Antigen–Antibody Interaction Limited by Mass Transport on an Optical Biosensor, *Biophysical Chemistry* 64, 127–137.

Nieba, L., Krebber, A. and Plückthun, A., 1996, Competition BIAcore for Measuring True Affinities: Large Differences from Values Determined from Binding Kinetics, *Analytical Biochemistry* 234, 155–165.

Nussenzweig, V. and Nussenzweig, R.S., 1989, Rationale for the Development of an Engineered Sporozoite Malaria Vaccine, *Advances in Immunology* 45, 283–334.

O'Shannessy, D.J., 1994, Determination of Kinetic Rate and Equilibrium Binding Constants for Macromolecular Interactions. A Critique of the Surface Plasmon Resonance Literature, *Current Opinion in Biotechnology* 5, 65–71.

Panayotou, G., Gish, G., End, P., Truong, O., Gout, I., Dhand, R., Fry, M.J., Hiles, I., Pawson, T. and Waterfield, M.D., 1993, Interactions Between SH2 Domains and Tyrosine-Phoshorylated Platelet-Derived Growth Factor *b*-Receptor Sequences: Analysis of Kinetic Parameters by a Novel Biosensor-Based Approach, *Molecular and Cellular Biology* 13, 3567–3576.

Pawlak, M., Grell, E., Schick, E., Anselmetti, D. and Ehrat, M., 1998, Functional Immobilization of Biomembrane Fragments on Planar Wave-Guides for the Investigation of Side-Directed Ligand-Binding by Surface-Confined Fluorescence, *Faraday Discussions* 273–288.

Piehler, J., Brecht, A., Giersch, T., Hock, B. and Gauglitz, G., 1997, Assessment of Affinity Constants by Rapid Solid Phase Detection of Equilibrium Binding in a Flow System, *Journal of Immunological Methods* 201, 189–206.

Sackmann, E., 1996, Supported Membranes: Scientific and Practical Applications, *Science* 271, 43–48.

Schuck, P., 1996, Kinetics of Ligand Binding to Receptor Immobilized in a Matrix, as Detected with an Evanescent Wave Biosensor. I. A Computer Simulation of the Influence of Mass Transport, *Biophysical Journal* 70, 1230–1249.

Shoup, D. and Szabo, A., 1982, Role of Diffusion in Ligand Binding to Macromolecules and Cell-Bound Receptors, *Biophysical Journal* 40, 33–39.

Spinke, J., Liley, M., Schmitt, F.-J., Guder, H.-J., Angermaier, L. and Knoll, W., 1993, Molecular Recognition at Self-Assembled Monolayers: Optimization of Surface Functionalization, *Journal of Chemical Physics* 99, 7012–7019.

Stankowski, S., 1983, Large Ligand Adsorption to Membranes: I. Linear Ligands as a Limiting Case, *Biochimica et Biophysica Acta* 735, 341–351.

Stenberg, E., Persson, B., Roos, H. and Urbaniczky, C., 1991, Quantitative Determination of Surface Concentration of Protein with Surface Plasmon Resonance Using Radiolabeled Proteins, *Journal of Colloid and Interface Science* 143, 513–526.

Szabo, A., Stolz, L. and Granzow, R., 1995, Surface Plasmon Resonance and its use in Biomolecular Interaction Analysis (BIA), *Current Opinion in Structural Biology* 5, 699–705.

Terrettaz, S., Stora, T., Duschl, C. and Vogel, H., 1993, Protein Binding to Supported Lipid Membranes: Investigation of the Cholera Toxin-Ganglioside Interaction by Simultaneous Impedance Spectroscopy and Surface Plasmon Resonance, *Langmuir* 9, 1361–1369.

Thompson, N.L., Burghardt, T.P. and Axelrod, D., 1981, Measuring Surface Dynamics of Biomolecules by Total Internal Reflection Fluorescence with Photobleaching Recovery or Correlation Spectroscopy, *Biophysical Journal* 33, 435–454.

Thompson, N.l., Drake, A.W., Chen, L. and Broek, W.V., 1997, Equilibrium, Kinetics, Diffusion and Self-Association of Proteins at Membrane Surfaces: Measurement By Total Internal Reflection Fluorescence Microscopy, *Photochemistry and Photobiology* 65, 39–46.

Part III
Transducer technology

6 Optical transducers

Martha Liley

Introduction

In the development of biosensors, optical transduction methods have, from the beginning, attracted a great deal of interest and research effort. The reasons for this become clear when one considers the constraints imposed by the biosensor format: biosensors use receptor molecules (often antibodies or DNA strands) attached to a solid transducer surface which is in contact with the sample to be analysed. The presence of analyte in the sample is detected by measuring the binding of analyte molecules to the surface-bound receptors. The number of receptor molecules that are bound to the transducer surface is normally very low (for a medium-sized protein such as streptavidin, 10^{12} molecules per square centimetre represents a good surface coverage), which implies that the method used to detect the binding of the analyte to the surface must be very sensitive. In addition, the detection technique should be surface specific – sensitive only to molecules at or very close to the surface – so that interference from the bulk solution is eliminated. This is particularly important if washing steps are to be reduced or eliminated from the assay. Finally, the choice of detection method is further limited by the requirement that measurements be performed in aqueous solution, and that the method be non-destructive so that real-time measurements giving kinetic information can be performed. Optical methods fulfil all these requirements and have the additional advantage of a long history of use in biochemical and chemical analysis. Examples such as UV-vis and infrared absorbance spectroscopy, fluorescence spectroscopy and Raman spectroscopy demonstrate the extent to which optical analysis methods are established and the wide range of information about molecular structure, conformation and environment that they can give.

In this chapter, two basic approaches to optical transducers will be described: direct detection and fluorescence detection. While other methods, such as optical absorbance or electrochemiluminescence, have been reported in the literature, research into these methods remains very limited. Direct detection is based on the difference in refractive index between water and biological molecules. Binding of analyte molecules to the transducer surface results in a minute change of the refractive index at the surface that can be detected either by evanescent wave techniques such as waveguide or surface plasmon resonance spectroscopy, or by reflection/interference techniques such as ellipsometry. In contrast, fluorescence assays use molecules labelled with fluorescent dyes and detection is based on measurements of the intensity of the light emitted by the labels. Fluorescence-based immunoassays are available in a multitude of forms. However, in the context of optical transducers we only consider heterogeneous formats in which the transducer is in direct contact with the sample. Descriptions of the many, highly successful homogeneous or pseudo-homogeneous fluorescence assays that have been developed can be found in a large number of excellent review articles or books (see, for example, Hemmilä 1991; Wood and Barnard 1997).

The field of optical transducers is too broad and fast-moving for this chapter to be an exhaustive review of the subject. Instead some general principles of optical transducers will be discussed and illustrated using examples from several different classes of transducers.

Optical transducers: general considerations

Assay formats

Three biological assay formats have been widely used in biosensors: the direct assay, the sandwich assay and the competitive assay (see Figure 6.1).

The direct assay is used in direct sensors. An antibody or other receptor molecule bound to the sensor surface. The binding of the analyte molecule to the antibody is measured in real time as an increase in the optical thickness of the biological layer on the transducer surface.

The sandwich assay involves two antibodies directed towards different binding sites ('epitopes') of the analyte and can be used in direct or fluorescence sensors. One antibody is immobilized on the sensor surface where it binds the analyte molecule. The second antibody (which is fluorescently labelled when used with fluorescence transducers) binds to the immobilized analyte. The binding of the second antibody is detected by the transducer, resulting in a signal whose intensity is proportional to the amount of bound analyte. The sandwich assay can only be used for the detection of large analyte molecules that have at least two distinct binding sites or epitopes, with separate antibodies against each epitope. Small analytes such as pesticides, drugs or antibiotics cannot normally be detected using this method.

The competitive assay is the assay format most commonly used for small analytes. It involves competition for a limited number of antibody binding sites. In one assay format, the antibody is immobilized on the sensor surface. A known quantity of labelled analyte analogue is added to the sample where it competes with the unlabelled analyte present in the sample for the antibody binding sites on the sensor surface. In an alternative assay format, an (unlabelled) analyte analogue is immobilized on the sensor surface and a known quantity of labelled antibody added to the sample. In this case the antibody binds either to the analyte in solution or to the analyte on the sensor surface. In both competitive assay formats, the resulting signal is highest for low concentrations of sample analyte. This has important implications for the sensitivity of these assays: in general, the highest sensitivity is obtained for competitive assays when the density of binding sites on the sensor surface is relatively low. This is in contrast to sandwich assays, where high sensitivity requires a high density of binding sites on the sensor surface.

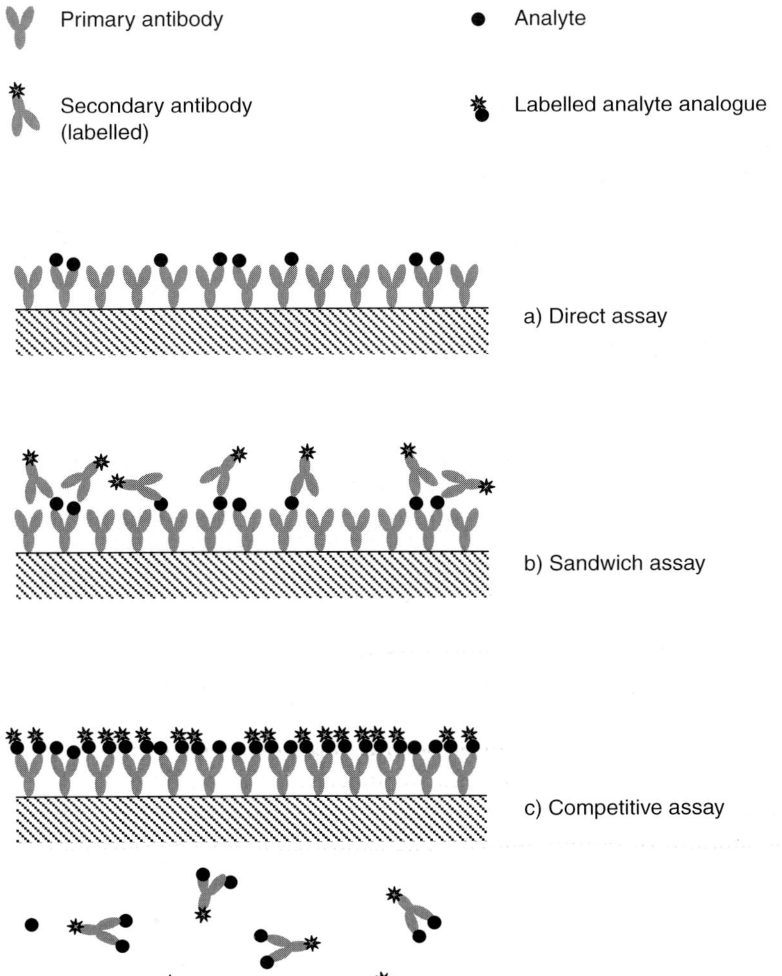

Figure 6.1 Assay formats for use with biosensors

a Direct assay. No labels are used. Binding of the analyte to the surface is measured directly

b Sandwich assay. After binding of the analyte to the sensor surface, a secondary antibody binds to surface-bound analyte molecules

c Competitive assay. Competition takes place for binding sites on the sensor surface. A low sensor signal is obtained for a high analyte concentration

Sensitivity and selectivity

Sensitivity

The sensitivity of a biosensor depends both on the transducer and on the biological element of the sensor. The transduction technique determines how many molecules must be present at the surface of the sensor in order to obtain a signal, while the biological element defines the relationship between the number of analyte molecules at the surface and the bulk concentration of analyte.

If an antibody (or other receptor molecule) is present at the surface of a sensor, then both the surface density of the antibody, σ (number of molecules per mm^2), and the binding constant, K, of the receptor, influence the number of antigen molecules bound to the surface. The surface density of antigen molecules on the sensor, Γ, and its dependence on the bulk concentration of antigen, $[Ag]$, is described by the Langmuir isotherm:

$$\Gamma([Ag]) = \frac{\Gamma_{max}K[Ag]}{(1 + K[Ag])} \tag{6.1}$$

This is shown graphically in Figure 6.2. The maximum surface density of antigen molecules, Γ_{max}, will be twice the surface density of antibodies (assuming that each antibody binds two antigen molecules). When the bulk concentration of analyte is equal to the reciprocal of the binding constant, half of the binding sites will be occupied. (Note: strictly speaking, the Langmuir isotherm describes the binding of

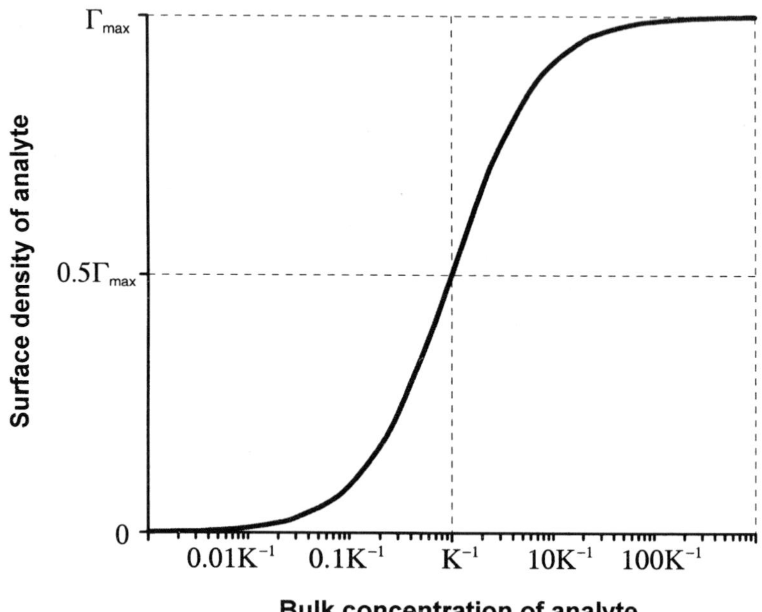

Figure 6.2 The Langmuir isotherm relates the density of molecules bound to a surface to their bulk concentration

molecules to surfaces under conditions that are almost certainly not fulfilled by antigen molecules binding to immobilized antibodies. However, the formula is an excellent approximation to the observed binding and has the important advantage of being very simple.)

A transducer produces a signal that depends on the number or surface density of molecules bound to its surface. In the case of direct sensing, the signal will be proportional to the mass of analyte bound, while for fluorescence techniques, the signal will be proportional to the number of fluorescent labels bound. For any transducer there will be minimum detectable surface coverage of antigen, Γ_{min}. The minimum detectable coverage of antigen defines the minimum detectable bulk concentration of antigen $[Ag]_{min}$.

$$\Gamma_{min} = \frac{\Gamma_{max}K[Ag]_{min}}{(1 + K[Ag]_{min})} \tag{6.2}$$

Solving for $[Ag]_{min}$ we obtain:

$$[Ag]_{min} = \frac{\Gamma_{min}}{K[\Gamma_{max} - \Gamma_{min})} \approx \frac{\Gamma_{min}}{K\Gamma_{max}} \tag{6.3}$$

From this we see that, for a given transducer and a given antibody/antigen pair, the minimum measurable concentration depends on the ratio of lowest detectable surface coverage to the maximum surface coverage of analyte and on the binding constant of the antibody/antigen pair.

Thus, if we wish to compare different transduction methods, quoting the sensitivity of a sensor in terms of bulk concentration – as is so often done in the literature – gives no useful information (unless, of course, the binding constant of the antibody/antigen pair is known). Transducer sensitivities must be given in terms of surface mass or number density of the analyte molecules. This point cannot be overemphasized.

When values for transducer sensitivity are given in terms of surface mass density or numbers of bound molecules, it is important that the same definition of sensitivity be used in all cases. For a rigorous definition of sensitivity, the minimum measurable surface density must be defined using the blank or null signal and its standard deviation (i.e. the noise). The lowest signal that can be reliably distinguished from the blank signal must be three standard deviations higher or lower than the blank signal. This corresponds to the minimum detectable surface density. If the surface coverage is to be quantified, the signal should be at least nine times the standard deviation observed on the baseline. Unfortunately, most authors, even when they quote transducer sensitivity in terms of surface coverage, do not specify their definition of sensitivity so that comparisons of different transducer techniques may still be difficult.

Selectivity

Unlike sensitivity, selectivity is not a well-defined concept in biosensor research. If a biosensor is exposed to a sample containing a mixture of different proteins, then a highly selective sensor will give a signal that depends only on the concentration of analyte, with little or no interference from other species present in the sample. In contrast, a poorly selective sensor will show marked effects (resulting in falsely high or low determinations of the analyte concentration) due to signals caused by other molecules in the sample.

Selectivity is particularly hard to quantify because the effects due to other molecules in a sample depend very strongly on which molecules are present: some molecules (such as albumin) are particularly surface active and bind very strongly to exposed surfaces. Other species, such as dextran, are very soluble and interact only weakly with surfaces. Thus the selectivity observed will depend on the sample measured.

When considering the selectivity of biosensors we must distinguish between two very different cases: direct transduction systems, and fluorescence transducers. Direct transducers measure mass coverage of the sensor surface: all molecules binding to the surface result in a transducer signal. Thus, for direct transducers, the selectivity of the sensor comes entirely from the biological component of the sensor, which defines which molecules will bind to the surface. In contrast, fluorescence transducers only detect labelled molecules: other molecules binding to or desorbing from the surface do not contribute to the sensor signal. Fluorescence sensors are, therefore, inherently more selective than direct transducers.

One of the key issues in biosensor research is the development of the biological layer so that only the chosen analyte will bind to the sensor surface and so that binding of other species to the surface ('non-specific binding') is suppressed. While this is clearly of the utmost importance for direct sensors, it is also important for fluorescence sensors where non-specific binding may block analyte-binding sites. For both direct and fluorescent transducers the selectivity of the sensor can be enhanced by the use of sandwich assays.

Reflection, refraction and interference of light

Optical transducers exploit surface-specific optical phenomena to interrogate the properties of surfaces while neglecting to a large extent bulk effects. This section summarizes some of the basic optics necessary to understand the transduction methods described later in this chapter: an excellent and much more detailed account can be found in the book *Optics* (Hecht 1998).

Travelling waves

Most of optical transducers use collimated light beams, which can, for the purposes of this chapter, be considered identical to a plane travelling wave. A plane travelling wave can be described by the equation:

$$\bar{E} = \bar{E}_0 \, exp \left[i(\bar{k}.\bar{r} - \varpi t + \epsilon) \right] \tag{6.4}$$

Where E is the electric field vector which defines both the polarization and amplitude of the wave; k, the wavevector; ω the angular frequency; and ϵ a phase constant.

These parameters are related to the more familiar parameters of light as follows:

$$|E| = I = \text{the intensity of the light}$$

$$|k| = \frac{2\pi}{\lambda} = \text{the momentum of the light}$$

$$\hat{k} = \text{the direction of propagation of the light}$$

$$v = \frac{\omega}{|k|} = \text{the velocity of the light}$$

The velocity of light, v, depends only on the material in which the light beam is propagating, and is determined by the index of refraction, n. The index of refraction is related to the velocity of light by the equation:

$$n = \frac{c}{v} \qquad (6.5)$$

where c is the speed of light in a vacuum.

Reflection

When light is incident on an interface between two media of different refractive indices, in general, part of the incident beam will be transmitted into the second medium, while part of the beam will be reflected (see Figure 6.3).

The law of reflection states that the angle of incidence is equal to the angle of reflection, that is:

$$\theta_i = \theta_r \qquad (6.6)$$

In this and the following formulae, the subscript i denotes angles of incidence, refractive indices and fields in the medium of the incident beam and t and r those in the medium of the transmitted and reflected beam, respectively.

The ratio of the reflected and incident light amplitudes (the amplitude reflection coefficient) depends on the refractive indices of the different media, on the angle of incidence and on the polarization of the incident light (see Figure 6.3) and is obtained from the Fresnel equations:

$$r_s \equiv \left(\frac{E_{0r}}{E_{0i}} \right) = \frac{n_i \cos\theta_i - n_t \cos\theta_t}{n_i \cos\theta_i + n_t \cos\theta_t} \qquad (6.7)$$

$$r_p \equiv \left(\frac{E_{0r}}{E_{0i}} \right) = \frac{n_t \cos\theta_i - n_i \cos\theta_t}{n_t \cos\theta_i + n_i \cos\theta_t} \qquad (6.8)$$

where r_s and r_p are the amplitude reflection coefficients for s polarized light (which has the electric field perpendicular to the plane of incidence) and p polarized light

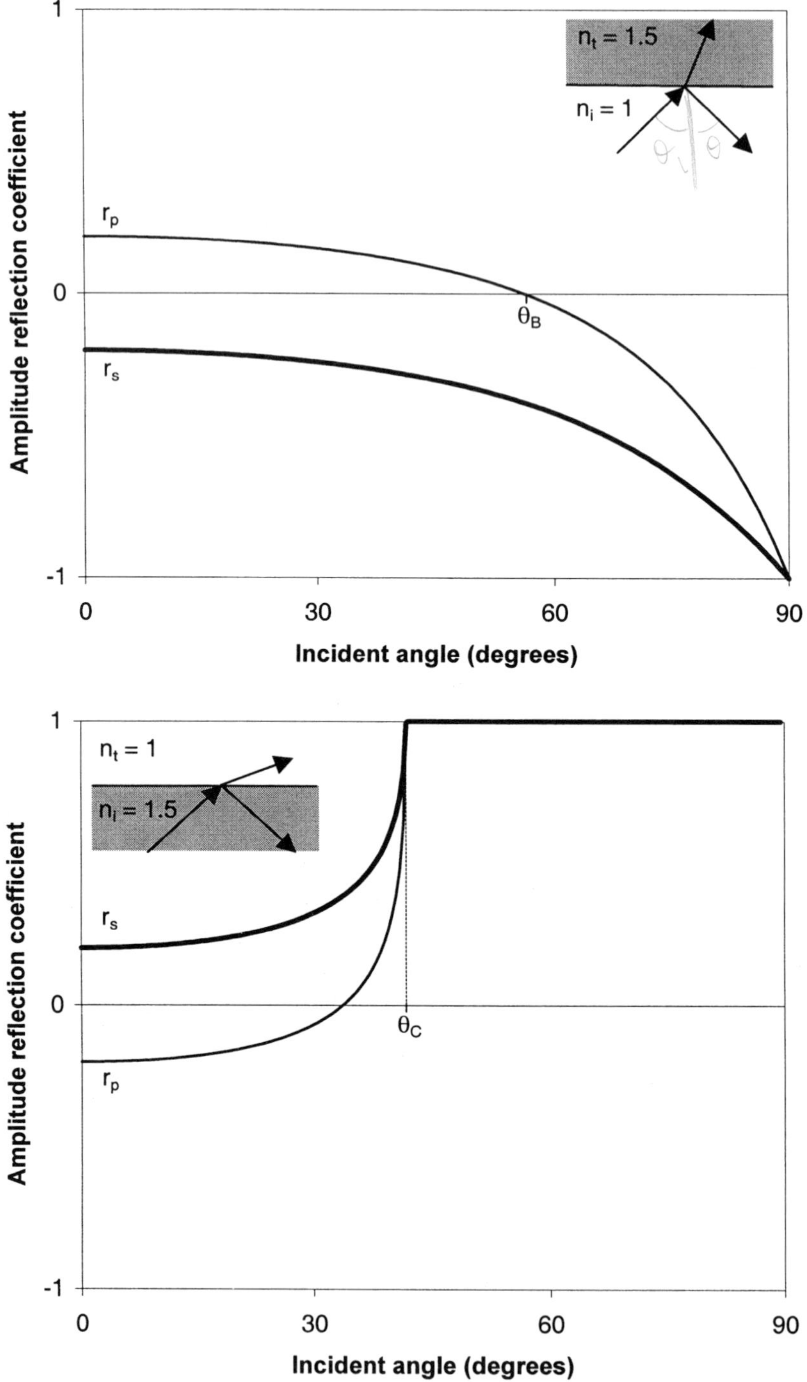

Figure 6.3 The amplitude coefficients of reflection as a function of angle for
    a   external reflection ($n_i = 1$; $n_t = 1.5$). θ_B is the Brewster angle
    b   internal reflection ($n_i = 1.5$, $n_t = 1.0$) θ_c is the critical angle

(electric field parallel to the plane of incidence) respectively. The reflection of light at an interface is also associated with a phase change. This phase change is 0° or 180° except for totally internally reflected light (see below).

Refraction

In addition to being partially reflected, light incident at an interface is refracted (or bent). The extent to which the light beam is bent is described by the law of refraction, or Snell's law, and depends on the refractive indices of the two media:

$$n_i \sin \theta_i = n_t \sin \theta_t \qquad (6.9)$$

When light goes from a low to a high refractive index medium, it is bent towards the normal to the interface. When it goes from a high to a low refractive index medium, it is bent away from the normal. At the critical angle, θ_c, the light leaves a high refractive index medium at 90° to the normal – at a tangent to the interface. At angles above the critical angle no transmitted beam can exist and all of the incident light is reflected back into the incident medium at the interface. This phenomenon is known as total internal reflection (see Figure 6.3) and can easily be observed using a glass prism.

In total internal reflection, the incident light is entirely reflected at the interface, but this does not mean that there is no electromagnetic wave in the second medium. In fact, an electromagnetic wave is obtained which has maximum intensity at the interface, and which decreases in intensity exponentially as it penetrates into the second medium.

$$E_t = E_{0t} \exp(-\beta z) \exp i \left(\frac{n_i k_f x \sin \theta_i}{n_t} \right) - \omega t \qquad (6.10)$$

This evanescent wave is the basis of many optical transducers since it is confined to the interfacial region and is therefore only sensitive to phenomena in this region. The depth of the interfacial region which is 'seen' by the evanescent wave is characterized by its exponential decay length, $1/\beta$, where β is given by:

$$\beta = k_t \left(\frac{n_i^2 \sin^2 \theta_i}{n_t^2} - 1 \right)^{1/2} \qquad (6.11)$$

Interference

If two (or more) electromagnetic waves are present in any region, then the resultant disturbance is a wave that is the algebraic sum of the individual waves – an effect known as interference.

If the two waves are of the same frequency:

$$E_1 = E_{01} \exp[i(\omega t + \alpha_1)] \qquad (6.12)$$

$$E_2 = E_{02} \exp[i(\omega t + \alpha_2)] \qquad (6.13)$$

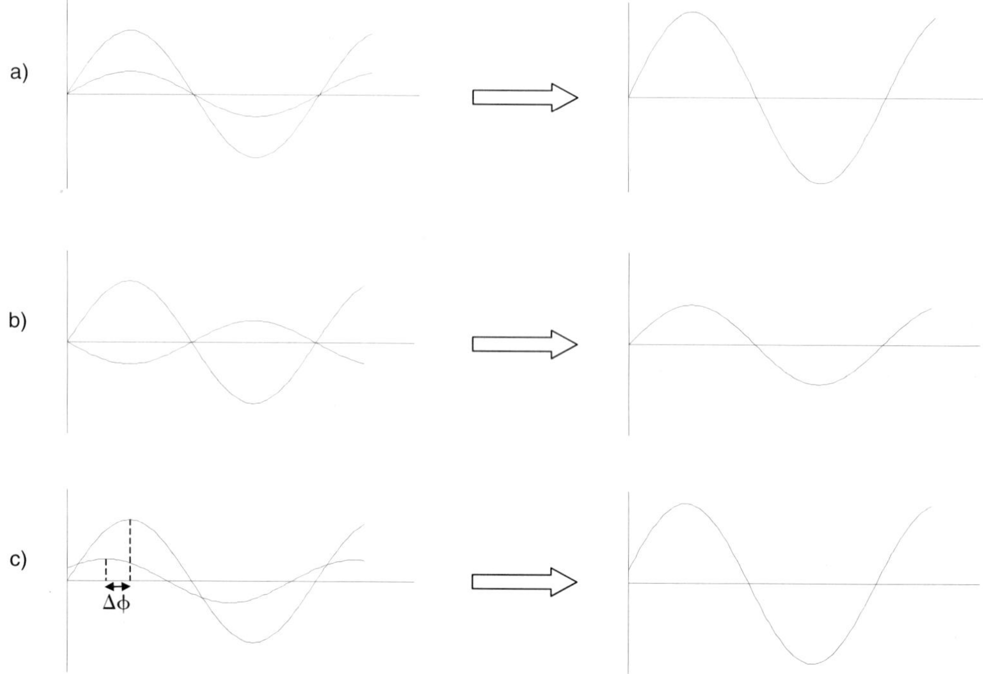

Figure 6.4 The interference of light
    a Two light waves of different amplitude, but in phase. Constructive interference
    b Two light waves 180° out of phase. Destructive interference
    c Two light waves out of phase by Δφ radians

Then the resultant wave is of the same frequency as the two components although its amplitude and phase are different (see Figure 6.4):

$$E_{1+2} = E_0 \exp\left[i(\varpi t + \alpha)\right] \tag{6.14}$$

with the amplitude given by

$$E_0^2 = E_{01}^2 + E_{02}^2 + 2E_{01}E_{02}\cos(\alpha_2 - \alpha_1) \tag{6.15}$$

and the phase by

$$\tan\alpha = \frac{E_{01}\sin(\alpha_1) + E_{02}\sin(\alpha_2)}{E_{01}\cos(\alpha_1) + E_{02}\cos(\alpha_2)} \tag{6.16}$$

The crucial factor in interference is the phase difference between the two waves:

$$\delta \equiv \alpha_2 - \alpha_1 \tag{6.17}$$

when $\delta = 0, 2\pi, 4\pi \ldots$ the resulting amplitude is a maximum
while $\delta = \pi, 3\pi, 5\pi \ldots$ gives a minimum amplitude

Interference effects are used in optical transducers to measure film thicknesses. In these applications both interfering waves come from the same light source (and thus have the same wavelength). The origin of the phase difference is the difference in the path length travelled by the two waves:

$$\delta = \frac{2\pi}{\lambda_0} n(x_2 - x_1) = \frac{2\pi}{\lambda_0} n\Delta x \tag{6.18}$$

where λ_0 is the wavelength in a vacuum, and $n\Delta x$ is the optical path difference.

Direct detection

Two fundamentally different detection principles have been developed for the direct detection of immunoreactions on surfaces. Both detection principles rely on the fact that biological molecules have a higher refractive index than water. This allows the binding and desorption of proteins and other molecules to be directly monitored in real time and without the use of labels – hence the term 'direct' detection.

The first, and most widely used principle is based on guided electromagnetic wave modes: a waveguide mode or a surface plasmon polariton mode. Binding of proteins or other biological molecules to the surface changes the effective refractive index of the guiding structure and this change is detected via the changed propagation properties of the guided wave. A large number of different sensor configurations have been developed based on this approach. In this chapter, the basic principles of the approach and some of the more important sensor geometries are described.

The second approach to direct detection is based on the interference of light reflected from planar multilayer structures. Ellipsometry and related reflectometric techniques developed for biological detection are described here. Reflectometric methods are less well developed than guided wave methods, although recent work by Brecht (Brecht *et al.* 1992; Brecht and Gauglitz 1997) and Ghadiri (Lin *et al.* 1997), amongst others, has demonstrated their potential as optical transducers.

Biological molecules and refractive index

The refractive index of pure water at room temperature, and at wavelengths close to 600 nm, is about 1.333. However, the exact value of the refractive index of an aqueous solution depends on a number of parameters, including the concentration of biological molecules, total salt concentration and temperature. While the first of these parameters is the basis of direct detection transducers, the other two are interfering effects, which must either be controlled or compensated for.

Investigations of the bulk refractive indices of protein solutions have shown that, at constant temperature and salt concentration, the refractive index is proportional to the mass concentration of dissolved protein (De Feijter *et al.* 1978). The increment in refractive index is given by:

$$\frac{dn}{dc} = 0.188\,\text{ml.g}^{-1} \tag{6.19}$$

where n is the refractive index of the solution and c is the protein concentration in grams per ml. Importantly, the same refractive index increment was observed (to a good approximation) for all the proteins investigated, with a linear dependence of refractive index on concentration up to very high protein concentrations (approximately 50 per cent protein w/w). It is general practice to extrapolate from these results and to apply this refractive index increment to protein films (where the protein concentration is usually higher than 50 per cent) and to all proteins under investigation.

The interpretation of results obtained from direct sensing transducers is often based upon an implicit model of the biomolecules bound to the sensing surface as a layer with a uniform refractive index and with smooth, well defined interfaces between the bulk media and the layer (see Figure 6.5). In fact, apart from some very rare exceptions, this is not the case: normally the roughness of the sensing surface is at least comparable to the biomolecule size (and thus to the layer thickness). In addition, molecules at the surface are usually bound in a variety of orientations, so that the protein volume fraction (or 'concentration', if one can think in these terms at these length scales) varies across the layer.

For a large number of direct detection transducers, the signal obtained on protein

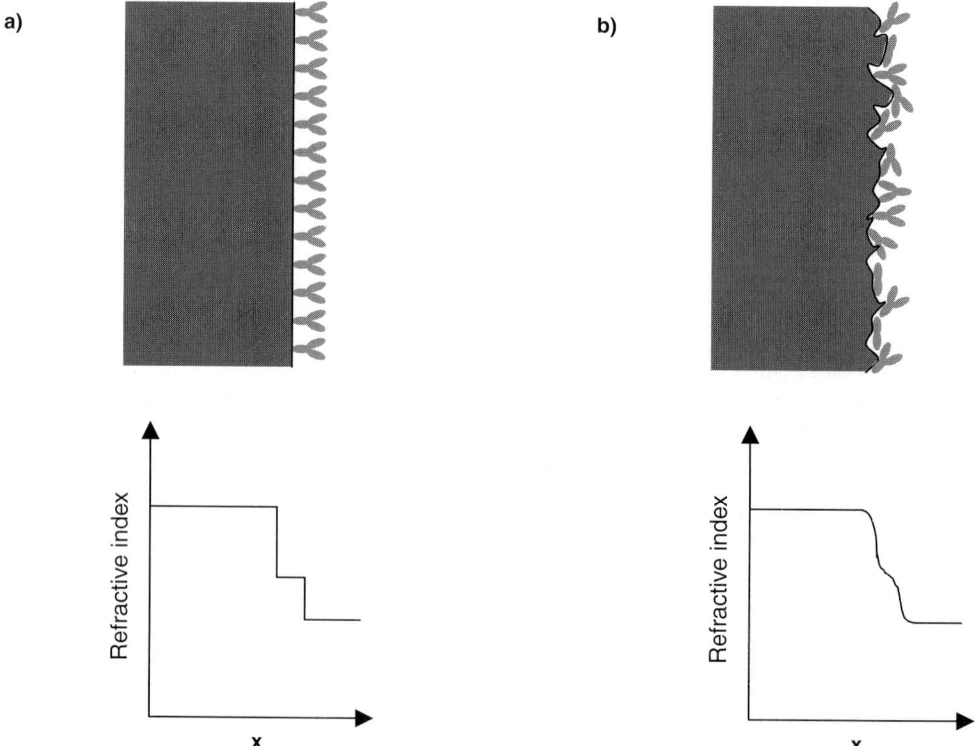

Figure 6.5 Two models for the adsorption of antibodies to a solid surface
   a The model implicitly used when separate values for biological layer thickness and refractive index are derived
   b A more realistic model in which the solid surface is rough, and the antibody is adsorbed in several different conformations

binding is a function of the optical thickness, $\Delta n.d$, of the protein layer (where Δn is the difference in refractive index between the layer and the aqueous solution, and d is the geometrical thickness of the protein layer). Since Δn depends linearly on the protein concentration, the optical thickness is directly proportional to the mass coverage of the surface:

$$\sigma \propto \Delta n.d \tag{6.20}$$

Some transducers give two parameters for the protein film, which are then used to separately determine Δn and d – thus distinguishing between, for example, a densely packed thin layer and a more loosely packed but thick protein layer. Clearly, this additional information can be valuable, particularly in a research environment. However, it is worth noting that, in general, two parameters are not sufficient to define the refractive index profile of the protein layer: the separation of optical thickness into geometric thickness and refractive index should be treated with some caution. In general, determination of the biomolecule mass coverage of the surface is sufficient for immunoassay applications.

Guided wave transducers are effectively highly sensitive refractometers that measure the refractive index of the solution close to the sensor surface (see below, page 139). These transducers can accurately determine the optical thickness of the bound protein layer only if the bulk refractive index of the solution remains constant (or is accurately known). Extremely small changes in bulk refractive index give a signal equivalent to significant mass densities of protein on the sensor surface. For this reason, great care must be taken to ensure that both the temperature and the salt concentration of the solution remain constant. Typically, if sensitivities in the pg/mm^2 range are desired, the temperature of sensor and sample must be stabilized to within a few hundredths of a degree Centigrade. An alternative approach is to independently monitor changes in bulk refractive index caused by temperature or salt concentration fluctuations and to correct the observed transducer signal for the bulk changes.

In contrast, reflectometric transducers measure the optical thickness of the layer of biomolecules by interferometry and are relatively insensitive to changes in the refractive index of the bulk sample.

Guided wave modes

Waveguide optics

Waveguiding is based on the phenomenon of total internal reflection described earlier (see page 131). The principles of waveguiding can be understood fairly easily by considering the case of a simple planar waveguide using a ray tracing analysis. The principles derived in this manner are also applicable to more complex planar waveguides and to waveguides with other geometries (such as optical fibres).

A simple planar waveguide is shown in Figure 6.6. The waveguide consists of a slab of high refractive index material between two low refractive index media. In general, a light beam travelling in the waveguide at angle θ, will be partly reflected from any interface it traverses. However, for angles bigger than the critical angle, θ_c, the light beam will be totally internally reflected. If internal reflection occurs at both

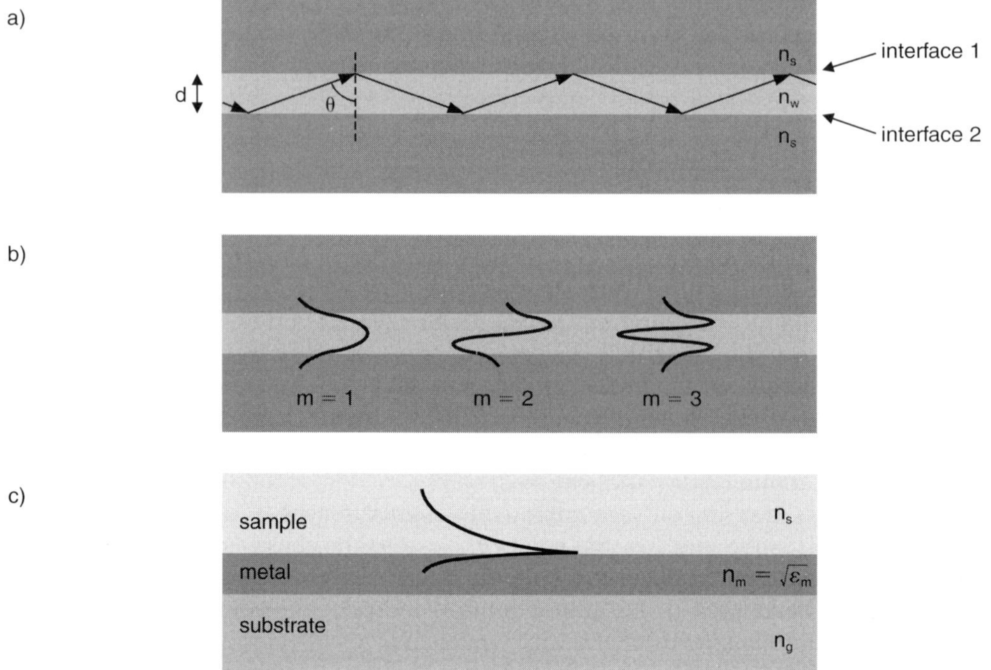

Figure 6.6 Guided wave modes

    a Waveguiding in a simple planar waveguide: a ray-tracing description

    b Waveguiding in a simple planar waveguide. The field amplitude distributions for three guided modes with $m = 1, 2, 3$

    c A surface plasmon polariton mode. Maximum field intensity occurs at the metal/sample interface. The field intensity decays exponentially into both sample and metal film

interfaces of the high refractive index slab, the light beam will travel inside the slab, which acts as a waveguide. As previously described, total internal reflection of the light beam at the interface of the waveguide does not imply that there are no electromagnetic fields outside the waveguide: an exponentially decaying evanescent field is present in both of the low refractive index media. The evanescent field is described by:

$$E_s = E_{0s} \exp(-\beta z) \tag{6.21}$$

where E_s is the electric field outside the waveguide and β is given by:

$$\beta = k_s \left(\frac{n_w^2 \sin^2 \theta}{n_s^2} - 1 \right)^{1/2} \tag{6.22}$$

The decay length of the evanescent field is $1/\beta$, which is typically around 100–300 nm for visible light.

    For ideal non-absorbing materials, the light beam will travel down the waveguide without loss of power as long as the multiple reflections of the light beam interfere to produce a stationary interference pattern i.e. if the phase change of the light beam is a multiple of 2π for the round trip from interface 1 to interface 2 and back again (see Figure 6.6). The phase change of the light beam depends on the optical path length and on the phase shift which occurs on reflection from the two interfaces so that we obtain waveguiding when:

$$m2\pi = \frac{2n_w k_0 d}{\cos\theta} - \phi_1 - \phi_2 \qquad (6.23)$$

where m is an integer, n_w the refractive index of the waveguide, k_0 the propagation number in a vacuum, and ϕ_1 and ϕ_2 the phase changes of the light on reflection from the respective interfaces. Each value of m is associated with a different waveguiding mode and each mode has a different, well-defined electromagnetic field distribution. The fraction of the optical power that propagates in the evanescent field is defined by this field distribution and is largest for the modes with the smallest values of m.

Since the phase change on reflection is different for s polarized light and for p polarized light, the conditions for waveguiding will be different for the two different polarizations: we obtain different waveguiding modes for s polarized light (TE modes) and for p polarized light (TM modes). In general, each mode is associated with a well-defined light angle, which allows us to define a propagation constant, κ, for the waveguide mode

$$\kappa = n_w k_0 \sin\theta \qquad (6.24)$$

and, by analogy, an effective refractive index, N_{eff}, of the mode:

$$N_{eff} = \frac{\kappa}{k_0} = n_w k_0 \sin\theta \qquad (6.25)$$

The effective refractive index depends on the polarization of the mode (TE or TM), on the mode number, m, and on the waveguide parameters: the thickness of the waveguide, and the refractive indices of substrate, film and sample. This effective refractive index of a waveguide and the changes to it that are caused by the binding of molecules to the waveguide surface are the central concepts of direct detection using waveguide sensors.

For light to couple into a waveguide, it is necessary that both the energy of the light (or ω) and the component of the wavevector along the propagation direction, k_x, be identical for both the waveguide mode and the incident light beam. The fact that light propagates in a waveguide by total internal reflection implies that externally incident light cannot couple directly into the waveguide. Three methods are commonly used to couple light into waveguiding structures (Figure 6.7):

End-fire coupling: light is incident on the end face of the waveguide and is transmitted directly into it. Efficient coupling of the light into the guide requires fine adjustment of the relative positions of beam and guide, and matching of the intensity profiles of both light beam and guided mode. This is a time-consuming and experimentally difficult method.

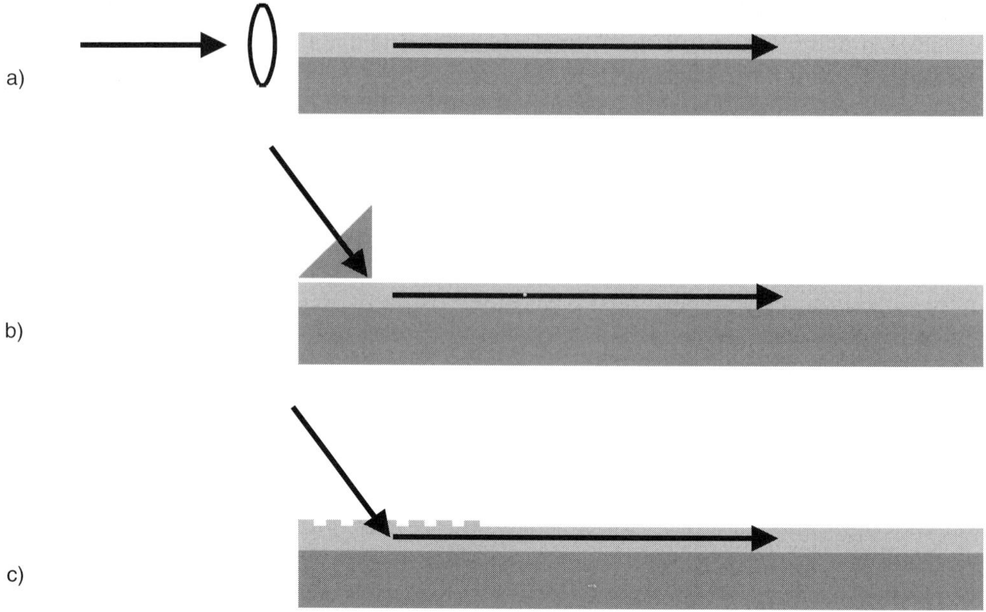

Figure 6.7 Methods for coupling light into planar waveguides
     a   End-fire coupling
     b   Prism coupling
     c   Grating coupling

Prism coupling: light is totally internally reflected at the base of a glass prism that is in close contact with the waveguide. At the coupling angle, θ_c, the component of the wavevector along the propagation direction, k_x is matched to the propagation number of the waveguide mode. If the evanescent field of the reflected light penetrates into the waveguide, coupling takes place. Efficient coupling requires prism and waveguide interfaces to be separated by a small, well-defined low refractive index layer (with a thickness on the order of μm).

Grating coupling: a grating at the waveguide surface can couple externally incident light into the waveguide by diffraction. The grating has a wavevector G in the propagation direction, and coupling takes place when:

$$k_x + mG = \kappa \tag{6.26}$$

where k_x is the x component of the externally incident wavevector and m is an integer.

    Grating coupling efficiencies are usually low.

Surface plasmon polariton modes

Surface plasmon polariton modes (commonly known as surface plasmon modes) are guided electromagnetic modes that resemble planar waveguide modes in many respects (see Figure 6.6).

Surface plasmon polaritons are *p*-polarized (TM) electromagnetic waves, which propagate at the interface between a metal and a dielectric. They exist when the real part of the complex dielectric constant of the metal, ϵ'_m, is negative and when $\epsilon'_m > n_s^2$. An effective propagation constant κ of the SP mode can be defined, with:

$$\kappa = \frac{2\pi}{\lambda_0}\left(\frac{\varepsilon'_m n_s^2}{\varepsilon'_m + n_s^2}\right)^{1/2} = k_0\left(\frac{\varepsilon'_m n_s^2}{\varepsilon'_m + n_s^2}\right)^{1/2} \tag{6.27}$$

where n_s is the refractive index of the sample, and ϵ'_m the real part of the dielectric constant of the metal layer.

The intensity of the electric field associated with the mode decays exponentially into both metal and dielectric media (see Figure 6.6) with the exponential decay in the dielectric medium (sample) given by:

$$E = E_0 \exp(-\beta z) \tag{6.28}$$

where β is defined by:

$$\beta = \left(\frac{2\pi}{\lambda_0}\right)\frac{n_s^2}{(\varepsilon'_m + n_s^2)^{1/2}} \tag{6.29}$$

Surface plasmons propagate at the surface of the metal, but are attenuated as they travel by absorption in the metal film. This attenuation is determined by the imaginary part of the complex dielectric constant of the film ϵ''_m and results in an intensity of the surface plasmons that decays exponentially with lateral distance according to the equation:

$$E = E_0 \exp(-\alpha x) \tag{6.30}$$

where α is given by:

$$\alpha = \frac{1}{2}\left(\frac{\lambda_0}{2\pi}\right)^2 \kappa^3 \tag{6.31}$$

Surface plasmons typically have propagation distances around $4\,\mu$m on gold surfaces and $25\,\mu$m on silver surfaces (for $\lambda = 633\,$nm). This rapid attenuation of surface plasmons contrasts strongly with the long propagation distances which can be attained by waveguide modes.

Surface plasmon modes, like waveguide modes, do not couple directly to incident light. Coupling of incident light into the mode can take place either via a grating coupler or a prism coupler.

Sensors based on guided waves

It is clear from the above descriptions that surface plasmons and planar waveguide modes are very similar. In both cases a propagating electromagnetic wave exists with an effective refractive index defined by the ratio of the guided mode propagation constant to k_0. An evanescent wave is generated in the sample.

The basis for the use of planar waveguide modes and surface plasmon modes as sensors is the effective refractive index: this is modified by changes in the refractive index of the sample close to the waveguide or metal surface (within the evanescent field). Thus if biological molecules arrive in the evanescent field, the local refractive index increases and the effective refractive index, N_{eff}, of the waveguide, or of the surface plasmon architecture also increases.

The sensitivity of a transducer based on guided modes thus depends principally on two factors: firstly, the dependence of the effective refractive index on the mass coverage of the surface,

$$\Delta N_{Eff} = \frac{dN}{dm} \Delta m \tag{6.32}$$

The second factor that determines the sensitivity of the transducer is the resolution with which changes in the effective refractive index, ΔN_{eff}, can be determined. In theory, the minimum resolvable change in effective refractive index, ΔN_{min}, is inversely proportional to the interaction length L_α over which the guided wave interacts with the sample.

$$\Delta N_{min} \propto \frac{1}{L_\alpha} \tag{6.33}$$

There are significant differences between surface plasmon and planar waveguide modes for both of these factors:

a   The fraction of guided power in the evanescent field is higher for the surface plasmon modes than for the guided wave modes. This results in higher values of *dN/dm* for surface plasmon transducers than for waveguide transducers.

   For waveguides, the highest *dN/dm* ratios are obtained for thin, high refractive index waveguides. For these waveguides, the amplitude of the electric field is very high at the sample/guide interface and the penetration depth of the evanescent field in the sample is small. For surface plasmon modes, *dN/dm* is inversely proportional to λ and to $\sqrt{\epsilon'_m}$ – it depends on the metal used.

b   The propagation lengths (L_a) of planar waveguide modes can be much longer than those obtained for surface plasmon modes. Input grating or prism couplers can have interaction lengths on the order of millimetres, while interferometers may have interaction lengths on the order of a centimetre or more. In contrast, the propagation distance (and thus the maximum interaction length) of surface plasmons is on the order of microns (approximately $4\,\mu$m for gold and $25\,\mu$m for silver at $\lambda = 633\,$nm).

When these two factors are taken into account, we find that waveguide transducers based on grating or prism couplers (see below, pages 141–143) should be about an order of magnitude more sensitive than surface plasmon transducers based on gold (Lukosz 1991). Waveguide transducers with exceptionally long interaction lengths such as interferometers should be even more sensitive.

One final difference between surface plasmon and waveguide sensors is that surface plasmon sensors exploit one mode only (a *p*-polarized mode) while waveguide sensors often exploit two guided wave modes. Since the two different guided waveguide modes have different penetration depths, they have different responses to changes in the waveguide architecture. The two values of ΔN_{eff} obtained for the different modes can be used to model the protein film as a layer with a uniform refractive index and thickness which can be determined independently (see the introduction to this section).

Waveguide sensor geometries

Input grating couplers

Input grating couplers have been commercially available for use as optical transducers for a number of years. They allow easy determination of the effective refractive index of the waveguide via the coupling condition:

$$k_x + mG = \kappa \tag{6.34}$$

which gives us:

$$N_{eff} = n \sin \theta + \frac{m\lambda_0}{\Lambda} \tag{6.35}$$

where N_{eff} is the effective refractive index of the waveguide, n, the refractive index of the medium from which the light is incident, θ, the external angle of incidence of the light, λ_0 is the wavelength, Λ the grating period and m is an integer number (the diffraction order).

In a conceptually simple approach, Lukosz and Tiefenthaler (Lukosz and Tiefenthaler 1988) coupled a collimated beam of light from a low-power laser into a waveguide via a grating coupler. The waveguide is repeatedly rotated through a fixed range of angles, and the power coupled in is monitored as a function of incident angle using a photodetector mounted at the end of the waveguide (see Figure 6.8a). The coupling angle, θ, found in this way, is then used to determine the effective refractive index, and thus the mass of analyte molecules bound to the waveguide. With a time for the angular scan of around one second, this method allows the binding of molecules to the waveguide surface to be followed with a time resolution of a few seconds. If waveguides supporting both TE_0 and TM_0 modes are used, both coupling angles can be determined and the information can be used to define both the refractive index and the geometrical thickness of the layer.

Simple input grating couplers of this kind have been shown to have very high sensitivities – comparable to those obtained using more complex transducer configurations. Kunz *et al.* demonstrated a resolution better than $2\,\mathrm{pg.mm^{-2}}$ over an hour for a simple input grating coupler (Kunz *et al.* 1994). However, this was only achieved under laboratory conditions of extremely stable temperature ($\Delta T < 0.024\,°C$), laser wavelength ($\Delta\lambda < 10^{-3}\,\mathrm{nm}$) and angle of incidence ($\Delta\theta < 5 \times 10^{-5°}$) (Dübendorfer *et al.* 1997).

An alternative approach to the use of grating couplers as optical transducers was described by Brandenburg *et al.* (Brandenburg *et al.* 1996). A light beam is focussed

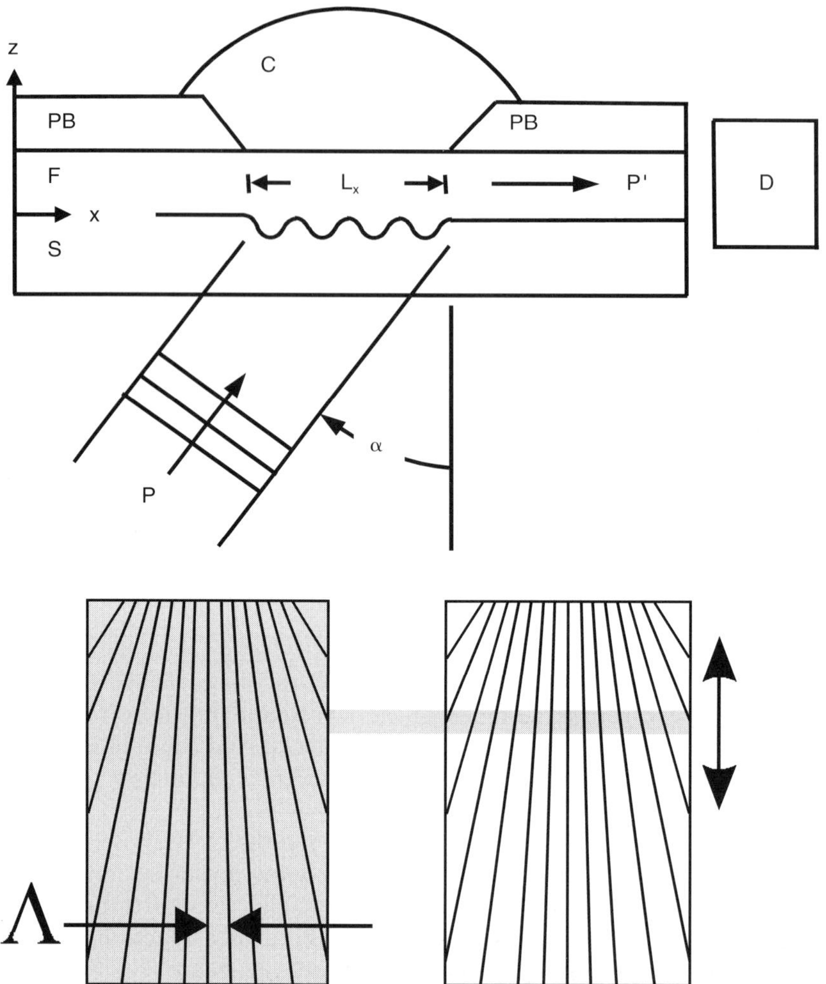

Figure 6.8 a   An input grating coupler with *F* the waveguiding film, *S* the substrate, *C*
the sample solution, and *D* the detector. α is the angle at which incident
light couples into the waveguide
(Reproduced from Lukosz and Tiefenthaler, *Sensors and Actuators* 15
(1988), 273. With permission)
b   The light pointer. Light is coupled into the waveguide at one position on
the first chirped grating. It travels along the waveguide to the second
grating where it is coupled out. Changes in the mass bound to the waveguide change the position of coupling

in a wedge shape onto the grating coupler and the reflected light distribution is
observed using a CCD array. At the incoupling angle, light falling on the grating
couples into the waveguide and a well-defined minimum is observed in the reflected
light. The angular position of this minimum is then used to derive the effective
refractive index of the waveguide and hence the mass of molecules bound to the
waveguide surface. This optical configuration has a number of advantages over the

approach of Lukosz and Tiefenthaler: most notably no moving parts are required, and the incoupling angle can be acquired very rapidly, allowing fast binding processes to be followed in detail.

Further possibilities of the grating coupler approach were demonstrated by Dübendorfer and Kunz (Dübendorfer and Kunz 1997; Dübendorfer and Kunz 1998) with the use of chirped grating couplers. This method (also known as the light pointer) uses a non-uniform (chirped) grating to couple light into the waveguide. The period of the grating varies from one end to the other so that, for a given angle of incidence, the coupling conditions are met at only one position on the grating (see Figure 6.8b). Binding of molecules to the waveguide surface results in changes in the coupling condition, and thus movement of the position of incoupling to a new grating period. The chirped gratings are illuminated uniformly and the position of incoupling can be observed using a second grating to couple the light back out of the waveguide and a CCD camera to record the position of the light. Once again, there are no moving parts in the transducer, and few optical components other than the laser head and detector. The sensitivity of the detector depends (among other things) on the chirp of the grating – in other words on the rate of change of grating periodicity: a large chirp gives a large measuring range but low sensitivity, while a small chirp gives high sensitivity but has a limited dynamic range.

Prism couplers

Prism couplers for waveguide sensors have been largely neglected with one important exception: the resonant mirror (RM), which is commercially available from IAsys (Cush *et al.* 1993). Since the RM is the subject of a separate chapter in his book, only a brief description will be given here.

The transduction principle of the RM is very similar to that of input grating couplers: monochromatic light is incident on the RM and the input conditions that allow the light to couple into the waveguide are determined. These coupling conditions are then used to determine the effective refractive index of the waveguide and, hence, the mass of analyte molecules bound to the waveguide surface. Repeated and rapid measurement of the coupling conditions allows biological recognition processes to be followed in real time.

The resonant mirror (Figure 6.9) consists of a high refractive index prism, a low-index spacer layer of about 500 nm thickness and a high refractive index waveguide layer. The waveguide properties (thickness and refractive index) have been chosen to allow it to support both TE_0 and TM_0 modes, so two coupling angles may be determined. There are two ways to interrogate the resonant mirror: either using a wedge-shaped incident beam with a CCD array to analyse the reflected beam, or, alternatively, using a collimated laser beam which is scanned rapidly through a range of incident angles. In general, light polarized at 45° to the plane of incidence (with *s* and *p* components of equal intensity) is incident on the spacer layer of the RM above the critical angle. When the coupling conditions are fulfilled for one of the polarizations, light with this polarization will couple into the waveguide and then directly out again, undergoing a phase change of π. (Below and above the coupling angle, the phase changes on reflection are 0 and 2π, respectively.) The phase change in one of the orthogonal polarizations is detected as a change of 90° in the polarization of the light beam.

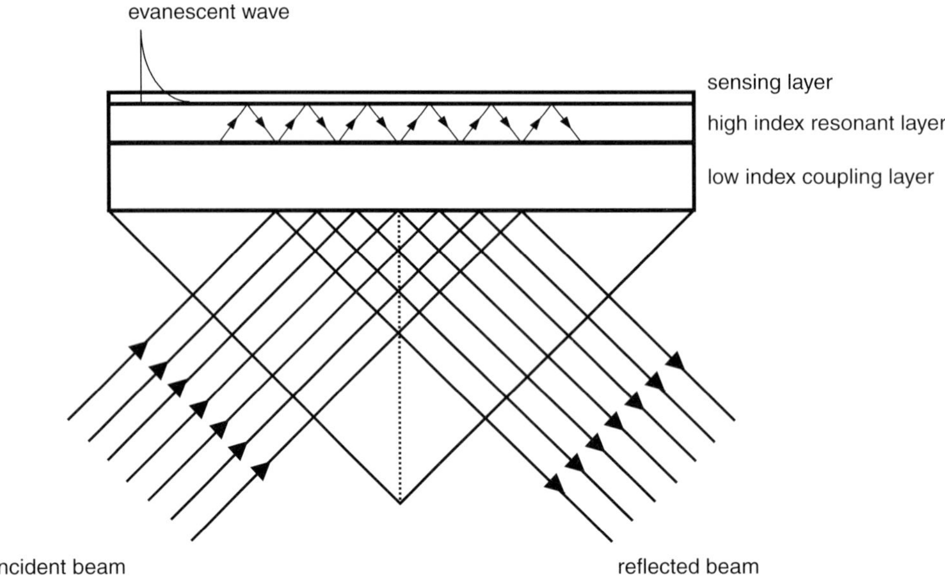

Figure 6.9 The resonant mirror transducer

(Reproduced from Cush *et al. Biosens. Bioelect.* 8 (1993) 347. With permission)

The fabrication of the RM is relatively simple compared to that of grating couplers, since no optical grating structures are required. The resolution of the resonant mirror has been reported to be better than $9\,\mathrm{pg.mm}^{-2}$ (Cush *et al.* 1993).

Waveguide interferometers

Waveguide interferometers have also been developed for use in immunosensing applications. Two different approaches have been developed: the Mach-Zehnder interferometer uses two spatially separated guided modes, one of which interacts with a sensing area, while the other is used as a reference. Alternatively, two different guided modes (often TE and TM) can be used to interrogate the same sensing area: this is the principle of polarimetric interferometry.

The operating principle of a Mach-Zehnder interferometer is shown in Figure 6.10. Light from a single source is split into two beams, which are directed into two separate waveguide arms. One of the waveguide arms (the measurement arm) is coated with receptor molecules: binding of the analyte to this surface changes the effective refractive index of the waveguide, N_{eff}, which produces a change of the propagation (phase) velocity of the guided wave. The other arm of the interferometer (the reference arm) is treated to minimize binding of the analyte to the surface. The guided waves from the two different measurement arms are recombined to give an output intensity which depends on their phase difference. From the phase difference between the two modes, the difference in N_{eff} and thus the mass of analyte bound to the measurement arm are derived.

If we assume that the change in refractive index at the surface Δn_c is small, then the phase change, $\Delta\phi$, obtained is:

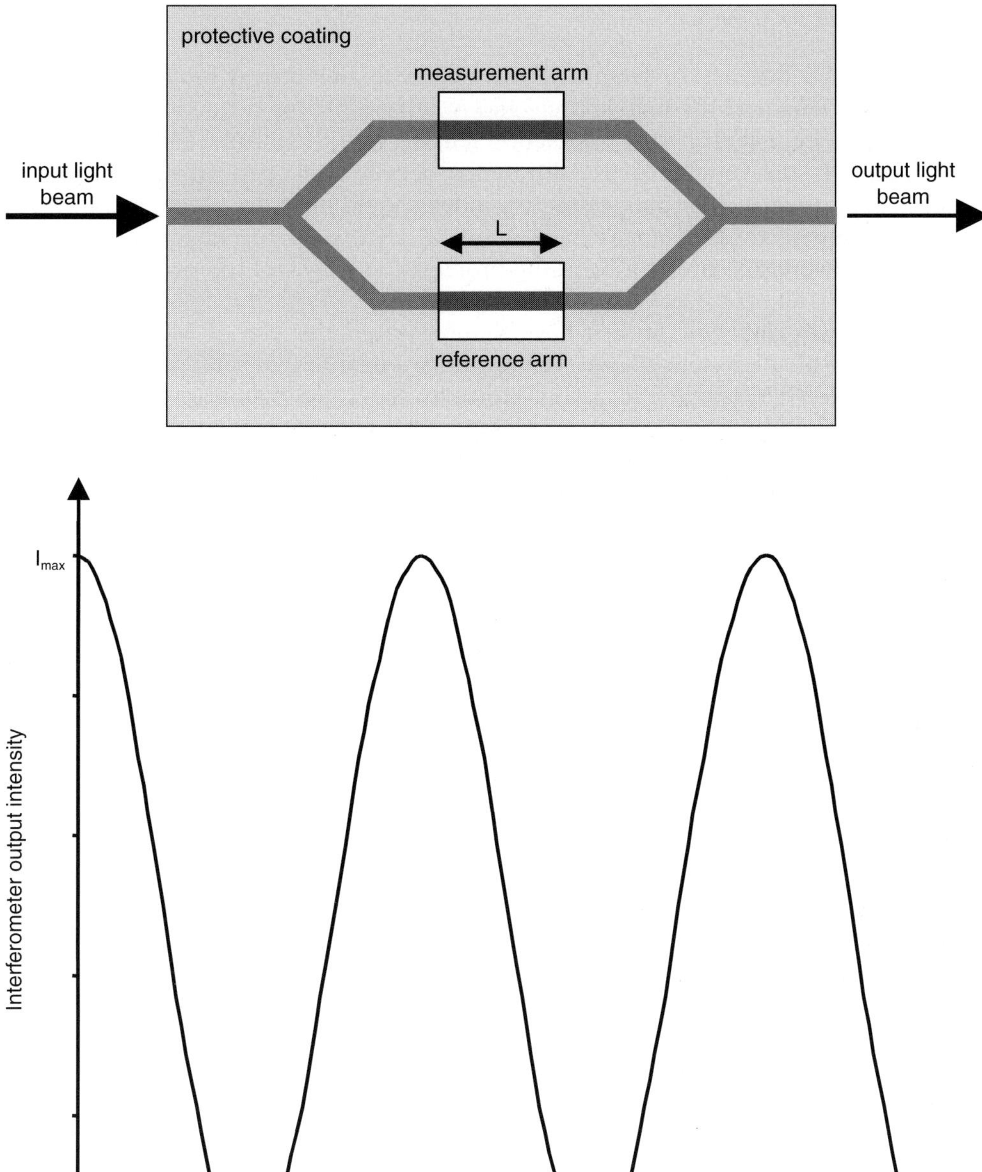

Figure 6.10 a A Mach-Zehnder interferometer. The light is split and passes through both measurement and reference arm. Binding of molecules to the measurement arm result in a phase shift of the light passing through this arm

b The output of the Mach-Zehnder interferometer. The output intensity varies sinusoidally as a function of the effective refractive index of the measuring arm

$$\Delta\phi = L\left(\frac{2\pi}{\lambda_0}\right)\left(\frac{\partial N_{eff}}{\partial n_c}\right)\Delta n_c \qquad\qquad (6.36)$$

where L is the interaction length of the arm with the bound molecules, λ_0, the vacuum wavelength of the light source, and N_{eff}, the effective refractive index of the waveguide mode. Clearly, long interaction lengths give the most sensitive transducers. Since only the relative phase difference between the two arms is measured, changes in temperature or bulk refractive index should have no effect on the sensor signal, as long as these changes have identical effects on the effective refractive index of the two arms, and as long as the propagation length of the two arms is identical.

A number of different groups have demonstrated the use of MZIs as optical transducers with high sensitivities (better than $1\,\mathrm{pg.mm}^{-2}$) (Schipper *et al.* 1995; Drapp *et al.* 1997; Weisser *et al.* 1999). However there are significant disadvantages to the configurations that have been demonstrated to date. Firstly, all the MZIs described require extremely accurate alignment, either because they use end-fire coupling of the light into the waveguide (Drapp *et al.* 1997), or because they require precise splitting of the original incoming beam into two (Lenferink *et al.* 1997). Secondly, adsorption of a protein layer to the surface of the waveguide will usually involve the output signal passing through a number of complete fringes. These fringes must be counted if the adsorption is to be quantified: this means that the adsorption process must be followed on line in its entirety. This is in contrast to other refractometric techniques for which a 'start' and 'end' measurement are sufficient to quantify binding processes taking place at the transducer surface.

In addition, a simple measurement of the intensity of the recombined beam is usually not sufficient to obtain an accurate value for the phase difference between the two arms. The intensity of the outcoupled beam depends on a number of factors, including the intensity of the light source, the efficiency of coupling into the waveguide, and the amount of light scattering from the waveguide surface. All of these factors may change during the measurement. Two methods have been shown to allow good determination of the phase difference. The first involves modulation of the path length of one of the arms using a rotating glass plate: this allows the position of the interference maxima and minima to be accurately determined (Lenferink *et al.* 1997). The second method uses an integrated 3×3 coupler structure (Drapp *et al.* 1997): the three different outputs obtained from this device all have a different phase, which allows unambiguous determination of the phase difference between the two arms. Of the two approaches, the 3×3 output coupler has the clear advantage of requiring simpler read-out instrumentation.

A very recent publication has demonstrated the fabrication of an integrated MZI using silicon technology, with end-coupling from optical fibres, phase modulation of the optical pathlengths, and beam splitting and recombination all taking place on-chip (Heideman and Lambeck 1999). The integration of the different components onto the optical chip results in simple readout instrumentation (although at the cost of a highly complex chip fabrication process), no alignment problems and an extremely stable interferometer. Although this MZI has not yet been used to monitor biological reactions, tests of its response to bulk refractive index changes show an improvement of one to two orders of magnitude in sensitivity compared to previously reported MZI sensors.

The polarimetric approach to waveguide interferometers – the use of different guided modes as measurement and reference beam – has been demonstrated in its simplest form by Stamm and Lukosz (the 'integrated optical difference interferometer') (Stamm and Lukosz 1993; Stamm and Lukosz 1996). A waveguide that can support two guided modes is used as the transducer and coated with receptor molecules. Polarized light is endcoupled into the waveguide so that both TE_0 and TM_0 modes are coherently excited. The two modes propagate together through the waveguide and, after end-face outcoupling, the phase difference between the two modes is analysed using polarimetry.

Binding of biological molecules to the surface of the waveguide results in a change of the effective refractive index, N_{eff}, of the waveguide. Since the distribution of the field strengths is different for the two guided modes, the change in the effective refractive index, ΔN_{eff} is also different for the two modes. This results in a change in the phase difference between the two modes, which is observed on outcoupling, and which is then used to derive the bound mass of analyte. The integrated optical difference interferometer has the advantages of a very simple chip design – a simple planar waveguide – but is experimentally difficult to use, requiring accurate optical alignment.

In contrast, another transducer based on polarimetry, the bidiffractive grating coupler, is designed to eliminate problems of optical alignment (Kubitschko *et al.* 1997; Spinke *et al.* 1997), being relatively insensitive to small lateral or angular displacements. This transducer consists of an oxide waveguide, which supports both TE_0 and TM_0 modes, on a plastic substrate with a bidiffractive grating (the superposition of two gratings, G_a and G_b) (Danielzik *et al.* 1997). The two gratings are chosen so that under measurement conditions the two guided modes will be coupled out of the waveguide – each by a different grating – at almost exactly the same angle. The outcoupled light passes through a polarizer and is focussed onto a CCD camera where the interference fringes from the two modes are imaged (see Figure 6.11). As in the previous example, binding to the surface of the waveguide results in an effective refractive index change which is different for the two modes. The outcoupling angles change by different amounts, and this relative angle change is observed as a change in the fringe spacing on the CCD camera. As can be seen in Figure 6.11, the bidiffractive grating is also used to couple the light into the waveguide. For each mode, one grating is used for incoupling and the other for outcoupling. The modes are outcoupled over a distance of about 1.5 mm – the interaction length of the mode with the sample. The bidiffractive grating coupler has been shown to have a short-term resolution of around $2 \, pg.mm^{-2}$ (over 10 minutes). However, over longer timescales, drifts caused by porosity of the waveguiding film were found to limit the sensitivity to $10–20 \, pg.mm^{-2}$ (Schlatter *et al.* 1993).

It is worth noting that, in polarimetric transducers, the different field distributions (e.g. penetration depths) of the two guided modes mean that the modes have different sensitivities both to binding of molecules to the waveguide surface and also to bulk refractive index changes. This means that in the bidiffractive grating coupler, and the integrated optical difference interferometer (and all similar devices) changes in the bulk refractive index of the solution caused by temperature changes or compositional changes do not cancel out. Instead, bulk refractive index changes result in a signal equivalent to the adsorption or desorption of biological molecules. This is in contrast to the MZIs, where bulk refractive index changes caused by

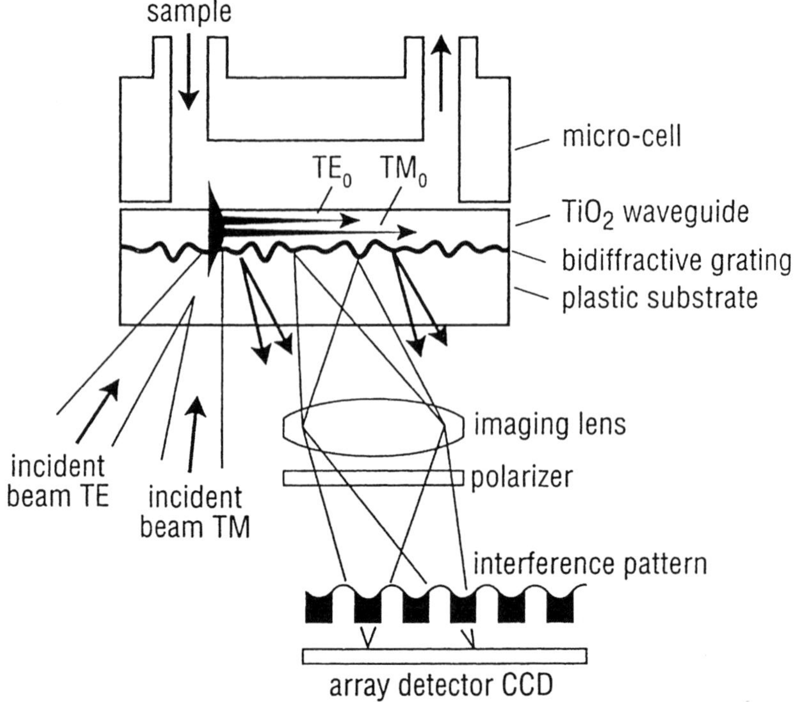

Figure 6.11 The bidiffractive grating coupler

(Reproduced from Spinke *et al. Sensors and Actuators B*, 38–39 (1997) 256. With permission)

temperature or concentration fluctuations have identical effects on the two arms and give no signal. The ability of MZIs to eliminate bulk refractive changes is limited by asymmetries between the two arms and by the spatial separation of the arms, which makes them sensitive to temperature gradients.

Surface plasmon resonance sensors

Prism couplers

The principles of surface plasmon resonance transducers are very similar to those of waveguide grating coupler transducers. In both cases an incident light beam is coupled into a confined mode (in one case a guided mode, in the other a surface plasmon mode) and the in-coupling conditions are modified by the binding of biological material to the transducer surface. Surface plasmon resonance transducers, however, are almost exclusively based on the use of prism couplers. Prism-based surface plasmon biosensors have been commercially available for a number of years, the most widely used instruments being the BIAcore instruments, developed and distributed by Pharmacia Biosensors. Since the BIAcore family of instruments is the subject of a separate chapter, the description of surface plasmon transducers in this section will be limited to the essential features.

Surface plasmon polaritons are similar to the waveguide modes described in the previous sections, in that light falling directly on the sensing surface cannot couple into the surface plasmon, since matching of both ω and k_x is not possible. For coupling to take place, the value of k_x of the incoming light must be increased: in the prism configuration this is done by having the incident light come through a high-refractive index prism onto the back of a gold layer of precisely controlled thickness. The light is totally reflected at the glass/gold interface, but the evanescent field traverses the gold layer to the gold/sample interface where it can excite a surface plasmon mode at a given angle of incidence – the resonance angle (Kretschmann 1971). The resonance condition is given by:

$$\kappa = \frac{2\pi}{\lambda_0}\, n_p \sin\theta_r = \frac{2\pi}{\lambda_0} \left(\frac{\varepsilon'_m n_s^2}{\varepsilon'_m + n_s^2} \right)^{1/2} \tag{6.37}$$

where κ is the propagation constant of the surface plasmon polariton, n_p and n_s the refractive indices of prism and sample respectively, and ϵ'_m the real part of the dielectric constant of the metal layer.

The optical configuration used in the Pharmacia BIAcore, shown in Figure 6.12, is very similar to the reflectance method for interrogating input grating couplers. A monochromatic light beam is focussed into a wedge shape, which falls on the gold/glass interface from which it is totally reflected. The reflected light beam is observed using a CCD array, and a pronounced minimum in the reflected intensity occurs at the resonance angle where coupling takes place between the incident light and the surface plasmon polariton (Karlsson *et al.* 1991). Binding of biological molecules to the transducer surface results in an increase in the angle of resonance, and the minimum in reflected intensity moves across the CCD array. (An alternative configuration uses a collimated monochromatic light beam to excite the surface plasmons: a scan through a range of angles of incidence allows the angle of minimum reflectance to be determined.) The excellent thermal and mechanical stability of the BIAcore instrument has resulted in reported resolutions on the order of $1\,\mathrm{pg.mm^{-2}}$ (Schuck 1997).

One of the great advantages of surface plasmon resonance transducers is clear from Figure 6.12: both the optical configuration and the sensor device are extremely simple. No optical gratings and no moving parts are required. In its simplest form, the sensor device consists of a glass prism with a 45 nm thick layer of gold deposited on its surface. For practical applications, the gold film is usually deposited on a glass slide, which is optically coupled to the prism with an index-matching fluid or polymer.

A less obvious advantage of surface plasmon sensors is the use of gold as the transducer surface. A number of metals or semiconductors can be used in surface plasmon sensors, depending on the size and sign of the real and imaginary parts of their dielectric constants at the wavelength to be used. Indeed, the use of silver in a surface plasmon transducer results in an increase in sensitivity of a factor 3–4 (Kooyman *et al.* 1988). However, gold has a number of desirable features which outweigh the loss of sensitivity associated with its use. Firstly, gold is a very stable metal, highly resistant to oxidation and degradation. In addition, thin layers of gold

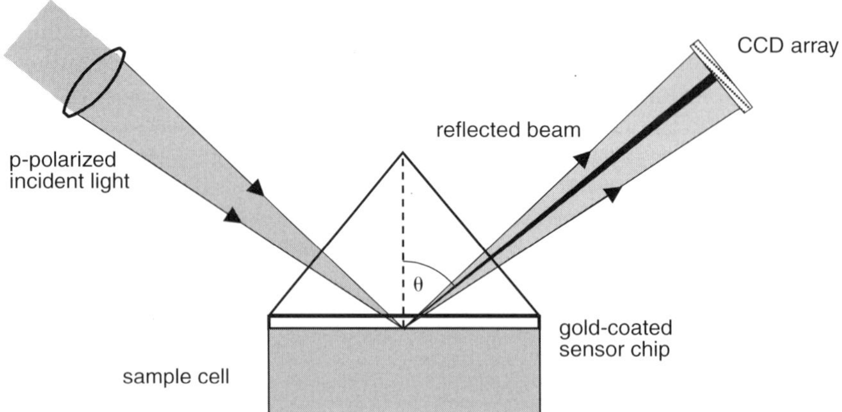

Figure 6.12 The surface plasmon resonance transducer: the optical configuration used in the BIAcore instrument

with a well-defined layer thickness can be easily and reproducibly fabricated by vacuum deposition. Finally, gold is the material of choice when chemical and biological modification of the surface is considered: simple solution methods using thiols allow the formation of highly reproducible surfaces with a wide variety of chemical and biological groups (Bain *et al.* 1989; Bain and Whitesides 1989). The use of gold as a transducer surface has undoubtedly made a significant contribution to the dominance of the biosensors market by surface plasmon sensors.

An alternative optical configuration for surface plasmon resonance transducers that has been investigated by a number of groups, uses a broadband light source that is incident on the gold/glass interface at a fixed angle (Mar *et al.* 1999; Stemmler *et al.* 1999). Light at the resonant wavelength couples into the surface plasmon mode. The reflected spectrum is monitored using a diode array spectrometer and a pronounced dip in the reflected intensity is observed at the resonant wavelength. Binding of biological molecules to the gold/sample interface changes the resonant wavelength.

Interest in spectral interrogation configurations of surface plasmon resonance transducers has grown in recent years thanks to the availability of low-cost diode array spectrometers that allow rapid and cheap analysis of the reflected spectrum. Replacing angular interrogation with spectral interrogation should increase the sensitivity of surface plasmon transducers (Homola *et al.* 1999). However, this has not yet been demonstrated.

One final and important aspect of surface plasmon resonance transducers is their inherent suitability for use as multichannel sensors. In most of the surface plasmon optical configurations described above, a monochromatic light beam is incident on the gold/glass interface and the angle of minimum reflectivity is used to measure the mass of bound molecules at the gold/sample surface. The value obtained in this configuration is the average mass bound over the illuminated area. However, the gold/glass surface can also be imaged, using a lens and a CCD camera to analyse the reflected light intensity at each point of the surface. This then gives the mass bound to the transducer surface for each pixel analysed (Rothenhäusler and Knoll 1988). Since the entire transducer surface can be imaged in this way, and since no additional optical structures, such as gratings, are required, the binding reactions taking place at each point on the gold surface can be analysed separately. In a surface plasmon resonance multichannel transducer, a measurement channel is defined solely by the area on which a certain antibody (or other receptor molecule) has been deposited: there is no need for additional gratings or other optical structures. This feature is already exploited in the BIAcore instrument, where four measurement channels are available on each gold-coated sensing chip. Further development should allow great improvements in terms of both the number and size of the measurement channels available.

Other surface plasmon resonance transducers

Grating couplers have been shown to be an alternative to prism couplers for surface plasmon resonance transducers (Cullen *et al.* 1987/88) (Cullen and Lowe 1990). In this configuration, a gold film is deposited on a grating on a glass or plastic substrate. Monochromatic light is incident on the gold/sample interface of the grating through the sample and couples into the surface plasmon polariton at the resonance angle θ_r, which is given by:

$$k_{spr} = \frac{2\pi}{\lambda_0} \, n_s \sin \theta_r + mG \qquad (6.38)$$

where k_{spr} is the momentum of the surface plasmon polariton, m is an integer and G is the wavevector of the grating.

Theoretical studies have shown that grating and prism couplers have similar sensitivities when used in angular-interrogation configurations (Homola *et al.* 1999). Grating couplers have advantages over prism couplers in that the thickness of the gold film, which must be controlled to within a few nanometres in prism couplers, is no longer critical, thus facilitating one of the fabrication steps of the transducer devices. However this advantage is more than offset by two disadvantages of grating couplers: firstly, grating couplers require the fabrication of optical gratings in the substrate. Secondly, and most importantly, in grating couplers the incident light beam passes through the sample to arrive at the grating surface. Turbid or opaque samples cannot be used with this transducer.

Several groups have investigated the combination of planar waveguides with surface plasmon resonance for direct sensing applications (Homola *et al.* 1997; Harris *et al.* 1999; Weisser *et al.* 1999). The transducer consists of a planar waveguide coated with a thin layer of gold: at resonance, the guided mode couples into the surface plasmon at the gold/sample interface, resulting in very strong attenuation of the guided mode. Off-resonance, the attenuation is a strong function of the effective refractive index of the waveguide. Thus, by monitoring the output intensity of the guided mode, the effective refractive index of the waveguide architecture, and hence the mass of bound protein at the surface of the waveguide, can be determined.

Transducers combining waveguides and surface plasmon resonance combine some of the advantages of the two approaches to direct sensing: the high stability and reproducible chemical functionalization of the gold surface found in surface plasmon sensors (Harris *et al.* 1999), together with the high degree of miniaturization and integration associated with waveguide sensors. Disadvantages of this approach include its lack of flexibility and its low dynamic range – the transducer must be accurately tuned to the bulk refractive index of the solution.

Referencing and mass labels in refractometric sensing

A number of analyses of the theoretical sensitivities of refractometric sensors: have indicated that waveguide sensors should be significantly more sensitive than surface plasmon sensors (see, for example, Lukosz 1991). However, there is little experimental data to support this. Cited experimental sensitivities of waveguide transducers are, almost without exception, well below the theoretically possible values and little progress has been made in transducer sensitivity in recent years. The actual sensitivities obtained are a function of the care and attention paid to such issues as mechanical and thermal stability of the instruments. At the moment, differences in sensitivities for different sensors are relatively small (Dübendorfer 1997), particularly when one bears in mind the uncertainties associated with the different evaluations of sensitivity given by different authors.

However, one very successful approach to increasing the sensitivity of direct transducers has been demonstrated in the past few years: it is the use of mass labels (Buckle *et al.* 1993; Kubitschko *et al.* 1997). Colloidal particles of gold, latex or oxides with diameters in the range of tens of nanometres are used as mass labels. Antibodies, antigens or other biological molecules are bound to the surface of the particles. When the molecule-coated particle binds to the transducer surface, the local refractive index change is much higher than when the molecule alone bonds. The resulting increase in sensitivity has been demonstrated for waveguide transducers: in one case, a sandwich assay using secondary antibodies bound to latex particles showed an increase in sensitivity of three orders of magnitude compared to the same assay with using unlabelled secondary antibodies (Kubitschko *et al.* 1997).

In addition to an increase in sensitivity, the use of a sandwich assay also improves the specificity of the signal, since the final binding step is performed in a buffer of known composition, in which non-specific binding can be minimized. Studies using the determination of thyroid stimulating hormone (TSH) in serum as a model assay, demonstrated good sensitivity and correlation with a commercial enzyme assay for TSH (Kubitschko *et al.* 1997) in blood plasma – a medium in which non-specific binding is often high. However, the use of a labelled secondary antibody in the assay poses significant questions as to the suitability of the choice of transduction method, since the advantages of direct sensing – label-free detection in real time – are no longer available. Under these circumstances, the use of a fluorescence transduction method and fluorescently labelled secondary antibodies would almost certainly result in further improvements in sensitivity.

Refractometric transducers are inherently sensitive (although to different extents) to the refractive index of the bulk sample: fluctuations caused by temperature changes and/or changes in total solute concentration can give spurious signals. One approach to this problem has been to concentrate on precise control of the temperature of sample and transducer elements (for example, in BIAcore instruments, the transducer temperature is stabilized to within 0.01 °C). An alternative approach is the use of reference transducing elements which are insensitive to the biological reaction to be measured, but which respond to bulk refractive index changes in an identical manner to the measurement element. Subtraction of the reference signal from the sensing channel signal results in a signal that is, ideally, due only to the binding of molecules to the sensing channel. Effective referencing requires that the sensing and reference elements be very close to each other, so that any temperature differences between the two are minimized. It also requires that the reference element be resistant to non-specific adsorption of solutes to the surface. Reference elements are inherent to MZI transducers. They have also been demonstrated for waveguide grating couplers (Dübendorfer *et al.* 1997; Dübendorfer and Kunz 1997) and for surface plasmon resonance transducers (O'Brien *et al.* 1999). In these studies the reference channel was made resistant to non-specific adsorption either by blocking with bovine serum albumin, or by the formation of a self-assembled layer terminated with oligo(ethylene glycol) moieties. A detailed analysis of the effectiveness of the referencing showed that the reference element reduced the sensitivity of the system to temperature fluctuations by one order of magnitude, and to wavelength and angle of incidence variations by two orders of magnitude (Dübendorfer *et al.* 1997).

Reflectometric techniques

An alternative technique for the direct detection of biological recognition reactions at surfaces is reflectometry: the thin biological film at the surface of the transducer is characterized by a combination of reflection and interference. A beam of light is incident on the thin film where it is reflected from both front and back interfaces of the film. The interference between the two reflected beams gives a resultant beam whose intensity, polarization or spectral composition depends on the optical thickness of the film. There are several different transduction methods based on reflectometry: the best known is ellipsometry.

Ellipsometry

Ellipsometry is a well-established technique for the characterization of thin films. It has been used very successfully for many years in the semiconductor industry for the characterization of hard thin films such as oxide layers and is based on the reflection and interference of s and p-polarized light at interfaces.

In general, when light is reflected from an interface, both the amplitude reflection coefficients and the phase shift are different for s and p-polarized light. Thus, reflection of light from an interface normally produces a change in the polarization state of the light. For a thin film, s and p-components of the light are reflected from both front and back interfaces of the film and the interference of the two reflected beams for each polarization component results in a well-defined amplitude for that component. Thus the polarization state of the reflected beam depends on the polarization of the incident beam and of the optical thickness of the thin film. The change in polarization state on reflection is defined by the complex reflectivity:

$$\frac{r_p}{r_s} = \frac{E_{rp}/E_{rs}}{E_{ip}/E_{is}} = \tan\Psi\exp(i\Delta) \tag{6.39}$$

where

$$\tan\Psi = \left|\frac{E_{rp}/E_{rs}}{E_{ip}/E_{is}}\right| \tag{6.40}$$

and

$$\Delta = (\delta_{rp} - \delta_{rs}) - (\delta_{ip} - \delta_{is}) \tag{6.41}$$

with $(\delta_{rp} - \delta_{rs})$ and $(\delta_{ip} - \delta_{is})$ the phase differences between p and s components in the reflected and incident beams respectively.

For well-defined interfaces between bulk media, both the thickness and the refractive index of a homogeneous, isotropic thin film can be derived from the change in complex reflectivity. Usually the reflectivity is determined by null ellipsometry, in which elliptically polarized light is incident on the surface, and the polarization of the incident beam is varied until a linearly polarized reflected beam is obtained. Calculation of the thin film parameters is based on a Fresnel model of a homogeneous layer with smooth interfaces.

Early studies of protein adsorption and thin organic films at solid/water and air/water interfaces demonstrated that ellipsometry is capable of sensitive measurement of surface loading. (Sensitivities on the order of hundreds of pg/mm^2 were obtained (De Feijter *et al.* 1978; Jönsson *et al.* 1985).) However, these studies also demonstrated the difficulty of independently determining both the geometric thickness and the refractive index of a thin biological layer. For biological films only a few nanometres thick, the variations in Ψ caused by changes of the film thickness are minute (below the resolution of currently available high precision ellipsometers). Only one good experimental parameter is available, the parameter Δ, which was used in these studies to determine adsorbed protein mass at the surface.

More recent work has shown that the use of spectral ellipsometry (i.e. simultaneous ellipsometric measurements at several wavelengths) can improve the reliability of the data (Spaeth *et al.* 1997). This allows both geometrical film thickness and refractive index to be determined independently for relatively thick (>20 nm) protein layers, although not for protein films below monolayer thickness or for monolayers of small molecules. It is important to note that the use of spectral ellipsometry does not allow the determination of additional parameters of the protein layer. The information obtained at the different wavelengths is not independent (to within the accuracy of the measurements) for the very thin, low-dispersion films obtained on protein binding. Measurements carried out at different wavelengths are effectively repeated determination of the same parameters, Ψ and Δ (Reiter *et al.* 1992).

One serious disadvantage of ellipsometry for biosensor applications is that the interrogating light beam is incident on the sensor surface through the sample solution (see Figure 6.13). This clearly excludes the use of ellipsometry for the investigation of turbid samples and also has important implications for non-turbid samples:

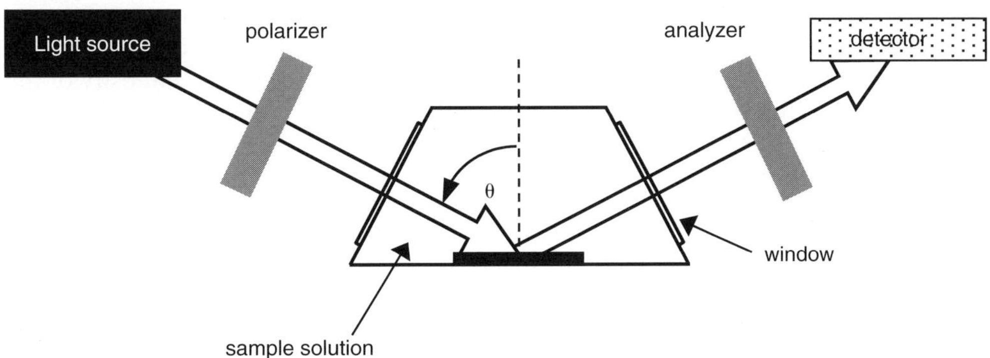

Figure 6.13 Ellipsometer configuration for biosensing applications

not only does the light beam traverse the solution, it also passes through two windows with two surfaces at which protein adsorption may occur. A great deal of care must be taken to avoid changes in the polarization state of the light caused by the windows. In particular, both incident and reflected light beams must be perpendicular to the windows to avoid phase changes caused by protein binding at the glass/water interface.

Given the expense and complexity of ellipsometric measurements, the difficulty of independent determination of two parameters for the protein layer, and the limited usefulness of the second parameter in biosensing applications, a large number of researchers have concentrated on the development of simplified reflectometric transducers. These methods measure one parameter of the thin film only and so are limited to the determination of the optical thickness (*n.d*) of the film (which is equivalent to the adsorbed mass of protein at the surface). This allows a reduction in instrumental complexity while not significantly changing the amount of useful information available about the biological layer.

Brewster angle reflectometry

One such simplified reflectometric approach is Brewster angle reflectometry: the measurement of the reflected intensity of *p*-polarized light at or near the Brewster angle.

At a clean smooth interface between two non-absorbing media, *p*-polarized light, incident at the Brewster angle is entirely transmitted: there is no reflection from the interface (see Figure 6.3). (In real systems, the reflectivity observed at the Brewster angle is determined by the smoothness of the interface: any roughness results in scattering of the incident light.) The Brewster angle, θ_B, is defined by the equation:

$$\tan \theta_B = \frac{n_t}{n_i} \tag{6.42}$$

where n_t and n_i are the refractive indices of the media in which the light is transmitted and incident, respectively.

The formation, at the interface, of a thin layer of biological molecules with a refractive index, n_l, different from n_t and n_i results in the Brewster conditions no longer being fulfilled, and light is reflected from the interface. The amount of protein adsorbed at the interface can be quantified by measuring the intensity of the reflected light. The change in reflectance on absorption of a thin biological layer is low in absolute terms but, since it is measured against a very low background, it can be determined with high sensitivity. The increase in reflectivity is not a linear function of adsorbed mass, and increased sensitivity to protein adsorption can be obtained by the use of silicon oxide 'spacer' layers on the substrate. The refractive index of silicon oxide, 1.5, is similar to that of the adsorbed protein layer, and the oxide and proteins layers can, to a first approximation, be modelled as a single layer. Oxide layers of around 10–20 nm thickness seem to offer the best compromise between increased reflectivity change on protein adsorption and increased background signal due to the oxide layer.

Brewster angle reflectometry has been used in optical biosensing (Welin *et al.* 1984; Arwin and Lundström 1985), giving sensitivities similar to those of ellipsometry. However, although certainly much simpler and cheaper than ellipsometry, it still suffers from ellipsometry's major disadvantage: the incident light beam passes through the sample solution.

Reflection-interference spectroscopy

Currently the most promising and well-developed of all the reflection–interference techniques is reflectometric interference spectroscopy or RIfS, which is based on white light interference in thin films (Brecht *et al.* 1992). An optic fibre is used to bring white light to a glass substrate normal to the glass surface. The light is partially reflected from both faces of a thin (500 nm) silica film on the opposite surface of the substrate and wavelength-dependent interference takes place between the two reflected light beams, as described by:

$$E_0^2 = E_{01}^2 + E_{02}^2 + 2E_{01}E_{02}\cos\left(\frac{2\pi}{\lambda_0}\Delta nd\right) \tag{6.43}$$

where Δnd is the optical path difference between the beams reflected from the two faces of the silica film. Thus, the spectrum of the reflected light depends on the optical thickness of the silica film.

The spectrum of the light reflected from the silica film can be used as a measure of the mass of biological molecules binding to its surface. Since the refractive indices of silica and proteins are similar, the binding of biological molecules can (as for Brewster angle reflectometry) be modelled as a simple increase in the thickness of the silica film. This produces a characteristic change in the spectrum of the reflected beam which is measured using a diode array spectrometer (see Figure 6.14). These low-cost spectrometers are the key to RIfS, since they allow the rapid acquisition of complete spectra with high wavelength resolution. A number of studies using RIfS as a transduction technique have demonstrated that the method is highly sensitive, allowing the detection of mass increases on the order of a few pg/mm^2. This is sufficient for the direct detection of small molecules such as biotin (molecular weight 244) at submonolayer coverages (Piehler *et al.* 1996).

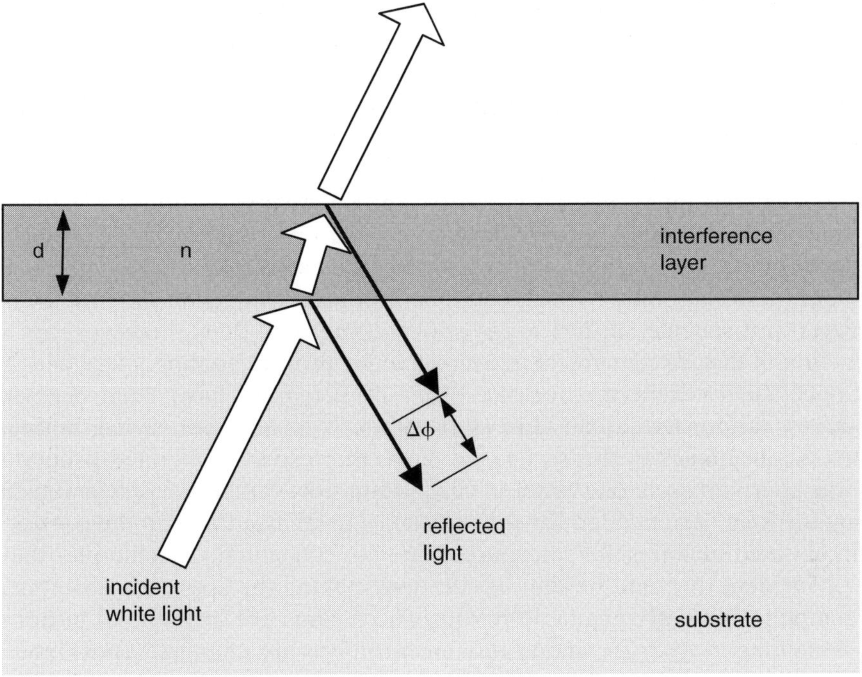

Figure 6.14 Reflection of white light at a weakly reflecting thin film: the optical principle of RIfS transducers

RIfS has important advantages over ellipsometry and Brewster angle reflectometry. The most significant advantage is that the light beam is not incident through the sample, but, instead, through the glass substrate. This eliminates all possible problems caused by solution turbidity and/or reflections from multiple interfaces. When compared with refractometric methods such as waveguide grating couplers or surface plasmon resonance, RIfS has the advantage of being only weakly sensitive to the refractive index of the bulk sample. This means that reference channels are not necessary to compensate for bulk refractive index changes in high precision measurements, or, alternatively, that the requirements for temperature stabilization are hugely reduced. (For example, the RIfS measurements of biotin binding were performed without temperature stabilization: similar sensitivity with surface plasmon resonance requires the temperature to be stabilized to within 0.01°.)

Sensitivity enhancement

A very interesting development in the field of reflectometric transducers is the use of oxidized porous silicon films to increase transducer sensitivity. In the reflectometric methods described above, proteins (or other biological molecules) bind to a planar surface which is coated with a thin inorganic film with a refractive index of around 1.4–1.5. The binding of the proteins increases the geometrical thickness of the coating by a few Angstroms and this increase is observed as a change in the reflectivity of the surface. However, if a relatively thick (200–300 nm) and highly porous film such as oxidized porous silicon, is used, a huge surface area is available for protein binding, which takes place throughout the film. This binding results in a change in the refractive index of the entire film, which thus gives a large increase in its optical thickness and, consequently, a large optical signal.

The use of oxidized porous silicon for direct measurement of biomolecule binding was reported by two different groups which used the porous layers as substrates in ellipsometry (Bjorklund *et al.* 1997) and a variant of RIfS (Lin *et al.* 1997). The porous silicon films were prepared by anodic etching of boron-doped silicon wafers and the hydrophobic porous layer obtained was then thermally oxidized to give a hydrophilic porous layer into which aqueous solutions can penetrate. Films some hundreds of nanometres thick could be produced, with pore sizes up to 200 nm in diameter and specific surface areas above $200 \, m^2/cm^3$. It is, however, not clear what fraction of this surface area was accessible for protein binding.

Extremely high sensitivities on the order of $10 \, fg/mm^2$ have been claimed for reflectometry on porous silicon (Lin *et al.* 1997). It is, however, worth noting that there are disadvantages to this technique. First, the response obtained using porous silicon was approximately one order of magnitude slower than for planar substrates, reflecting the long time needed for the molecules to diffuse through the porous film. In addition, since silicon is not transparent, in both transduction techniques the light beam was incident through the sample solution, making the approach unsuitable for turbid samples. Currently available porous silicon films are unstable in buffer solutions, giving long-term drifts when salt concentrations are changed (Bjorklund *et al.* 1997). Finally, the underlying mechanism of the refractive index change in the porous layer is not clear: a refractive index increase due to the binding of molecules does take place, but the dominant effect may be due to changes in interfacial capacity reducing the carrier density of the silicon (Lin *et al.* 1997).

Fluorescence transducers

The fundamentals of fluorescence

Fluorescence is a three-stage process, which results in light emission. It occurs in certain molecules, which are known as fluorescent dyes, fluorophores, or fluorochromes (Cantor and Schimmel 1980; Wood and Barnard 1997). The process is often illustrated by a simple electronic state (or Jablonski) diagram (see Figure 6.15).

Stage 1 is the absorption of a photon of energy hv_{ex} by the fluorophore. This absorption creates an unstable excited electronic state (S') in the molecule.

Stage 2. The excited electronic state has a lifetime of typically 1–10 nanoseconds. During this time some of the energy of state S' is dissipated as heat to give a relaxed state S, from which light emission occurs.

Stage 3 is the emission of a photon of energy hv_{em} by the fluorophore, which returns to the ground state. Since energy was dissipated between S' and S, the energy of the emitted photon is lower than that of the excitation photon. This difference ($hv_{ex}-hv_{em}$) is known as the Stokes shift. The Stokes shift is fundamental to fluorescence assays since emitted light can be separated from excitation light on the basis of wavelength. This allows low numbers of emitted photons to be measured against a low background to give a very sensitive assay format.

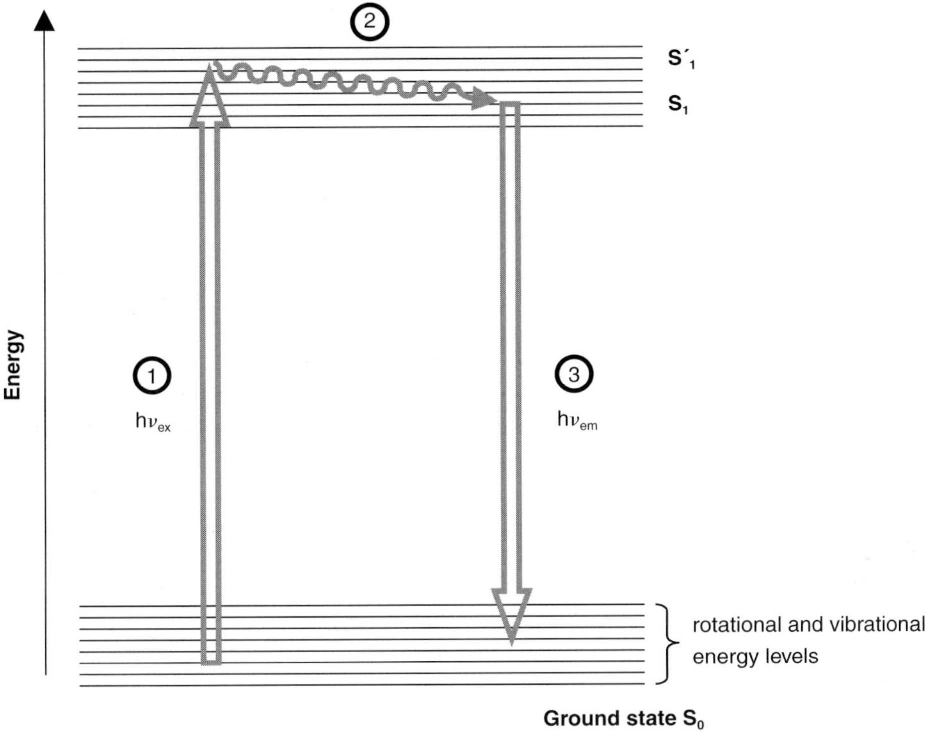

Figure 6.15 Jablonski diagram illustrating optical excitation and fluorescence emission of a molecule

Excitation spectra

The probability of excitation of a fluorophore depends on the wavelength of the incident light – as shown in Figure 6.16. (In general the excitation spectrum is the same as the absorption spectrum of the fluorophore.) The wide excitation peak is a consequence of the fact that a molecule in the ground electronic state can be in any one of a number of vibrational and rotational states. Similarly a number of vibrational and rotational energy levels are associated with the excited electronic state. Thus a relatively wide range of photon energies can promote the electron from ground to excited state.

The extinction coefficient of the fluorophore is a measure of the amount of light it absorbs at a given wavelength. The molar extinction coefficient (ϵ) is defined as the optical density of a 1 M solution with a 1 cm path length. For commercially available fluorophores, ϵ ranges from about 5000 to 200,000 cm^{-1} M^{-1}.

Emission spectra

The emission spectrum of a fluorophore plots the intensity of the emitted light as a function of wavelength. It is always red-shifted compared to the excitation spectrum because of the Stokes shift. The emission spectrum is independent of the excitation wavelength.

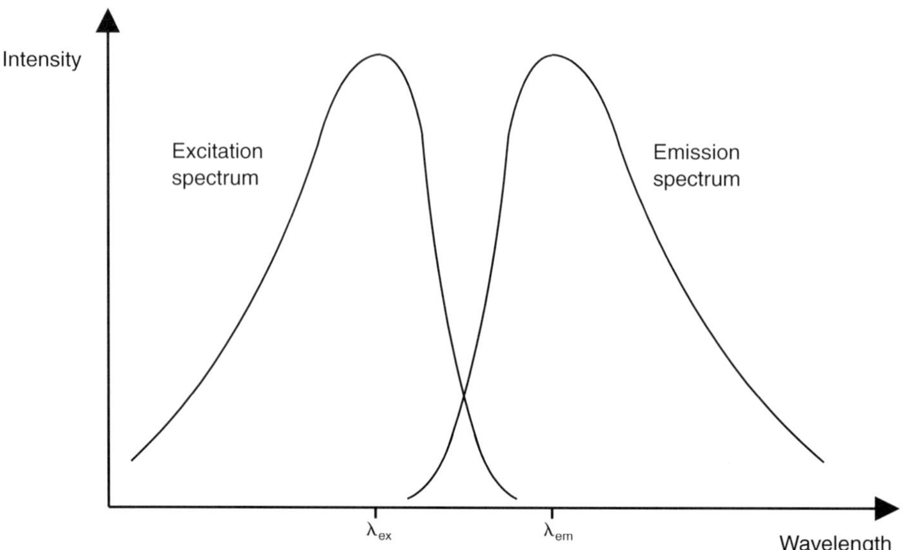

Figure 6.16 Schematic excitation and emission spectra of a fluorophore

Fluorescence intensity

The intensity of light emitted from a fluorophore is proportional to its molar extinction coefficient, ϵ, (which defines the efficiency of photon absorption) and its quantum efficiency ϕ:

$$\phi = \text{number of photons emitted/number of photons absorbed} \tag{6.44}$$

The quantum efficiency is determined by the interactions of the fluorophore with its environment during the lifetime of the excited state. These interactions may depopulate the excited state (e.g. by collisions or fluorescence energy transfer) so that the fluorophore returns to the ground state without fluorescence emission.

A third factor which influences the fluorescence intensity of a sample is the number of times a fluorophore can absorb and emit a photon. At high light intensities photobleaching, the irreversible destruction of the excited state fluorophore, plays an important role and reduces the observed fluorescence.

Fluorescence sensing

A number of formats exist for the detection and measurement of analytes using fluorescence methods. Notable progress has been made in recent years in the development of techniques for homogeneous assays or pseudo-homogeneous assays (e.g. assays based on the use of latex particles). These developments have been reviewed elsewhere (see, for example, Hemmilä 1991; Wood and Barnard 1997), but are outside the scope of this chapter, which is restricted to heterogeneous detection based on the use of the evanescent wave.

Like many refractometric sensors, fluorescence sensors rely on a combination of biological recognition and an evanescent wave at a surface to obtain specificity: receptor molecules, such as antibodies, bind the analyte to the surface where it can be detected via the evanescent wave, while bulk signals are largely eliminated. However, fluorescence sensors have an extra degree of specificity in that only fluorescent (or fluorescently labelled) molecules are detected. This minimizes false signals caused by non-specific binding of other sample components to the transducer surface: non-specific binding is only a problem if the non-specifically bound protein interferes with the specific antibody/antigen recognition process (for example, by covering specific binding sites on the sensor surface).

A second advantage of fluorescence sensing is that of sensitivity. The optical detection and even characterization of single fluorophores has been well established for many years (see, for example, Weiss 1999). While the instrumental costs of these techniques exclude their use in biosensing, it is clear that fluorescence has the potential to be orders of magnitude more sensitive than direct sensing, particularly for small molecules.

The evanescent wave can be used to detect surface-bound fluorescent labels in a number of different configurations (see Figure 6.17) all of which are used in current biosensor developments:

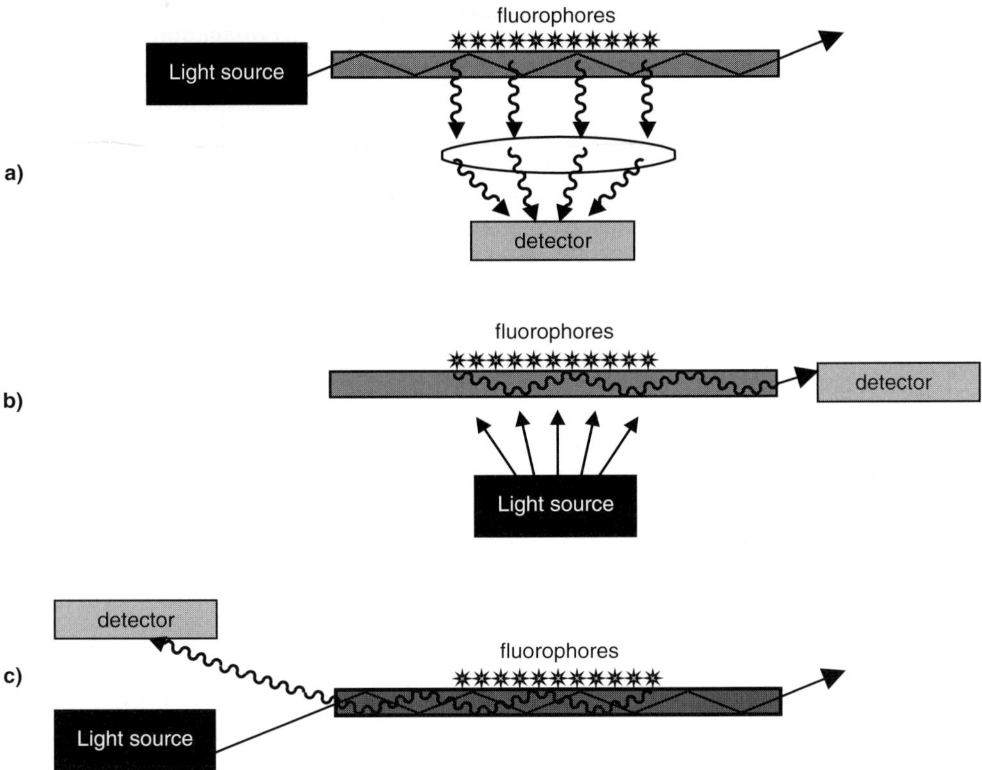

Figure 6.17 Three different fluorescence transducer configurations:
  a Excitation of the fluorophores by the evanescent wave
  b Collection of the fluorescently emitted light by the optical waveguide
  c Fluorecence excitation and collection of the emitted light via the waveguide

a   The evanescent wave is used to excite the fluorophores: light at the excitation wavelength is totally internally reflected at the surface to which the fluorescent labels are bound. Emitted light is collected in the bulk medium.

b   The evanescent wave is used to collect fluorescent emission at the surface. Excitation takes place by light incident from the bulk.

c   Both excitation and collection of emitted light exploit the evanescent wave. This configuration differs from the previous two in that the contribution of the bulk sample to the fluorescence signal is smaller. Since both excitation and coupling of emitted light back into the waveguide are proportional to $\exp(-d/\lambda)$ the depth dependence of the fluorescence signal in this configuration is proportional to $\exp(-2d/\lambda)$.

Fluorescence labelling for biosensors

Apart from a very few assays for naturally fluorescent analytes, fluorescence biosensors depend exclusively on the use of proteins or other biological molecules, which have been modified by the addition of a fluorophore. The correct choice of this fluorophore is central to the development of a sensitive transducer and depends on a number of factors: high quantum efficiencies and molar extinction coefficients are required to give high intensity fluorescence emission; a large Stokes shift allows easy separation of excitation and emission signals; insensitivity to quenching and photobleaching is desirable; excitation and emission wavelengths should be appropriate for inexpensive light sources and detectors; the fluorophore should have suitable functional groups to allow covalent attachment to the molecule to be labelled. In addition, labelling of the biological molecule with the fluorophore should result in only minimal perturbation of its biological function. This is a particular problem for small molecules, where the addition of a fluorophore – often a large, hydrophobic molecule – may significantly alter, for example, solubilities and binding constants.

Background signals in fluorescence assays are largely due to light scattering and natural fluorescence in both sample and hardware (cuvettes, lenses, etc.). Two approaches are very promising for the reduction of these background signals.

The first approach is the use of near infrared fluorophores (Patonay and Antoine 1991). Absorption, scattering and fluorescence from biological and organic molecules are most intense between 350 and 600 nm. Only very few classes of organic molecule exhibit significant absorption or fluorescence between 600 and 1000 nm. The development of inexpensive near-IR laser diodes and semi-conductor photodiodes in recent years has stimulated the development of suitable fluorophores for this spectral region. The most commonly used are the cyanine dyes, such as Cy5, which has an excitation maximum of 650 nm and an emission maximum of 668 nm.

The second approach is the use of long-lifetime fluorophores (Diamandis 1988). Background and signal fluorescence are then distinguished on the basis of fluorescence lifetime. Background fluorescence generally has a decay time constant on the order of nanoseconds, while fluorescent chelates of the rare earth metals, Sm(III), Eu(III), Tb(III), and Dy(III) may have lifetimes in the millisecond time regime. Time-resolved measurements allow effective exclusion of background. However the cost of time-resolved fluorimeters has restricted its use in biosensor applications.

Fluorescence transducers based on planar waveguides

While early research into fluorescence biosensors concentrated on the use of fibre optic transducers (see below, pages 167–170), in recent years there has been a very marked move towards the development of planar waveguide sensing systems. Planar waveguides have a number of advantages over optical fibres including simplified manufacturing processes, easier integration of fluidics systems, the spatial separation of the fluorescently emitted light from the excitation light, and a convenient format for the creation of multianalyte transducers.

One of the earliest and simplest planar waveguide devices to be developed is the fluorescence capillary fill device (Badley *et al.* 1987; Robinson *et al.* 1993) (FCFD) shown in Figure 6.18. The FCFD consists of two plates of glass separated by a narrow gap, typically of 100 microns. The lower glass plate is coated with capture antibodies and acts as a multimode waveguide. The upper plate has a layer of fluorescently labelled reagents (secondary antibody or competitive ligand) on its surface. The sample is drawn into the FCFD by capillary action (which ensures reproducible sample size) and dissolves the fluorescent reagents from the upper plate. The fluorescent reagents bind to the lower plate and are excited by illumination of the flat face of the CFD with filtered white light. Discrimination between fluorescent labels bound to the waveguide surface and those in solution is based on the angular separation of the two components: fluorescence emitted light from labels in solution may enter the waveguide and be propagated along it by multiple reflections, but only at angles lower than the critical angle. In contrast, light emitted by labels bound to the waveguide surface can couple into the waveguide via the evanescent field and propagate along the waveguide at angles greater than the critical angle. At the end of the glass plate the light is transmitted into a photodetector, which measures the fluorescence intensity, while a spatial filter is used to exclude fluorescent light from molecules in the bulk phase. The use of bulk excitation and

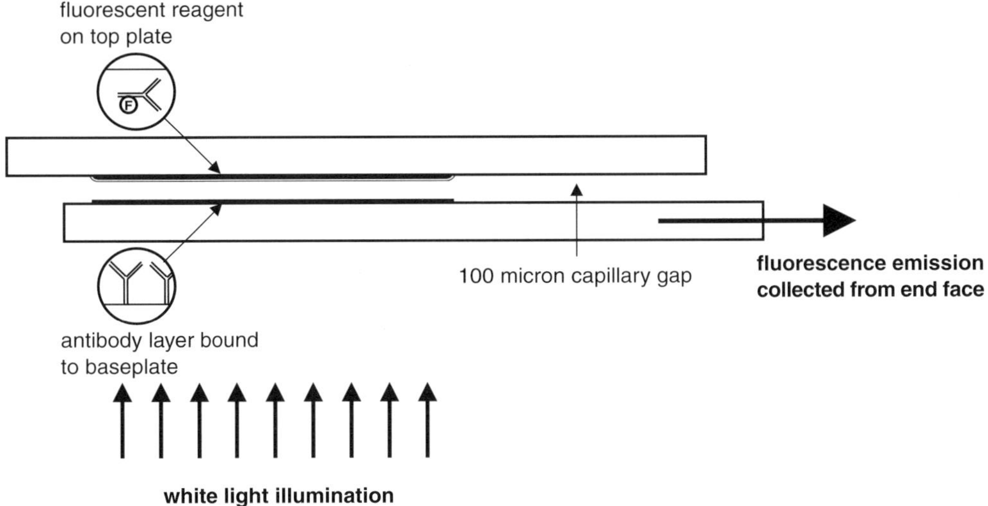

Figure 6.18 The fluorescence capillary fill device

(After a figure from Deacon *et al. Biosens. Bioelect.* 6 (1991) 193. With permission)

evanescent detection relaxes the constraints on waveguide alignment, which can be very stringent, for example, for end-face coupling into waveguides.

A very different approach to planar waveguide fluorescence sensing was taken by Duveneck *et al.* (Duveneck *et al.* 1997; Duveneck and Abel 1999) who developed a sensor system based on single mode waveguides and evanescent wave excitation. The waveguides used in this configuration have a very high refractive index ($n = 2.2$) and are formed by the evaporation of 100 –150 nm of TaO_5 onto glass substrates. Light is coupled into the waveguide using a grating coupler etched into the glass substrate. As the light travels through the waveguide the evanescent wave interacts with the bound labels and excites fluorescence. The light emitted by the fluorophores can be detected in one of two ways (Figure 6.19): light passing approximately perpendicularly through the glass substrate may be collected and measured ('volume detection'); emitted light which couples back into the waveguide may be detected using a second grating to couple it out as a collimated beam which is directed onto a photodetector ('grating detection'). The Stokes shift between excitation and emitted light results in a small angular separation between the two outcoupled beams. Quantitative comparison of volume and grating detection showed that the fluorescence signals and signal-to-noise ratios obtained in the two configurations were similar (Duveneck *et al.* 1996). The choice of which sensing configuration is most suitable for a given assay depends more on such factors as the necessity for multianalyte detection (see below, page 166) or on the bulk fluorescence of the sample. A significant difference between bulk detection and grating detection is the sensitivity of the two configurations to the bulk: the z dependency of the fluorescence signal in volume detection is given by $\exp(-z/\lambda)$ while for grating

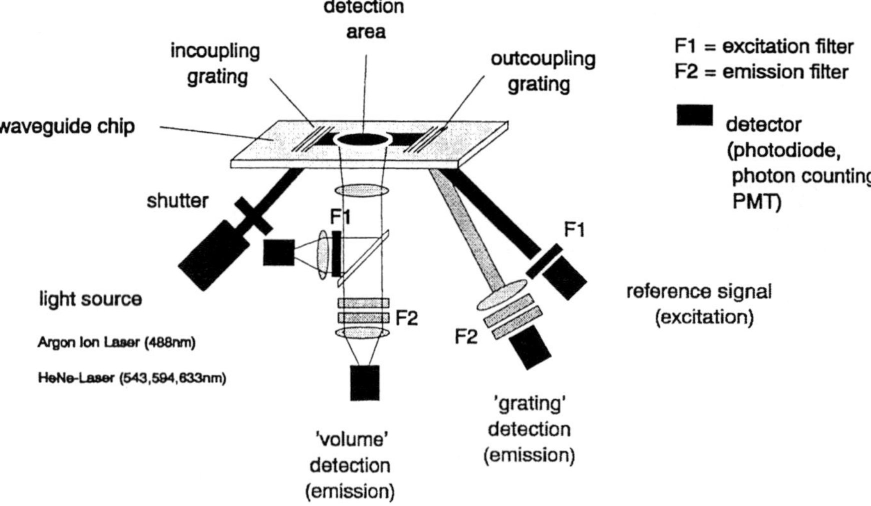

Figure 6.19 Fluorescence transduction based on single mode waveguides. Excitation of the fluorophores via the evanescent wave, with 'grating detection' and 'volume detection'

(From Duveneck *et al. Proc. SPIE*, 3858 (1999) 1. With permission)

detection the equivalent term is $\exp(-2z/\lambda)$. Thus for highly fluorescent samples, grating detection should have reduced background fluorescence from the bulk when compared to volume detection. The sensitivities reported for both of these configurations is on the order of 10^4–10^5 labels bound to a total area of $20\,\text{mm}^2$.

Other authors have reported similar configurations for fluorescence sensing, usually using gratings to couple into and out of the waveguides, but sometimes using prism couplers (Plowman *et al.* 1996). Perhaps the most interesting of the grating configurations are those which use only one grating both for coupling the excitation light into the waveguide and for coupling the emitted light out again. Early versions of this configuration placed the grating next to the waveguide area coated with capture molecules and exploit the fact that the fluorescently emitted light which couples into the waveguide propagates equally in all directions (Voirin and Kunz 1998) (see Figure 6.20). However, in more recent publications, the coupling grating itself was coated with antibodies (Kunz *et al.* 1999). Excitation of the fluorescent labels was performed using a TM guided mode while only the TE emission was collected at the photodetector. Since the coupling conditions of the two waveguide modes are different, their coupling angles also differ and this configuration produces a much larger angular separation of the excitation and emission beams. In addition to the increased angular separation, the change of polarization between excitation and collected light significantly reduces the background signal caused by diffuse reflection and scattering of the excitation beam at the grating.

Monomode high refractive index waveguides as used by Duveneck *et al.* have a significant advantage over multimode guides in that the fraction of the guided mode power in the evanescent wave is much higher. Thus, the same intensity of guided light will result in a higher total intensity of excitation light (and thus fluorescence) for a monomode guide than for a multimode guide, all other things being equal. The high refractive index of the waveguide also results in a small penetration depth, reducing background from unbound label or bulk fluorescence in the bulk sample.

A recent study with multimode waveguides has shown that improved field intensity distributions (high guided power in the evanescent wave and short penetration depth) can also be obtained in multimode waveguides if a thin high refractive index

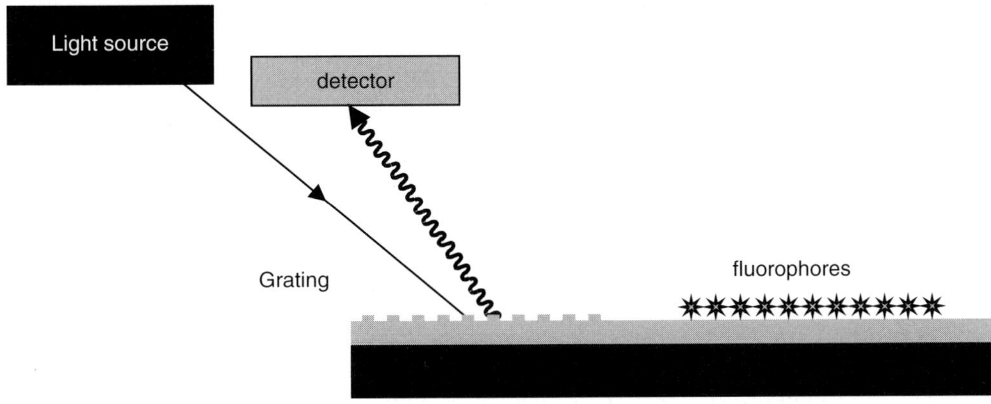

Figure 6.20 Fluorescence transduction based on single mode waveguides. Excitation and 'grating detection' of the emitted fluorescence using a single grating coupler

layer is deposited on the top of the waveguide (Klotz *et al.* 1999). In a comparative study of multimode waveguides with and without high refractive index overlayers, fluorescence intensity differences of a factor of 5–7 were obtained, together with signal-to-noise improvements of up to a factor of 10 on addition of the high refractive layer.

In contrast to most direct sensing techniques, fluorescence transducers use intensity measurements to quantify the amount of analyte bound to the sensing surface. Thus the efficiency of light coupling into and out of the waveguide must be highly reproducible from device to device. Failing this, some calibration mechanism must exist. Since the efficiency of coupling into thin film waveguides depends strongly on such parameters as incident light angle and grating profile, most sensor configurations use the intensity of guided excitation light to calibrate the fluorescent signal. For example, Duveneck *et al.* (1996) measured the guided excitation light either coupled out at the end of the waveguide and collected by optic fibres, or coupled out by a second grating. In contrast, Plowman *et al.* (1996) used volume detection and separately measured the excitation light scattered into the bulk together with the fluorescently emitted light. Alternatively, in array sensors (see below) reference channels, where the quantity of fluorescence label on the surface is known, may be used to calibrate the signals obtained (Robinson *et al.* 1993).

Recent developments in fluorescence sensing have concentrated on multianalyte detection, in which several different analytes can be simultaneously detected in one measurement. Planar waveguides lend themselves particularly well to this approach since arrays of different capture antibodies can be readily deposited on the planar waveguide surface. The different areas of the waveguide – and thus the response of the sensor to the different analytes – can be interrogated using the volume detection method and imaging optics. This approach has been demonstrated for multimode waveguides (Rowe *et al.* 1999) using a glass slide as a waveguide with end-coupling of light from a diode laser into the slide. Use of a gradient refractive index lens array to focus the light produced a compact and lightweight instrument (Wadkins *et al.* 1998), which has been shown to allow simultaneous detection of three very different antigens in whole blood (Rowe *et al.* 1999). Similar results were obtained with monomode waveguides (Budach *et al.* 1999) with grating couplers being use to couple excitation light into the guides. Quantitation of the results obtained from these multianalyte arrays remains difficult, however, since the intensity of the guided excitation light varies across the array. This is the case both perpendicular to the direction of travel of the mode – due to the Gaussian intensity distribution of the incoming light beam – and also parallel to the direction of travel, due to attenuation of the guided mode. As the attenuation of the mode may also depend on the presence of fluorescent labels on the waveguide surface (Duveneck *et al.* 1996), the fluorescent intensities obtained for the different analytes are not entirely independent, and cannot be analysed separately. An alternative approach – to use an array of small gratings, with a separate grating for each analyte – described by Kunz *et al.* (1999) may have advantages in this respect, since the excitation of the guided mode is independent for each analyte. However, the use of separate gratings for each analyte has not yet been demonstrated for a large number of analytes, nor has any attempt been made to use this configuration to quantitate the analyte bound to the surface.

Fluorescence transducers based on optical fibres

The alternative to the use of planar waveguides in fluorescence biosensors is the use of optical fibres – cylindrical waveguides. Optical fibres have certain advantages over planar waveguides, most notably for *in situ* applications where their small size and the fact that the sensing surface may be some distance from the optical instrumentation may be essential.

The most common and very successful approach to the use of optical fibres as transducers involves the removal (or 'stripping') of the cladding from a section of fibre, which then acts as the sensing surface (see Figure 6.21). Light at the excitation wavelength is focussed onto the end of the optical fibre and couples in. At the stripped section of the fibre, fluorescence labels bound to the core surface are excited and a fraction of the emitted fluorescence is coupled back into the fibre core. This emitted light travels along the fibre, both in the direction of propagation of the excitation light, and in the opposite direction. The fluorescence signal can be monitored either at the opposite end of the fibre from the light in-coupling, or at the same end. In both cases, filters must be used to separate the excitation wavelength from the emitted wavelength. However, when the same fibre end is used both for in and out coupling of the light, as shown in Figure 6.21, the problem of separation of the two wavelengths is much reduced, since only scattered and reflected excitation light is coupled out, as opposed to the directly transmitted beam at the other end of the fibre. Often the outcoupled excitation light is used as a reference to correct for any variation either in the output of the light source or in the efficiency of light coupling into the optical fibre.

As described in the previous section, fluorescence sensors use intensity measurements to quantify the binding of analyte to the sensing surface. This implies that the efficiency of coupling of excitation and emission light into and out of the fibre must be highly reproducible from fibre to fibre. Where this is not the case the signal response of each fibre must be independently quantified. Sources of variation in the signal response of optic fibre transducers include: small differences in launch angle of light into the fibre; defects in the fibre itself; surface defects of the unclad core; minor differences in sensing surface geometry; and variations in the coating of the

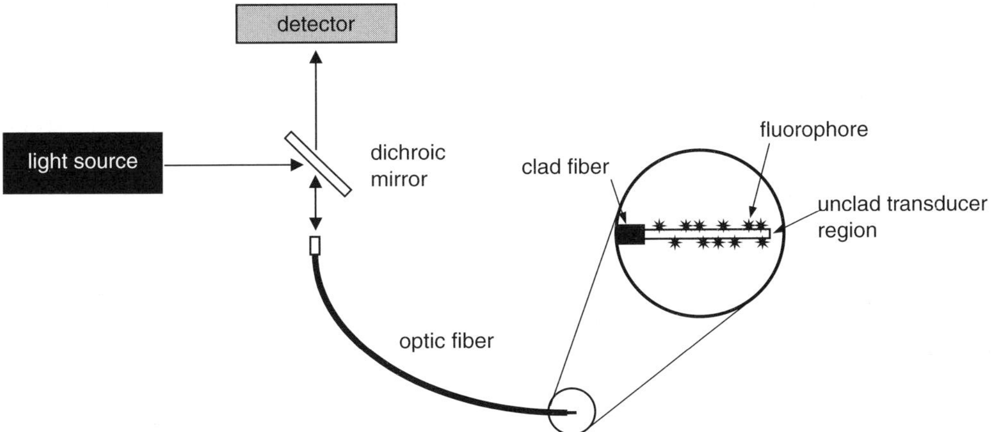

Figure 6.21 A fiber optic transducer

fibre with capture antibody. Since it is extremely difficult to eliminate these sources of variation, one promising approach, analogous to the use of reference channels in planar sensors, is the use of an internal reference fluorescence system for calibration of the sensor response. An elegant solution has been proposed by Ligler *et al.* (Wadkins *et al.* 1995), which uses two fluorescence labels that can be excited by the same wavelength. Capture antibodies immobilized on the sensing surface of the fibre are labelled with the cyanine dye Cy5.5, while the assay is based on competition between unlabelled antigen and antigen labelled with the dye Cy5. Both Cy5 and Cy5.5 can be excited by light at 635 nm, while the emission signals from the two dyes, which have maxima at 668 and 696 respectively, can be separately quantified using filters. The Cy5.5 intensity is used as a measure of the fibre's signal response: the ratio of the Cy5 signal to the Cy5.5 signal being used to quantify the analyte bound to the surface.

The use of two dyes that can be excited by the same wavelength is essential for the elimination of signal variation in this technique since this allows the use of one light source only. The use of two light sources would inevitably imply minor differences in the alignment of the incident light beams, which would result in significant differences in coupling into the optic fibre.

The optical fibres used in biosensing applications are generally multimode fibres, with the light guided simultaneously in a number of different modes. The number of modes available (the mode capacity) depends on the waveguide parameter or *V*-number:

$$V = \frac{2\pi r_{co}}{\lambda_0} \left(n_{co} - n_{cl} \right)^{1/2} \tag{6.45}$$

where r_{co} is the radius of the fibre core, n_{co} and n_{cl} the refractive indices of core and cladding, and λ the wavelength of the guided light. For each mode, the fraction of the electromagnetic field in the sample solution is different, which leads to different coupling efficiencies between guided light and the fluorescence labels. When the cladding of the fibre (n_{cl} typically 1.41) is removed and the fibre immersed in an aqueous solution (n_{sl} typically 1.33) the *V*-number increases and higher order modes are available for guiding. These higher order modes have an increased fraction of the propagated power in the evanescent wave that probes the sample. Since, however, the excitation light is propagated in the lower order modes, the intensity of excitation light in the evanescent field is less than optimal. In addition, fluorescence emission from molecules immobilized on the unclad fibre surface couples predominantly into the higher order modes that are not propagated in the clad fibre. These two effects together result in a significant loss of fluorescent intensity at the detector and thus of sensitivity in the transducer.

In order to avoid fluorescence loss by *V*-number mismatch, the core diameter, r_{co}, in the unclad fibre section can be reduced to give the same *V*-number as the clad fibre. Two methods have been investigated: the step-etch, where the core diameter changes abruptly at the end of the cladding; and the tapered probe, where the core diameter is reduced gradually (see Figure 6.22). Step-etched fibres are simple to produce and give a rapid transition to the optimal fibre diameter. However, this transition can be inefficient, with some of the excitation light being transmitted into the bulk solution at the step, resulting in loss of power in the evanescent field and

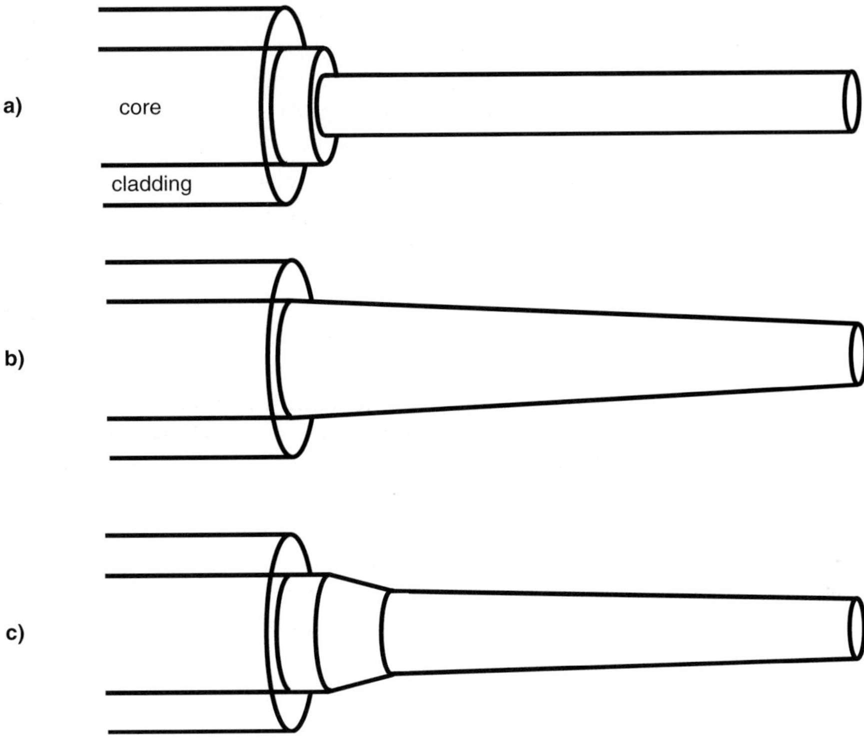

Figure 6.22 Etched optic fibres:
a   step etch
b   continuous taper
c   combination taper

(After a figure from Anderson *et al. Biosens. Bioelect.* 8 (1993) 249. With permission)

increased background fluorescence from the bulk solution. In contrast, in tapered probes, total internal reflection is conserved and the taper acts as a mode converter, converting lower order modes in the clad fibre into higher order modes in the unclad fibre. Tapered probes, therefore, have a higher proportion of the light in the evanescent field and consequently a better excitation efficiency for the immobilized labels on the sensing surface. Increases in sensitivity due to V-number matching have been found to be of the order of a factor 20 for step etching and 50 for simple tapered probes. Further increases in sensitivity have been obtained using combination tapered fibres, which have a steep taper to give a rapid transition to a suitable fibre diameter, followed by a shallow taper, which optimizes the fluorescence signal acquisition along the length of the unclad fibre.

Fibre optic transducers based on 'distal sensing' have also been developed by a number of authors (e.g. Hanbury *et al.* 1996; Hanbury *et al.* 1997). In these sensors, the biological layer is immobilized on the end of the fibre. Distal sensing has been less popular in recent years due to limited sensitivity caused by the small surface area available for sensing. The use of, for example, a gel to increase the number of antibody/antigen pairs interrogated by the transducer, can to some extent compensate for the small surface area (Hanbury *et al.* 1997). Recent developments have shown, however, that distal sensing has potential for the development of multi-analyte transducers based on optical fibres.

The use of stripped fibres for multichannel sensing has been reported (Golden *et al.* 1997), but it is clear that this approach is technically complicated: each sensing channel requires a separate fibre, which must be stripped and individually incorporated in a flow cell. In contrast, the combination of distal sensing with imaging optic fibres, which typically have 3000–100,000 individual pixels in a total diameter of 0.2 –1.0 mm, offers a route to simple multichannel transducers (Walt 1997). If the distal end of the imaging fibre is suitably etched it forms an array of wells, each of which is at the end of a separate fibre core. If the fibre is then exposed to a solution containing latex microspheres of a suitable size, the microspheres position themselves randomly in the individual wells (see Figure 6.23). Each microsphere can then be used as a separate sensing element. Different populations of microspheres are produced, each population having a unique sensing biochemistry (for example a monoclonal antibody) and also a unique fluorescent label. The different populations are then mixed and put on the end of an imaging fibre. Individual microspheres bind to different wells where each microsphere can be identified by its label. Subsequent reactions of the microspheres such as the binding of fluorescently labelled antigen or antibody binding can be individually monitored. What is attractive about this approach is the simplicity of the formation of the array and the fact that several of each kind of microsphere bind to the fibre, with the consequent redundancy providing increased confidence in the measurements. While much remains to be done in the development of multichannel sensors based on imaging fibre optics, particularly in the field of immunoassays, it is clear that they show great potential.

Conclusions and outlook

It is clear from the large number of new developments in optical transducers in the past few years that this continues to be a highly active research field. The last few years have produced some new and very original developments such as the use of porous silicon for reflectometric direct detection (Lin *et al.* 1997) or the use of etched imaging optic fibres for multianalyte fluorescence sensing (Walt 1997).

In direct detection, although numerous new transducer designs and configurations have been reported, little progress has been made in the improvement of transducer sensitivity. The superior sensitivity of waveguide transducers, when

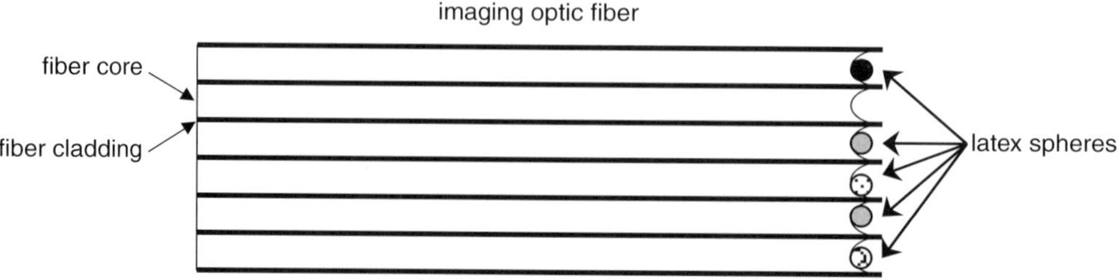

Figure 6.23 An imaging optic fiber transducer. Each latex particle can be identified by its fluorescent label and specifically binds one analyte

compared to surface plasmon resonance detection, has not yet been unambiguously demonstrated – although the recently reported integrated Mach-Zehnder interferometer (MZI) (Heideman and Lambeck 1999) seems likely to do this in the very near future. Indeed the sensitivity of an optical transducer is generally limited by instrumental factors such as mechanical, thermal and wavelength stability rather than by any theoretical limits. Optimization of these instrumental aspects is surely the reason for the exceptional sensitivity of the BIAcore range of sensors.

Refractometric sensors such as BIAcore or IAsys are now well established in the R&D laboratory where they are routinely used to investigate, for example, antibody/antigen interactions. However the step from the laboratory environment with its controlled and well-defined solutions to the analysis of 'real' samples such as blood, plasma or food extract is not an easy one. It remains to be seen how soon these sensors will gain acceptance as analytical methods for quality control and diagnostics.

In contrast to direct sensors, fluorescence sensors are not widely commercially available – neither for R&D applications, nor for diagnostic or quality control applications – although this will probably change in the very near future. In fluorescence detection, recent years have seen a strong move towards multianalyte sensors. This has almost certainly been an important factor in the move away from optic fibres and towards planar waveguide formats, which are much more suitable for the fabrication of sensing arrays. (The exception to the move towards planar formats is the use of imaging optic fibres for multianalyte sensing: an original approach that will certainly influence the field in the next few years.) Multichannel detection is important, not only for the simultaneous detection of several analytes – a necessary feature for certain medical diagnostics and food quality control applications – but also for the incorporation of features to improve the reliability of measurements: the use of more than one measurement channel for the same analyte ('redundancy'), the use of reference channels to identify spurious non-specific effects and the use of pattern recognition are all important approaches to increase the reliability of optical transducers.

New developments in materials science and instrumentation continue to strongly influence transducer development. This can clearly be seen in the development of transducers based on porous silicon (Lin *et al.* 1997): a material which was first reported (in a form suitable for optical transduction) in 1991. Similarly, the development of reflection-interference spectroscopy was dependent on the commercial availability of high quality, low-cost diode spectrometers. In the field of fluorescence detection, the development of new long-wavelength dyes has been an important factor in the improvement of assay sensitivity. Important progress continues to be made in the field of fluorescent labels, for example in the use of fluorescent nanoparticles or quantum dots as labels in biological assays. These particles, which have highly advantageous spectral properties and are resistant to bleaching, will certainly influence future developments in fluorescence sensors.

In summary, research in optical transducers remains exciting and innovative. New trends towards multichannel sensors and simpler readout systems can be observed. Recent developments in instrumentation and materials science contribute strongly to the continuing progress in this field.

Acknowledgements

I would like to thank the many colleagues with whom I have worked in the field of optical biosensors. Special thanks to Claus Duschl and Juergen Spinke, from whom I have learnt so much. Thanks also to Rino Kunz, Guy Voirin and Christoph Stamm for help with this chapter. Finally, thanks to AM for being so patient during the preparation of the manuscript.

References

Arwin, H. and I. Lundström, (1985) A reflectance method for quantification of immunological reactions on surfaces. *Analytical Biochemistry* 145: 106–112.

Badley, R.A., R.A.L. Drake, Shanks, I.A., A.M. Smith, and P.R Stephenson (1987) Optical biosensors for immunoassays: the fluorescence capillary fill device. *Phil. Trans. R. Soc. Lond. B* 316: 143–160.

Bain, C.D., E.B. Troughton, Y.T. Tao, J. Evall, R.G. Whitesides and R.G. Nuzzo (1989) Formation of monolayer films by the spontaneous assembly of organic thiols from solution onto gold. *J. Am. Chem. Soc.* 111: 321–335.

Bain, C.D. and G.M. Whitesides (1989) Formation of monolayers by the coadsorption of thiols on gold: variation in the headgroup, tailgroup and solvent. *J. Am. Chem. Soc.* 111: 7155–7164.

Bjorklund, R.B., S. Zangooie and H. Arwin (1997) Adsorption of surfactants in porous silicon films. *Langmuir* 13: 1440–1445.

Brandenburg, A., R. Polzius, F. Bier, U. Bilitewski and E. Wagner (1996) Direct observation of affinity reactions by reflected-mode operation of integrated optical grating coupler. *Sensors and Actuators B* 30: 55–59.

Brecht, A. and G. Gauglitz (1997) Recent developments in optical transducers for chemical or biochemical applications. *Sensors and Actuators B* 38–39: 1–7.

Brecht, A., J. Ingenhoff and G. Gauglitz (1992) Direct monitoring of antigen–antibody interactions by spectral interferometry. *Sensors and Actuators B* 6: 96–100.

Buckle, P.E., R.J. Davies, T. Kinning, D. Yeung, P.R. Edwards and D. Pollard-Knight (1993) The resonant mirror: a novel optical sensor for direct sensing of biomolecular interactions. Part II: Applications. *Biosensors and Bioelectronics* 8: 355–363.

Budach, W., A.P. Abel, A.E. Bruno and D. Neuschäfer (1999) Planar waveguides as high-performance sensing platforms for fluorescence-based multiplexed oligonucleotide hybridization assays. *Anal. Chem.* 71: 3347–3355.

Cantor, C.R. and P.R. Schimmel (1980) *Biophysical Chemistry Part 2*. San Francisco, CA, W.H. Freeman.

Cullen, D.C., R.G.W. Brown and C.R. Lowe (1987/88) Detection of immuno-complex formation via surface plasmon resonance on gold-coated diffraction gratings. *Biosensors* 3: 211–225.

Cullen, D.C. and C.R. Lowe (1990) A direct surface plasmon-polariton immunosensor: preliminary investigation of the non-specific adsorption of serum components to the sensor surface. *Sensors and Actuators* B1: 576–579.

Cush, R., J.M. Cronin, W.J. Stewart, C.H. Maule, J. Molloy and N.J. Goddard (1993) The resonant mirror: a novel optical biosensor for direct sensing of biomolecular interactions. Part I: Principle of operation and associated instrumentation. *Biosensors and Bioelectronics* 8: 347–353.

Danielzik, B., W. Ehrfeld, C. Fattinger, M. Heming, H. Löwe, A. Michel, F. Michel, N. Oranth and J. Spinke (1997) A universal transducer for optical interface analytics: transducer design and concepts for an economical mass production. In: *Thin Films on Glass*, H. Bach and D. Krause (eds). Berlin, Springer-Verlag.

De Feijter, J.A., J. Benjamins and F.A. Veer (1978) Ellipsometry to study the adsorption behavior of synthetic and biopolymers at the air-water interface. *Biopolymers* 17: 1759–1772.

Diamandis, E.P. (1988) Immunoassays with time-resolved fluorescence spectroscopy: principles and applications. *Clinical Biochemistry* 21: 139–150.

Drapp, B., J. Piehler, A. Brecht, G. Gauglitz, B.J. Luff, J.S. Wilkinson and J. Ingenhoff (1997) Integrated optical Mach-Zehnder interferometers as simazine immunoprobes. *Sensors and Actuators B* 38–39: 277–282.

Dübendorfer, J. (1997) *Replicated Integrated Optical Sensors.* University of Fribourg.

Dübendorfer, J. and R.E. Kunz (1997) Reference pads for miniature integrated optical sensors. *Sensors and Actuators B* 38–39: 116–121.

Dübendorfer, J. and R. Kunz (1998) Compact integrated optical immunosensor using replicated chirped grating coupler sensor chips. *Applied Optics*-OT 37: 18–90.

Dübendorfer, J., R.E. Kunz, E. Schuermann, G.L. Duveneck and M. Ehrat (1997) Sensing and reference pads for integrated optical immunosensors. *J. Biomed. Optics* 2: 391–400.

Duveneck, G.L. and A.P. Abel (1999) Review on fluorescence-based planar waveguide biosensors. *Proc. SPIE* 3858 Advanced materials and optical systems for chemical and biological detection: 1–13.

Duveneck, G.L., M. Pawlak, D. Neuschäfer, E. Bär, W. Budach, U. Pieles and M. Ehrat (1997) Novel bioaffinity sensors for trace analysis based on luminescence excitation by planar waveguides. *Sensors and Actuators B* 38–39: 88–95.

Duveneck, G.L., M. Pawlak, D. Neuschäfer, W. Budach and M. Ehrat (1996) A novel generation of luminescence-based biosensors: single-mode planar waveguide sensors. *Proc. SPIE* 2928 Biomedical systems and technologies: 98–109.

Golden, J.P., E.W. Saaski, L.C. Shriver-Lake, G.P. Anderson and F.S. Ligler (1997) Portable multichannel fiber optic biosensor for field detection. *Opt. Eng.* 36(4): 1008–1013.

Hanbury, C.M., W.G. Miller and R.B. Harris (1996) Antibody characteristics for a continuous response fiber optic immunosensor for theophylline. *Biosensors and Bioelectronics* 11: 1129–1138.

Hanbury, C.M., G.M. Miller and R.B. Harris (1997) Fiber-optic immunosensor for measurement of myoglobin. *Clinical Chemistry* 43: 2128–2136.

Harris, R.D., B.J. Luff, J.S. Wilkinson, J. Piehler, A. Brecht, G. Gauglitz and R.A. Abuknesha (1999) Integrated optical surface plasmon resonance immunoprobe for simazine detection. *Biosensors and Bioelectronics* 14: 377–386.

Hecht, E. (1998) *Optics,* Reading, MA, Addison Wesley Longman Inc.

Heideman, R.G. and P.V. Lambeck (1999) Remote opto-chemical sensing with extreme sensitivity: design, fabrication and performance of a pigtailed integrated optical phase-modulated Mach-Zehnder interferometer system. *Sensors and Actuators B* 61: 100–127.

Hemmilä, I.A. (1991) *Applications of Fluorescence in Immunoassays.* New York, NY, John Wiley and Sons.

Homola, J., J. Ctyroky, M. Skalky, J. Hradilovka and P. Kolarova (1997) A surface plasmon resonance based integrated optical sensor. *Sensors and Actuators B* 38–39: 286–290.

Homola, J., I. Koudela and S.S. Yee (1999) Surface plasmon resonance sensors based on diffraction gratings and prism couplers: sensitivity comparison. *Sensors and Actuators B* 54: 16–24.

Jönsson, U., M. Malmquist and I. Rönnberg (1985) Adsorption of IgG, protein A and fibronectin in the submonolayer region evaluated by a combined study of ellipsometry and radiotracer techniques. *Journal of Colloid and Interface Science* 103: 360–372.

Karlsson, R., A. Michaelsson and L. Mattsson (1991) Kinetic analysis of monoclonal antibody–antigen interactions with a new biosensor based analytical system. *J. Immunological Methods* 145: 229–240.

Klotz, A., C. Barzen, A. Brecht, R.D. Harris, G.R. Quigley, J.S. Wilkinson and G. Gauglitz

(1999) Sensitivity enhancement of transducers for total internal reflection fluorescence. *Proc. SPIE* 3620 Integrated Optics Devices III: 345–354.

Kooyman, R.P.H., H. Kolkman, J. Van Gent and J. Greve (1988) Surface plasmon resonance immunosensors: sensitivity considerations. *Analytica Chimica Acta* 213: 35–45.

Kretschmann, E. (1971) Die Bestimmung optischer Konstanten von Metallen durch Anregung von Oberflächen plasmonenschwingungen. *Zeitschrift Physik* 241: 313–324.

Kubitschko, S., J. Spinke, T. Brückner, S. Pohl and N. Oranth (1997) Sensitivity enhancement of optical immunosensors with nanoparticles. *Analytical Biochemistry* 253: 112–122.

Kunz, R.E., G. Duveneck and M. Ehrat (1994) Sensing pads for hybrid and monolithic integrated optical immunosensing. *Proc SPIE* 2331 Medical Sensors II and fiber optic sensors: 2–17.

Kunz, R.E., P.N. Zeller and M. Wiki (1999) Integrated optical sensor and method for integrated-optically sensing a substance. Europe, CSEM centre suisse d'electronique et de microtechnique, Rue Jaquet-Droz 1, CH 2007 Neuchatel.

Lenferink, A.T.M., E.F. Schipper and R.P.H. Kooyman (1997) Improved detection method for evanescent waveguide interferometric chemical sensing. *Rev. Sci. Instrum.* 68: 1583–1586.

Lin, V.S.-Y., K. Motesharei, K.-P.S. Dancil, M.J. Sailor and M.R. Ghadiri (1997) A porous silicon-based optical interferometric biosensor. *Science* 278: 840–843.

Lukosz, W. (1991) Principles and sensitivities of integrated optical and surface plasmon sensors for direct affinity sensing and immunosensing. *Biosensors and Bioelectronics* 6: 215–225.

Lukosz, K. and K. Tiefenthaler (1988) Sensitivity of integrated optical grating and prism couplers as (bio)chemical sensors. *Sensors and Actuators.* 15: 273–284.

Mar, M.N., B.D. Ratner and S.S. Yee (1999) An intrinsically protein-resistant surface plasmon resonance biosensor based upon an RF-plasma-deposited thin film. *Sensors and Actuators B* 54: 125–131.

O'Brien, M.J., S.R.J. Brueck, V.H. Perez-Luna, L.M. Tender and G.P. Lopez (1999) SPR biosensors: simultaneously removing thermal and bulk-composition effects. *Biosensors and Bioelectronics* 14: 145–154.

Patonay, G. and M.D. Antoine (1991) Near-infrared fluorogenic labels: new approach to an old problem. *Analytical Chemistry* 63: 321A–327A.

Piehler, J., A. Brecht and G. Gauglitz (1996) Affinity detection of low molecular weight analytes. *Anal. Chem.* 68: 139–143.

Plowman, T.E., W.M. Reichert, C.R. Peters, H.K. Wang, D.A. Christensen and J.N. Herron (1996) Femtomolar sensitivity using a channel-etched thin film waveguide fluoroimmunosensor. *Biosensors and Bioelectronics* 11: 149–160.

Reiter, R., H. Motschmann, H. Orendi, A. Nemetz and W. Knoll (1992) Ellipsometric microscopy: imaging monomolecular surfactant layers at the air–water interface. *Langmuir* 8: 1784–1788.

Robinson, G.A., J.W. Attridge, J.K. Deacon, A.M. Thomson, C.A. Love, S. Whiteley, M. Pugh and P.B. Daniels (1993) The calibration of an optical immunosensor. *Biosensors and Bioelectronics* 8: 371–376.

Robinson, G.A., J.W. Attridge, J.K. Deacon and S.C. Whiteley (1993) The fluorescent capillary fill device. *Sensors and Actuators B* 11: 235–238.

Rothenhäusler, B. and W. Knoll (1988) *Nature* 332: 615–617.

Rowe, C.A., L.M. Tender, M.J. Feldstein, J.P. Golden, S.B. Scruggs, B.D. MacCraith, J.J. Cras and F.S. Ligler (1999) Array biosensor for simultaneous identification of bacterial, viral and protein analytes. *Analytical Chemistry* 71: 3846–3852.

Schipper, E.F., R.P.H. Kooyman, R.G. Heideman and J. Greve (1995) Feasibility of optical waveguide immunosensors for pesticide detection: physical aspects. *Sensors and Actuators B* 24–25: 90–93.

Schlatter, D., R. Barner, C. Fattinger, W. Huber, J. Hübscher, J. Hurst, H. Koller, C.

Mangold and F. Müller (1993) The difference interferometer: application as a direct affinity sensor. *Biosensors and Bioelectronics* 8: 109–116.

Schuck, P. (1997) Use of surface plasmon resonance to probe the equilibrium and dynamic aspects of interactions between biological macromolecules. *Annual Review Biophy. Biomol Struct.* 26: 541–566.

Spaeth, K., A. Brecht and G. Gauglitz (1997) Studies on the Biotin-Avidin multilayer adsorption by spectroscopic ellipsometry. *Journal of Colloid and Interface Science* 196: 128–135.

Spinke, J., N. Oranth, C. Fattinger, H. Koller, C. Mangold and D. Voegelin (1997) The bidiffractive grating coupler: application to immunosensing. *Sensors and Actuators B* 38–39: 256–260.

Stamm, C. and W. Lukosz (1993) Integrated optical difference interferometer as refractometer and chemical sensor. *Sensors and Actuators B* 11: 177–181.

Stamm, C. and W. Lukosz (1996) Integrated optical difference interferometer as immunosensor. *Sensors and Actuators B* 31: 203–207.

Stemmler, I., A. Brecht and G. Gauglitz (1999) Compact surface plasmon resonance-transducers with spectral readout for biosensing applications. *Sensors and Actuators B* 54: 98–105.

Voirin, G. and R. Kunz (1998) *Capteur optique utilisant une réaction immunologique et un marqueur fluorescent.* France, CSEm centre suisse d'electronique et de microtechnique.

Wadkins, R.M., J.P. Golden and F.S. Ligler (1995) Calibration of biosensor response using simultaneous evanescent wave excitation of cyanine-labeled capture antibodies and antigens. *Analytical Biochemistry* 232: 73–78.

Wadkins, R.M., J.P. Golden, L.M. Pritsiolas and F.S. Ligler (1998) Detection of multiple toxic agents using a planar array immunosensor. *Biosensors and Bioelectronics* 13: 407–415.

Walt, D.R. (1997) Fiber optic imaging sensors. *Acc. Chem. Res.* 5: 267–278.

Weiss, S. (1999) Fluorescence spectroscopy of single molecules. *Science* 283: 1676–1683.

Weisser, M., B. Menges and S. Mittler-Neyer (1999a) Refractive index and thickness determination of monolayers by multi-mode waveguide coupled surface plasmons. *Sensors and Actuators B* 56: 189–197.

Weisser, M., G. Tovar, S. Mittler-Neyer, W. Knoll, F. Brosinger, H. Freimuth, M. Lacher and W. Ehrfeld (1999b) Specific bio-recognition reactions observed with an integrated Mach-Zehnder interferometer. *Biosensor and Bioelectronics* 14: 405–411.

Welin, S., H. Elwing, H. Arwin and I. Lundström (1984) Reflectometry in kinetic studies of immunological and enzymatic reactions on solid surfaces. *Analytica Chimica Acta* 163: 263–267.

Wood, P. and G. Barnard (1997) Fluoroimmunoassay. In: *Principles and Practice of Immunoassay*, C.P. Prize and D.J. Newman (eds). London, Macmillan Reference.

7   Acoustic transducers

Electra Gizeli

Introduction

Acoustic waves have been extensively investigated and have had a practical significance in many different contexts. Seismology was one of the first areas where they were utilized and the substantial seismological literature, extending back to the nineteenth century, established many of the fundamental properties of acoustics. A second, but equally important application was the introduction of acoustic waves into the electronics field in the 1920s and the subsequent fabrication of components such as resonators, filters and delay lines. Today, the utilization of acoustic waves in signal processing has become commonplace and recent advances in transduction technology enable the manufacture of devices covering the frequency range from about 10 KHz to a few GHz. Following developments in the electronics field, the first attempt to utilize acoustic devices in analytical chemistry was made in the mid-1960s. This work involved the utilization of a bulk acoustic wave device as a sensitive mass-sensor for gas chromatography. Since then, acoustic sensors have been developed to detect not only gas molecules in the ppm and ppb range, but also sub-monolayers of proteins deposited on the device surface in presence of liquid. Today, despite considerable advances in the development of new acoustic sensors, their application in biological analysis is still limited compared to optical and electro-chemical devices. However, current research in the area strongly suggests that acoustic wave devices can become competitive sensors and many indications exist that their use will expand significantly in the future.

Elastic waves in solids

An acoustic wave propagating in a solid is defined as a form of disturbance involving deformation of the material (Morgan 1985). Deformation occurs when the motions of individual atoms are such that the distances between them change; this is accompanied by internal restoring forces which tend to return the material to its equilibrium position. If the deformation is time-variant, the motion of each atom is determined by these restoring forces and by inertial effects giving rise to propagating wave motion with each atom oscillating about its equilibrium position. In most materials the restoring forces are proportional to the amount of deformation, provided the latter is small, which can be assumed for most practical purposes. The material is then described as *elastic* and the waves are called *elastic* or *acoustic waves*. In an "ideal" elastic material, acoustic waves can propagate with no attenuation.

The simplest acoustic waves are the *plane waves*, which can propagate in an infinite homogeneous medium. There are two types of plane waves according to the polarization of the wave: the *longitudinal waves*, in which the atoms vibrate in the propagation direction, and the *shear waves*, in which the atoms vibrate in the plane normal to the propagation direction. Waves are also divided into three main categories, according to the characteristics of the propagation medium. If the propagation medium is bounded, the boundary conditions can substantially alter the character of the wave, and the propagating mode is normally described as a *surface acoustic wave*. On the other hand, a medium with dimensions much larger than the wavelength can support waves with characteristics similar to those of waves in an infinite medium, and these waves are described as *bulk waves*. Finally, the *surface generated bulk waves* are generated on the surface of a boundary medium but propagate and behave with characteristics similar to those of bulk waves. The above three types of waves will be described in more details in the following sections.

Bulk acoustic waves

The term "bulk wave" is generally used to describe waves which are not bound to a surface. Bulk elastic waves can propagate in any direction in any solid, whether elastically isotropic or anisotropic. In an infinite medium and for a given direction of propagation (defined by the wave vector k), three independent bulk waves can propagate: a pure longitudinal bulk wave and two degenerate pure shear bulk waves of arbitrary polarization in the plane perpendicular to k. Given appropriate crystal symmetry and orientation, either a shear or a longitudinal wave will be produced. Bulk waves propagating in a bounded medium must satisfy some boundary conditions; standing shear or longitudinal bulk waves will be generated in a bounded solid if all stresses normal to the surface are zero and the thickness of the solid is an integral number of half wavelengths (Bastiaans 1988).

Surface acoustic waves

The phenomenon of surface acoustic waves (SAW) first became of interest in the field of geology where acoustic energy released by earthquakes moves on the earth's crust. It is not surprising, therefore, that complex waves such as surface waves were discovered by physicists who had studied seismology: Rayleigh, Love and Stoneley. Generally, the term "surface acoustic wave" is used to describe two-dimensional waves propagating only on a surface or an interface separating two media. The amplitude of these waves decays exponentially with increasing distance from the surface or interface (Figure 7.1). The phase velocity of a SAW has the same general form as that for a bulk wave the only difference being that the surface–wave velocity is lower than the bulk velocities in the same medium. Detailed reviews on the different types of SAWs and their propagating characteristics can be found in a number of reviews and articles published in the literature (White 1970; Uberall 1972; Dieulesaint and Royer 1980; Gulyaev 1998).

The existence of a surface wave on the free plane surface of an elastic half-space was first described by Lord Rayleigh (Rayleigh 1885) and, hence, this wave has become known as the *Rayleigh wave*. The Rayleigh wave can exist on the surface of every solid. In the simple case of an isotropic, semi-infinite elastic solid bounded by

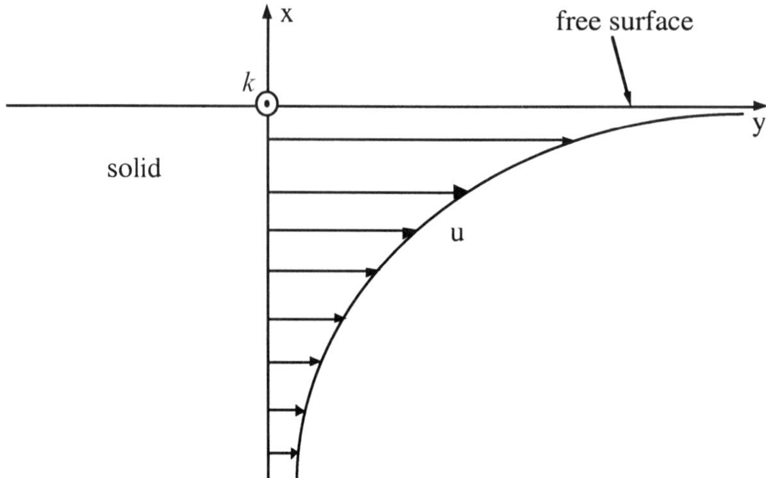

Figure 7.1 Exponential decay of the amplitude of a surface wave as the distance from the surface increases. In this example a shear-horizontal wave is illustrated where the wave polarization *u* is parallel to the surface and perpendicular to the wave vector *k*

a vacuum, it consists of a longitudinal component and a shear $\pi/2$-phase shifted component. At a distance of more than few wavelengths below the surface, the particle displacement becomes negligible.

The Rayleigh wave described in the preceding section is modified if the vacuum bounding the plane elastic half-space is replaced by a liquid or by another solid. In the case of a liquid layer, the wave corresponding to the Rayleigh wave on a free surface becomes exponentially damped in its direction of propagation as a result of leakage of energy into the other medium and other acoustic modes (known as *leaky surface waves*) are generated. In the case where a semi-infinite elastic solid is overlaid by an elastically different layer of finite thickness, *b*, two independent types of waves can exist, often referred to as *generalized Rayleigh-type waves* (Dieulesaint and Royer 1980):

1   A horizontally-polarized shear wave guided in the top elastic layer (Figure 7.2) known as the *Love wave* (Love 1911). In order for the wave to exist, the shear velocity in the layer must be less than the shear velocity in the elastic solid. Since the Love wave is a surface wave, the propagating energy is located in the layer and in that part of the substrate that is close to the interface.

2   A wave polarized in the sagittal plane, similar to the Rayleigh wave. This wave was analyzed by Stoneley (Stoneley 1924) in an attempt to specify more accurately how the energy of an earthquake is released and transmitted and is known as the *Stoneley wave*. Stoneley waves are non-radiating waves, since they remain confined to the interface and travel at a speed less than that of shear waves in either unbounded media. The main condition for the existence of Stoneley waves is that the velocity of shear waves for the two media is nearly equal.

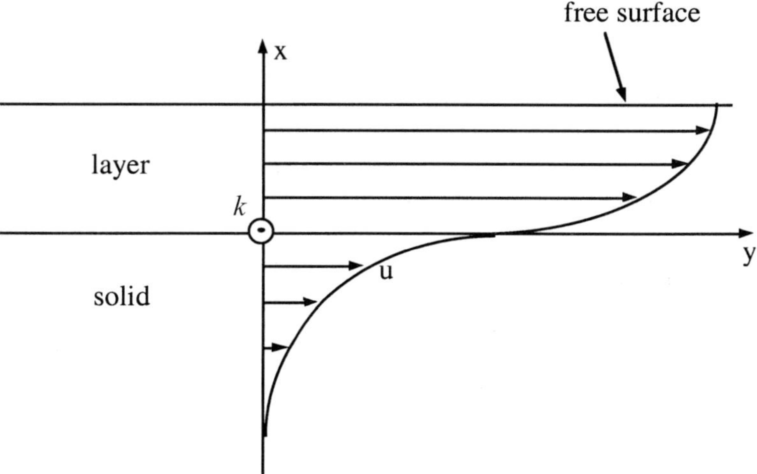

Figure 7.2 Love wave particle displacement for the first operating mode. The Love wave is guided predominantly in a layer of a dielectric material of thickness b deposited on top of an anisotropic elastic solid. The wave polarization u is parallel to the surface and perpendicular to the wave vector k

In practice, surface waves are generated and detected electrically on the surface of a piezoelectric medium. Individual modes of SAWs are excited by controlling the physical properties of the piezoelectric substrate involved, the dimensions of the acoustic media and the method of wave excitation. A detailed description of the SAW device can be found on page 182.

Surface generated bulk acoustic waves

These waves have been recently identified as a result of the development of SAW devices. In general, acoustic waves generated on a solid surface are accompanied by bulk waves which often degrade or interfere with the surface mode. However, within the last few years, there has been an increased interest in utilizing these otherwise spurious waves for practical applications. SAW devices were intentionally developed where bulk waves are used as the primary acoustic mode and surface waves are minimized. These newly developed waves include the *surface skimming bulk wave* (SSBW) and the *acoustic plate mode* (APM), also known as the *reflected bulk wave* (RBW). The SSBW, identified in 1977 (Browning and Lewis 1977), is a shear-horizontal wave where particle displacement is predominantly parallel to the device surface and normal to the direction of propagation. The wave propagates very close to, but underneath, the surface and penetrates slowly into the bulk of the device.

Acoustic plate modes, identified more recently by Lewis (1982), are shear-horizontal waves which propagate in the bulk of a piezoelectric medium. Like SSBWs, APMs can exist in piezoelectric substrates, where no surface wave excitation occurs. Thin piezoelectric plates, which act as acoustic waveguides, confine acoustic energy in the bulk of the plate as the wave propagates through multiple reflections between the top and bottom surface and generates a particle displacement at both surfaces. For certain substrate thicknesses the upper and lower faces impose a transverse resonance condition, which results in each APM having maximum displacement at the top and bottom surfaces, with sinusoidal variation between them. The number of distinct SH plate modes which can propagate inside the guide depends on the geometry, i.e. thickness, of the device (Martin *et al.* 1989). A schematic representation of the SSBW and APM devices is shown in Figure 7.3.

Acoustic wave devices

Acoustic wave devices utilize piezoelectric materials for the excitation and detection of the wave. These materials exhibit an electrical potential between deformed surfaces when pressure is exerted on them and conversely the application of a voltage induces physical distortions (Cady 1964). In other words, piezoelectric materials can be used to convert electrical signals to elastic (acoustic) waves and vice versa. Piezo-

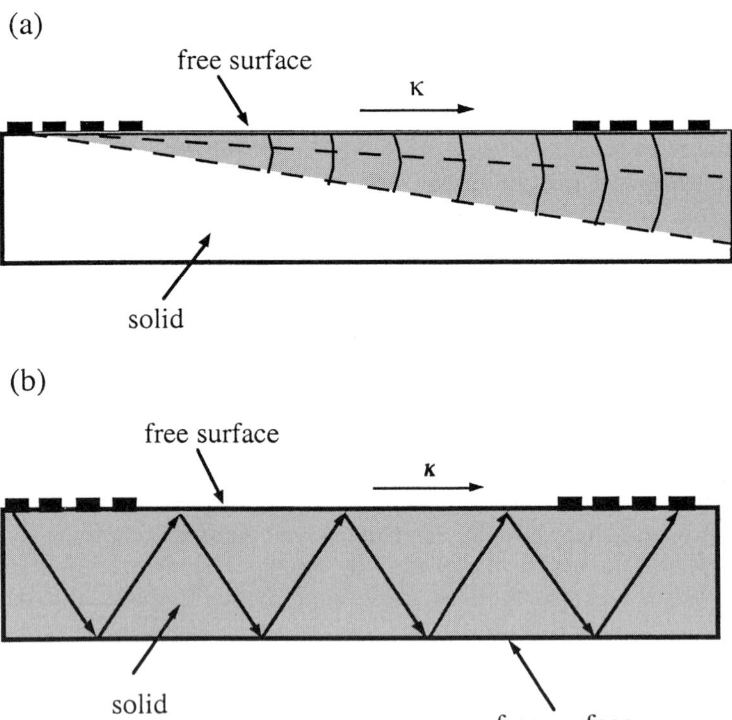

Figure 7.3 (a) Surface generated, shear-horizontal surface skimming bulk wave (SSBW) propagating in a SAW device. The wave is localised underneath the surface (grey area) and penetrates slowly into the bulk of the crystal; the k-vector is parallel to the surface. (b) Surface generated acoustic plate mode (APM) wave travelling in a SAW device through multiple reflections between the top and bottom surface of the crystal with the k-vector being parallel to the surface

electric materials are crystalline anisotropic solids and have no center of symmetry in their crystal structure. Twenty of the overall thirty-two crystal classes exhibit piezoelectric effects. Piezoelectric materials include cadmium sulphide (CdS), zinc oxide (ZnO), gallium arsenide (GaAs), lithium tantalate (LiTaO$_3$), bismuth and germanium oxide (Bi$_{13}$GeO$_{20}$), lithium niobate (LiNbO$_3$) and quartz (SiO$_2$) (Dieulesaint and Royer 1980). Though weakly piezoelectric, quartz is one of the most common choices for piezoelectric transducers, mainly due to its chemical stability in aqueous solutions and resistance to high temperatures with no loss of piezoelectric properties. Furthermore, different cuts of quartz, defined by the cut angle of the crystal with respect to its optical axis, give rise to acoustic devices with different properties, making each device suitable for a different application. Lithium niobate is also extensively employed, because of its good piezoelectric coupling, for several pure mode directions.

Acoustic wave devices can be divided in two categories based on the general wave classification; the *bulk acoustic wave* (BAW) device, which is used to generate bulk waves and the planar or *surface acoustic wave* (SAW) device, which is used to generate surface waves and surface shear-horizontal bulk waves.

Bulk acoustic wave device

The most common bulk acoustic wave device is the thickness-shear mode (TSM) resonator, illustrated diagrammatically in Figure 7.4, which has been used in electronics since 1920 (Pierce 1923; Cady 1924). This device consists of a parallel-sided plate of crystalline quartz with electrodes on both sides. If the thickness of the plate is small with respect to the major dimensions then application of a time-varying rf potential causes the plate to resonate at a frequency such that the wavelength is equal to twice the thickness of the plate. The natural resonance frequency (f_o) of the TSM resonator is given by:

$$f_0 = \frac{V_B}{2d} \tag{7.1}$$

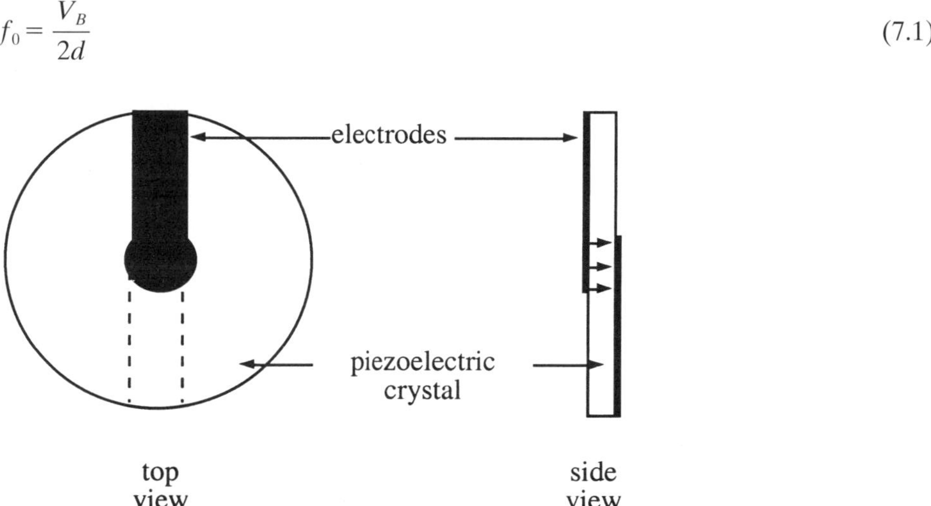

Figure 7.4 The thickness shear mode resonator from the top and side view (dimensions are not in scale). The diameter of a BAW device can vary between 9–15 mm and the thickness between 0.25–1 mm

where d is the plate thickness and V_B is the bulk wave-propagation velocity. The plate will also resonate at harmonics of this frequency

Since quartz is piezoelectric, the acoustic resonances can be excited electrically and an electrical oscillator circuit can be used to control its frequency. Given appropriate crystal symmetry and orientation, either a shear or a longitudinal wave, propagating along the thickness dimension will be produced. Crystal resonators are usually designed for shear wave operation and the crystal cut is either AT or BT quartz. Frequencies up to about 50 MHz are obtainable, the limitation being that higher frequencies require thinner, more fragile crystals. Because acoustic waves at specific frequencies can only efficiently propagate between the faces of bulk wave crystals, the crystal resonator is used in bandpass filters, in order to pass signals in some specific band of frequencies and reject signals at other frequencies.

Surface acoustic wave devices

Surface acoustic waves became of significant interest to the field of electrical engineering in the 1960s, when a simple and efficient method of exciting the waves on piezoelectric materials was developed (White and Voltmer 1965; White 1970). In this now standard method, metal electrodes are deposited on the surface of a piezoelectric solid in the form of thin interdigitated fingers. Alternating fingers are attached to two different connection bars. When an alternating voltage is applied across the two connectors, oscillating electrical fields are induced between each finger pair. The electrical fields created by this interdigital transducer (IDT) induce particle displacement within the solid in directions determined by the piezoelectric coupling coefficients. Given the correct choice of piezoelectric material and orientation and that the input signal is confined to a frequency band in which the transducers are effective, it is possible to excite a variety of non-dispersive elastic waves with the particle displacements corresponding to that of the allowed surface wave mode. This wave can later be detected on the surface with a second IDT through the reverse mechanism. Devices that consist of one or more IDTs on the surface of a crystal are known as planar or surface acoustic wave devices and have been extensively developed during the last few decades. The SAW device is shown schematically in Figure 7.5.

The spacing between the fingers of the IDT determines the operating frequency of a SAW device, since efficient transduction requires that adjacent fingers, connected to the same bar fingers should be one acoustic wavelength (λ) apart. Frequency f_o is related to finger spacing, which is equal to λ, by:

$$f_0 = \frac{V_S}{\lambda} \tag{7.2}$$

where V_S is the velocity of the surface wave. According to Equation 7.2, for a specific V_s only signals at a certain frequency will be efficiently transmitted.

Interdigital transducers can be also used to generate bulk acoustic waves such as SSBW and APMs in a SAW device. A bulk wave travelling at an angle θ to the surface should satisfy the following condition (Morgan 1985):

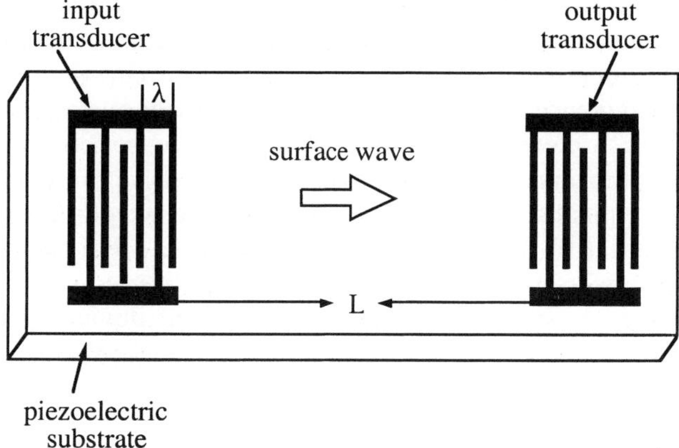

Figure 7.5 Schematic representation of a surface acoustic wave (SAW) device (dimensions are not in scale). The surface wave is generated by the output and detected by the input interdigital transducer (IDT). Typical SAW devices have a propagation length L of ~10 mm and a substrate thickness of ~1 mm. Interdigitated fingers width can vary between 25–50 μm

$$f_0 = \frac{V_B(\theta)}{\lambda \cos \theta} \tag{7.3}$$

where $V_B(\theta)$ is the bulk wave velocity. Equation 7.3 refers to the fundamental response; there will, however, also be harmonic responses satisfying the relation $M\lambda_b = L\cos\theta$, where M is an odd integer, λ_b is the bulk wave wavelength and L is the propagation length. Equation 7.3 shows that coherent bulk wave generation can only occur above a cut-off frequency,

$$f_o = \left(\frac{V_B(0)}{L} \right) \tag{7.4}$$

where $\theta = 0$ and the bulk wave travels parallel to the surface. At lower frequencies the equation has no real solution, but coherent bulk waves can be generated at all frequencies above f_c, with the angle θ increasing with frequency and approaching a limit of $\pi/2$ at infinite frequency. Generally, there will be several bulk waves involved and for each of these the velocity $V_B(0)$ parallel to the surface will be greater than the surface wave velocity V_S. The cut-off frequencies are therefore all greater than the center frequency of the surface wave response, which is equal to V_S/L.

Acoustic wave sensors for studying biomolecular interactions

Acoustic sensors first appeared in the field of analytical sensing in 1964 (King 1964); since then, their high mass sensitivity and large dynamic range has led to their use in a wide range of sensing applications. The simplest application of acoustic wave sensors is based on the following operating principle: mass deposited on the surface of a device in the presence of air will oscillate synchronously with the device surface under the influence of the propagating wave and will result in an acoustic signal change. The exact mathematical relationship between the acoustic change and the deposited mass depends on the type of wave and device used and will be examined separately for each system in the following sections. A common feature of all acoustic sensors is that the higher the operating frequency, the larger the acoustic response.

Biomolecular interactions typically occur in liquid so, for the development of acoustic biosensors, it is essential to use devices that can operate efficiently in a liquid environment. In the case where the predominant mode of the acoustic wave is a shear wave, it is possible to apply a liquid sample on the device surface without damping the wave since SH waves can propagate in low-viscosity liquids with insignificant attenuation. Instead, the liquid in contact with the device will oscillate with the device surface, giving rise to an acoustic evanescent field at the solid/liquid interface. The effective liquid thickness δ coupled to the acoustic wave is given by:

$$\delta = \sqrt{\frac{2\eta_l}{\rho_l \omega}} \tag{7.4}$$

where η_l and ρ_l are the liquid's shear viscosity and density, respectively, and ω the angular frequency (Figure 7.6). Changes in the properties of the liquid entrained within δ will result in a number of wave/matter interactions and, subsequently, affect the propagation characteristics of the wave. These interactions will result from: (1) mass deposited on the device surface; (2) changes in the viscosity of the entrained liquid; and (3) changes in the electric properties, i.e. permittivity and conductivity, of the interface. The latter is related to the piezoelectric nature of the oscillating substrate which generates an acoustoelectric field at the device surface. When the device is in contact with a solution an evanescent electric field will be also generated and will extend into the adjacent liquid coupling to ions and dipoles in solution.

The above interactions relate the acoustic signal to changes in the bulk properties of the solution within δ. However, because the wave propagates at the solid/liquid interface, it is also affected by the roughness and wettability of the solid surface and the presence of any interfacial layers adjacent to the solid which may have different viscoelastic and electric properties than the rest of the liquid within δ. The above interfacial properties are very difficult to control experimentally and can lead occasionally to inconsistent results. In particular, rough hydrophobic surfaces may affect the continuity of the acoustic field from the solid surface to the liquid and lead to slip-effects. The application of a reference device is normally a very effective way to minimize interferences and simplify the acoustic wave measurements.

The most important devices for biosensing applications will be examined in the following sections. More information on acoustic wave devices and their applications can be found in a number of books and review articles (Bastiaans 1988; Ballantine *et al.* 1997; Cavic *et al.* 1999; Janshoff *et al.* 2000).

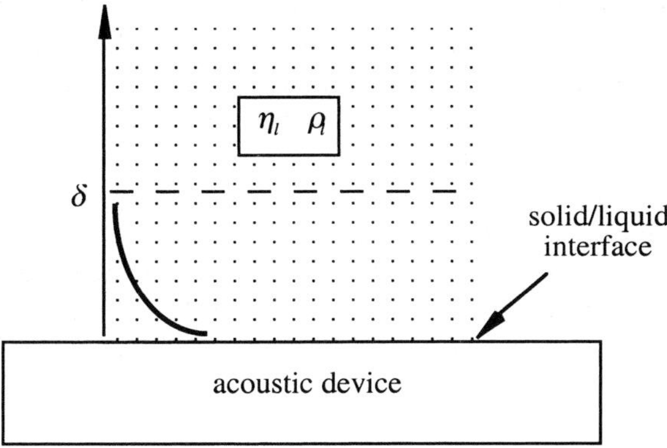

Figure 7.6 Schematic representation of the solid/liquid interface. The effective thickness of the liquid δ coupled to the acoustic wave and the acoustic evanescent field in liquid are also shown

Thickness-shear mode resonator

The TSM resonator, often referred to as the Quartz-Crystal Microbalance (QCM) is the most popular acoustic sensor used for biomolecular studies mainly due to its commercial availability and simple electronics required for data acquisition.

Theory

As explained on page 181, the frequency of operation of the TSM resonator depends on the crystal thickness of the device. Deposition of foreign material on the quartz crystal surface will result in an increase of the plate thickness, thus causing a change in the resonant frequency. A mathematical relationship between the mass of material placed upon a bulk wave resonator and frequency shift was first derived by Sauerbrey (1959):

$$\Delta f = -2f_0^2 \frac{\Delta m}{A\sqrt{\mu_q \rho_q}} \tag{7.5}$$

where Δf is the change in frequency due to mass deposition, f_o is the operating frequency of the uncoated device, Δm is the deposited mass, A is the coated area, ρ_q is the density of the substrate and μ_q the shear modulus of quartz. The above equation applies only to AT-cut crystals vibrating in the thickness shear mode, and assumes that the resulting frequency change is caused only by a change of the thickness of the crystal and that the crystal operates in a vacuum. Equation 7.5 provides a basic mathematical expression for the system but ignores the viscoelastic properties of the deposited material and the difference between its density and that of the crystal. According to Equation 7.5 the sensitivity of the TSM is proportional to the square of the operating frequency of the device, implying that for a constant crystal area (A) sensitivity increases with an increase in the frequency of oscillation. TSM sensors have been shown to exhibit a radial sensitivity distribution across the electrode surface with maximum sensitivity occurring at the center of the electrode (Carey and Kowalski 1986).

When the device is immersed in a liquid sample, energy is lost through viscoelastic coupling. The most general mathematical treatment of the surface loading effect on the bulk shear wave resonator was presented by Kanazawa and Gordon (1985). In this work, resonance is treated as arising from the matching of the shear waves in the quartz and in the overlayer to specific boundary conditions, assuming that the deposited material has both elastic and viscoelastic properties. This approach results in a quite complex mathematical model, but simplified relationships can be derived for the extremes of purely elastic and purely viscous behavior. In the case where a viscous medium is applied to the surface of a TSM resonator, the acoustic wave extends into the viscous layer and is accompanied by some damping of the wave. For a pure viscous medium the frequency shift Δf is given by:

$$\Delta f = -f_0^{3/2} \sqrt{\frac{\rho_l \eta_l}{\pi \rho_q \mu_q}} \tag{7.6}$$

where η_l and ρ_l are the viscosity and density of the viscous layer, respectively, and ρ_q and μ_q the density and shear modulus of the quartz crystal, respectively. Physically, the above equation predicts that only a thin layer of the viscous liquid will undergo displacement at the surface of the resonator and that device response will be a function of both the viscosity and mass of this layer.

The above analysis considers the effect of the change of the interfacial properties on the mechanical resonance of the TSM resonator. Because the device uses a piezoelectric substrate it is possible to relate changes in the resonance frequency of the device to the electrical characteristics of the resonator. This can be done by using an equivalent-circuit model to describe the electrical behavior of the device, i.e. the electrical admittance (the reciprocal of impedance) which is defined as the ratio of current flow to applied voltage. This model can be used to derive information about the energy stored and the power dissipated in both the TSM-resonator and the perturbing media. The equivalent circuit typically used to describe the unperturbed TSM resonator is the Butterworth-Van Dyke (BVD) circuit shown in Figure 7.7, in which a few lumped elements are used to simulate the electrical characteristics of the TSM-resonator over a range of frequencies around resonance. A "static" capacitance C_o arises between the electrodes located on opposite sides of the insulating quartz. In addition, because the quartz is piezoelectric, electro-mechanical coupling gives rise to a "motional" contribution (L_q, C_q, R_q) in parallel with the static capacitance. In the case where the TSM-resonator is loaded by a thin mass layer and a semi-infinite Newtonian liquid, it has been shown that the BVD equivalent circuit can be used to describe the loaded TSM-resonator and to relate the equivalent circuit elements to physical properties of the device and contacting media (Martin *et al.* 1991; Bandey *et al.* 1999). It is interesting to note that for small mass and liquid perturbations, the frequency behavior described by the equivalent circuit model is in good agreement with equations 7.5 and 7.6. A detailed description of the equivalent circuit model analysis can be found in the following book and review articles: Ballantine *et al.* 1997; Cavic *et al.* 1999; Janshoff *et al.* 2000.

In practice, when the Sauerbrey equation is applied the device is connected to an oscillator circuit and is continuously oscillating at a frequency controlled by the crystal itself. Changes in the series resonant frequency of the TSM-resonator (Δf) are measured with a standard frequency counter. However, a number of problems are associated with this method. Series resonant frequency is affected by both changes in mass and liquid properties at the interface, and with one single measurement it is not possible to differentiate between the two. Additionally, the resonant frequency depends on the components of the oscillator circuit and the oscillation does not function in certain cases such as heavy mass loading and highly viscous damping (Yang and Thompson 1993). When the equivalent circuit model analysis is applied, the device is connected externally to a network analyzer which provides an alternating voltage over a range of frequencies and at the same time measures the frequency response and impedance of the device/circuit. This passive method of measurement can characterize the device more completely by determining the values of the magnitude and phase of the impedance of the TSM-resonator. The above values can be further used to calculate the mass, density and shear modulus of the sensing overlayer (Yang and Thompson 1993; Yang *et al.* 1993). An inexpensive way to replace the Network Analyzer but still provide frequency and dissipation information is based on the use of the open-circuit-decay method (Rodahl and Kasemo 1996). This method and their applications will be discussed extensively in Part IV, Chapter 12.

Applications

The TSM resonator has been used extensively for studies of biological interactions such as the hybridization of DNA and RNA molecules, formation of biotin–avidin complexes, adsorption of biomolecules and cells to a solid surface, binding of ligands to phospholipid layers and antibody/antigen binding.

The application of the TSM resonator as a DNA sensor involves, initially, the immobilization of an oligonucleotide molecule on the gold electrode. Two approaches have been mainly used: the first involves the self-assembly of the thiolated single or double stranded DNA on the gold surface and the second the binding of the biotinylated DNA to the avidin or neutravidin-modified surface. The single stranded DNA-modified surface has been further applied to the study of the binding

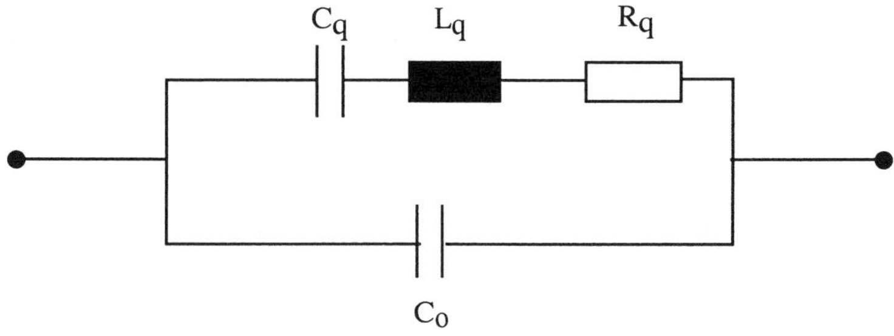

Figure 7.7 Butterworth-Van Dyke (BVD) equivalent circuit model. The lumped-elements used to describe the near-resonant electrical characteristics of the TSM-resonator include the static capacitance C_o, the inductance L_q, the resistance R_q and the capacitance C_q

of the complementary nucleotides during hybridization (Okahata *et al.* 1992; Wang *et al.* 1997), the *in vitro* selection of DNA from a library (Furuzawa *et al.* 1998; 1999) and the monitoring of DNA polymerase reaction (Niikura *et al.* 1998). The double stranded DNA-modified surface has been used to monitor the binding of sequence-specific peptides and ligands such as doxorubicin (Okahata 1998a; 1998b; Niikura 1999). Regardless of the method used for the immobilization of DNA, a common problem appears to be the poor coverage of the surface, which is in the range of 10 to 30 percent. In all the above works the device was connected to an oscillator circuit and the Sauerbrey expression was applied. However, the frequency decrease following DNA binding was a few times larger than theoretically predicted indicating that DNA is not rigidly bound to the surface and cannot be treated as an elastic mass layer according to the Sauerbrey equation. More insight on the mechanism of interfacial DNA binding was obtained when the device was connected to a Network Analyzer and the equivalent circuit model was used. It was shown that the acoustic response was related mainly to the surface hydrophilicity and interfacial viscoelastic properties while there was little or no correlation with mass-effects (Su and Thompson 1995; Su *et al.* 1996; Furtado *et al.* 1999). The above results suggest that DNA layers may exhibit some unusual time-dependent viscoelastic properties during their hybridization on a gold surface.

The variation of the acoustic response during probing of biological interactions was also shown in studies involving biotin–avidin layers. In one of these works, an impedance analyzer was used to monitor the change in the resonance frequency of the device caused by protein multilayers built on the device surface and extending over several hundred nanometers into the solution (Ricket *et al.* 1997). The protein multilayer structure consisted of layers of biotinylated bovine serum albumin coupled together through layers of streptavidin. Results showed that there was no change in the device sensitivity with an increase in the number of protein layers. The authors suggested that the above response indicated rigid coupling of the hydrated protein layers to the device surface and small contribution due to interfacial visco-elastic effects. The larger frequency response observed in relation to what was expected from Sauerbrey's equation was attributed to the additional water found to be rigidly bound to the protein. In a similar study performed by Ijiro *et al.* (1998), a frequency counter was used to measure the frequency change observed during the formation of a multilayer structure consisting of alternative layers of biotinylated double-stranded DNA and streptavidin. Langmuir-Blodgett transferred monolayers of biotinylated lipids deposited on the device surface were used to attach strepta-vidin, followed by the successive addition of biotinylated DNA and streptavidin. Figure 7.8 shows a schematic representation of the interface and the frequency change observed during the *in situ* monitoring of the binding events. Interestingly, it was found that by using Sauerbrey's equation, each binding step detected by the TSM-resonator was in good agreement with mass changes predicted based on the mean area of crystallizing streptavidin on the biotin lipid and the expected biotin/streptavidin molar ratios. Finally, the acoustic network analysis was applied to the study of various avidin- and neutravidin-biotin conjugate samples (Ghafouri and Thompson 1999). In this work, the series resonance frequency and motional resistance were measured during the binding of biotinylated molecules of various masses, i.e. dextran (Mw = 70,000), bovine serum albumin (Mw = 69,000) and insulin (Mw = 11,670) to surface-bound avidin or neutravidin. Figure 7.9 shows that the fre-

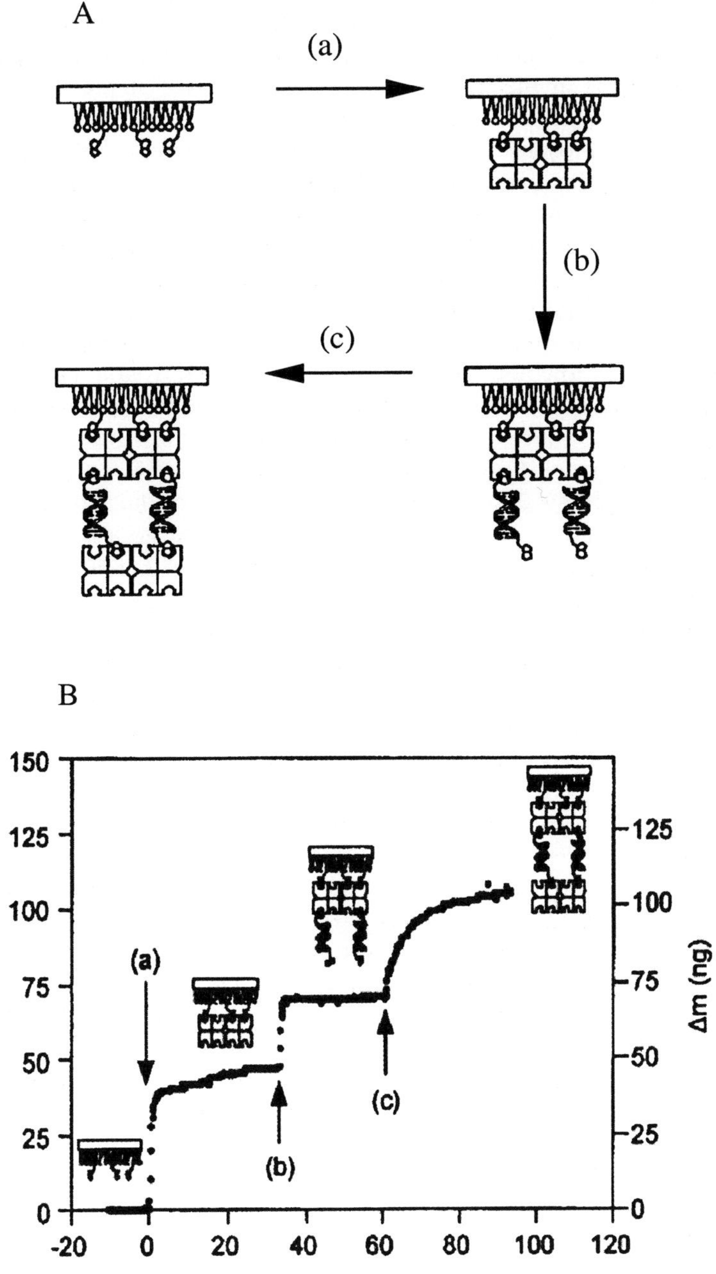

Figure 7.8 (A) Schematic representation of the multilayered interface: (a) addition of streptavidin to the biotinylated TSM surface, (b) binding of biotinylated DNA and (c) addition of a second streptavidin layer. (B) The frequency change during the formation of the above layers is also shown, as this was recorded in real time (Ijiro *et al.* 1998)

quency response cannot be related to mass since dextran has a similar molecular weight to that of albumin but gives a smaller drop in frequency and insulin, although smaller than albumin gives about the same Δf_s. It was concluded that the acoustic response was mainly reflecting differences in the viscoelastic properties of the proteins and/or variations in the acoustic coupling processes at the protein/liquid interface.

The fact that the TSM biosensor does not respond only to mass deposited in presence of liquid and that the Sauerbrey equation cannot describe fully the TSM-behavior was re-emphasized during experiments involving the interaction of biomolecules with functionalized surfaces. For example, the TSM-resonator combined with a special oscillator circuit was used to differentiate between frequency shift due to mass loading and frequency shift due to viscoelastic changes (Snellings *et al.* 2000). The above analysis was used to show that the viscoelastic properties of a film of low, high and very low density lipoproteins (LDL, HDL and VLDL, respectively) adsorbed on clean gold were very different. The unique rigid nature of LDL films in comparison to the more viscous HDL and VLDL films was related to the effect of the former lipoproteins on arteriosclerosis.

A good biological model system to study the dependence of the acoustic signal on interfacial properties is multilayers of lipid films which undergo a phase transition from the solid to the liquid crystalline state (Okahata and Ebato 1989). Figure 7.10 shows the frequency change of multi-bilayers consisting of the synthetic phospholipid 1,3-dihexadecyl-glycerol-2-phosphoethanolamine ($2C_{16}PE$) and the polymeric bilayer-forming amphiphile dioctadecyldimethylammonium poly(styrene-sulfonate) ($2C_{18}N^+2C_2/PSS^-$) as a function of temperature in both water and air. An abrupt frequency increase at a temperature that corresponds to the transition temperature T_c from the solid to the liquid crystalline state is only observed in water and not in air. This change is attributed in part to the change of the viscoelastic properties of the lipid layer during phase transition but mainly to slipping occurring when water penetrates between the fluid hydrophilic lipid layers.

In those cases where the properties of the surface, i.e. hydrophilicity and roughness, and of the medium, i.e. density, viscosity and temperature, do not change significantly with time, it is possible to use the TSM-resonator and Sauerbrey's equation to study binding events such as the interaction of ligands with lipid layers. The binding of lectins such as wheat and peanut agglutinin to a supported lipid layer doped with ganglioside molecules was investigated as a function of both the ganglioside and lectin concentration and the binding constant of the reaction was measured (Steinem *et al.* 1997; Sato *et al.* 1998). In addition, the inhibition of the reaction in the presence of different saccharides was also investigated. Similar studies were successfully performed during the binding of toxins and viruses to ganglioside-containing supported membranes (Sato *et al.* 1996; Steinem *et al.* 1997).

The TSM-resonator has been used extensively as an acoustic immunosensor. Although the exact nature of the acoustic response may not be clear, frequency change appears to be always concentration-dependent. The potential of the TSM-resonator as a liquid-phase immunosensor was analyzed in an early paper where the importance of the structure of the interface was examined in relation to the time- and mass-dependent frequency response of the device (Thompson *et al.* 1987). In practice, the standard strategy in developing an immunosensor is to attach an antibody on the device surface by using an immobilization method that has minimal

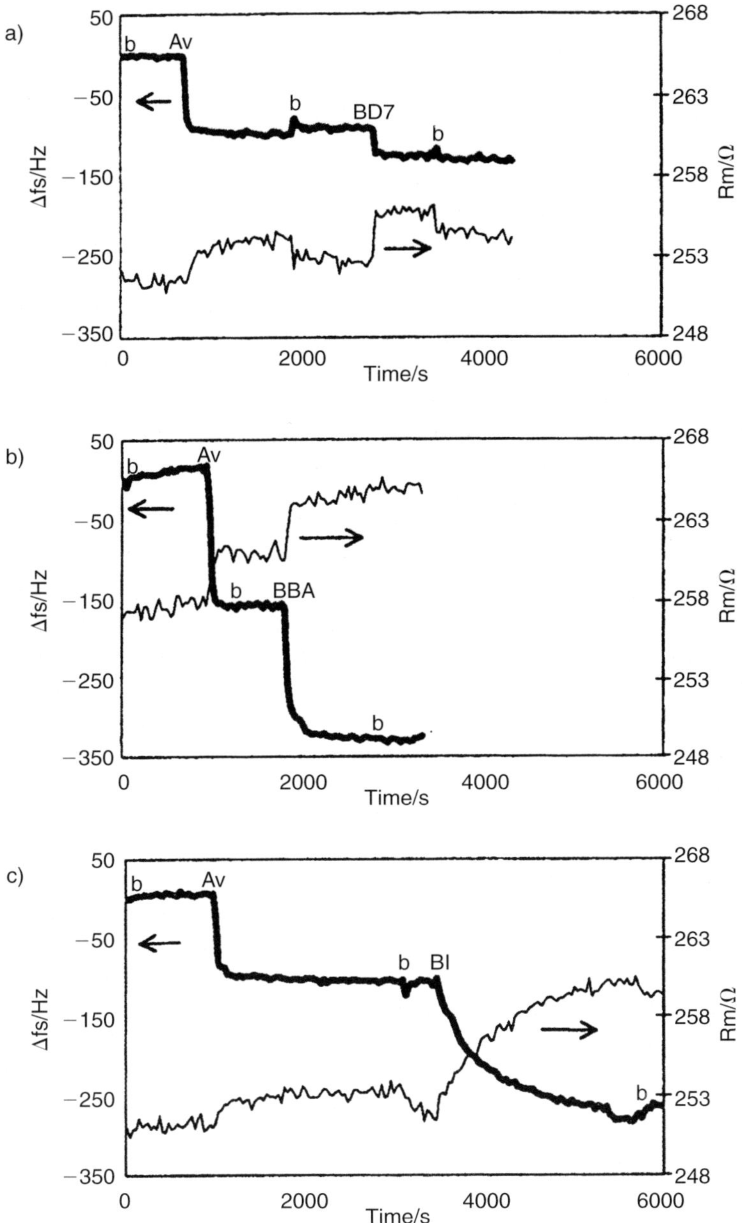

Figure 7.9 Response of the TSM device to coupling of avidin with biotin conjugates. The graphs show the frequency change Δf_s and the motional resistance R_m associated with the introduction of avidin followed by (a) biotinylated dextran 70,000, (b) biotinylated bovine albumin and (c) biotinylated insulin (Ghafouri and Thompson 1999)

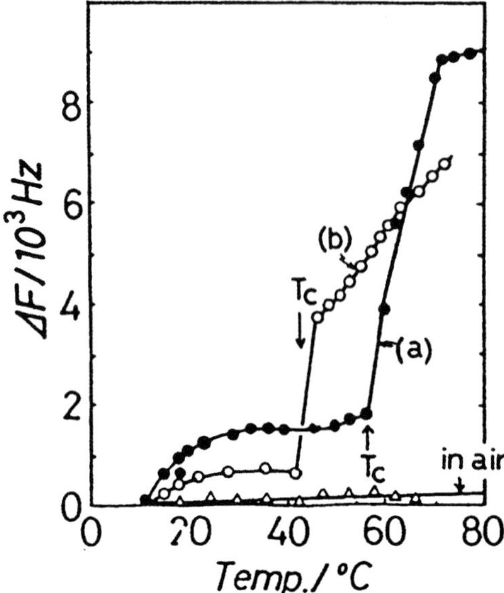

Figure 7.10 Frequency increase at T_c of a multilayer of (a) $2C_{16}PE$ and (b) $2C_{18}N^{+}2Cl/PSS^{-}$ in water and air as a function of temperature (Okahata and Ebato 1989)

effect on antibody activity. The following immobilization techniques and applications have been reported: coating of the piezoelectric crystal with protein A for the detection of immunoglobulins (Muramatsu *et al.*. 1987; Davis and Leary 1989; Dubrovski *et al.* 1995; Raman Suri *et al.* 1995), immobilization of synthetic peptides for the detection of anti-HIV antibodies (Köβlinger *et al.* 1992), silanization of the surface for the covalent attachment of antibodies or lectins in order to detect erythrocytes and sugars, respectively (Barnes *et al.* 1991; König and Grätzel 1993), attachment of haptens to the surface for the detection of antibody-coated liposomes (Yun *et al.* 1998) and self-assembly of a cystamine thiol layer on the gold electrode for the subsequent covalent binding of antibodies (Ben-Dov *et al.* 1997). In general, a linear correlation between frequency shift and biomolecule concentration is observed. The reported detection limits are $1.6 \times 10^{-9} M$ of human serum albumin (Muratsugu *et al.* 1993), 1 ng/ml of insulin (Raman Suri *et al.* 1995), $6.6 \times 10^{-11} M$ of IgM (Raman Suri *et al.* 1994), 5×10^3 cells (König and Grätzel 1993) and 260 ng/ml of bacteria (Ben-Dov *et al.* 1997). Apart from the above direct immunoassays, other applications of the TSM-resonator include an amplified mass immunoadsorbent assay for the detection of adenosine 5′-phosphosulphate (APS) reductase and human chorionic gonadotropin (hCG) (Ebersole and Ward 1988), human serum albumin (Sakai *et al.* 1995) and ochratoxin A (Hauck *et al.* 1998).

Finally, the TSM-resonator has been shown to be a useful analytical tool in combination with other techniques for the elucidation of the mechanism of biomolecular interactions. Ha *et al.* (2000) combined acoustic measurements with acoustic force microscopy (AFM) during the study of the mechanism of the binding of an antimicrobial peptide to a lipid bilayer. Based on the acoustic measurements it was confirmed that the interaction of the peptide indolicidin with DPPA layers occurs rapidly and at a ratio of indolidin molecules to lipids of 1 to 5.7.

Shear-horizontal surface acoustic wave sensors

Surface acoustic wave (SAW) devices operating in the shear mode produce waves where the particle displacement is parallel to the sensing surface. These devices have been used in liquid-sensing applications and, in general, promise higher sensitivity to surface-perturbations than TSM sensors due to SAWs' higher frequency of operation. However, they are still not as widely used as the TSM-resonator, mainly because SH-SAW devices are not commercially available like the TSM-resonator and their fabrication involves sophisticated and expensive methods.

During sensing applications the frequency or velocity of the wave and attenuation or insertion loss of the device are normally measured by using a network analyzer. The former group is related to changes in the propagation velocity as a result of mass, viscoelastic and electric changes in the interface layer. The latter is related to acoustic losses through energy coupling or decoupling from the device surface to the adjacent liquid and is affected by changes in the viscoelastic and electric properties of the interface. In most biosensing applications it is desirable to minimize changes related to the electrical properties of the interface. This can be achieved by depositing and grounding a thin metal layer (typically in the order of 100 Å) on the device surface between the transducers. The intended effect of this metal layer is to decouple ions in solution from the APM acoustoelectric field without affecting mass sensitivity.

The most popular SH-devices are the APM and the Love wave devices. The theoretical aspects of the SH-wave/matter interaction and some of the most important applications in biological sensing are presented below.

Acoustic plate mode sensors

APM waves propagate through multiple reflections between the top and bottom surface of the device, exhibiting the same sensitivity on the two surfaces (see Figure 7.3(b)). This has the advantage that the bottom surface can be used for sample localization, which is more convenient than utilizing the top surface where the transducers are deposited.

THEORY

The different types of interactions that can affect the propagation of APMs have been theoretically analyzed by Martin *et al.* (1989). The interaction of the wave with mass deposited on the device surface was modeled by applying perturbation theory. An elastic mass layer deposited at the surface of the APM device will result in a decrease in the propagation velocity (Δv) and frequency (Δf). The latter is given by the following expression:

$$\frac{\Delta v}{v_0} = \frac{\Delta f}{f_0} = c_f \rho \tag{7.7}$$

where c_f is the frequency sensitivity to surface mass and depends on the density and thickness of the quartz substrate ($c_f = 1/\rho_s b$, where ρ_s is the density of the substrate and the thickness), f_o is the operating frequency and ρ is the surface mass density ($\rho = m/A$).

The above equation predicts that frequency will decrease linearly with accumulated mass density.

Loading of the device surface with a viscous liquid results in both attenuation and change in the propagation velocity of the plate mode. Perturbation analysis is used to estimate the above changes, which are given by the following equations:

$$\frac{\Delta v}{v_0} = \frac{c_f n}{2\omega} \, \mathrm{Im}\!\left(\frac{\xi}{1 + j\omega\tau}\right) \tag{7.8}$$

$$\frac{\Delta \alpha}{\kappa} = \frac{c_f n}{2v_0} \, \mathrm{Re}\!\left(\frac{\xi}{1 + j\omega\tau}\right) \tag{7.9}$$

in which

$$\xi^2 = \left(\kappa^2 - \frac{\omega^2 \rho}{\mu}\right) + j\,\frac{\omega\rho}{\mu} \tag{7.10}$$

where τ is the relaxation time associated with the transition from viscous to elastic behavior in a Maxwellian liquid ($\tau = \eta/\mu$).

A more detailed description of the above analysis can be found in Ballantine *et al.* (1997).

APPLICATIONS

The application of the APM device to biomolecular studies has been mainly qualitative, while emphasis has been given on the optimization of the device performance for biosensing applications. The latter involves the study of the effect of the operating frequency and piezoelectric substrate on the device mass sensitivity. When ZX-LiNbO$_3$ was used as a substrate, it was shown that, for a given plate, thickness (0.5 mm), mass sensitivity was linearly dependent on frequency for devices operating at a frequency above 60 MHz (Josse *et al.* 1996). It was also emphasized that sensitivity depends not only on the operating frequency but also on the particular acoustic mode that corresponds to that frequency (Renken *et al.* 1996). The latter was clearly shown when a frequency peak at 152.6 MHz appearing very close to the third harmonic of the APM at 151.1 MHz gave much lower mass sensitivity than the third harmonic. The above behavior was largely dependent on the LiNbO$_3$ substrate used, which does not support a pure APM at the operating frequency. An additional problem with the above substrate was found to be its high acoustoelectric coupling. As a result, in a biosensing application involving protein adsorption, only 55 percent of the change was attributed to mass deposition while the remaining 45 percent was related to acoustoelectric interference (Bender *et al.* 1997). In order to overcome the above problems, quartz was used as an alternative piezoelectric substrate for APM devices (Bender *et al.* 1999). Y-cut quartz plates were shown to be a promising alternative to $LiNbO_3$ since they not only support a single APM series over a wider range of frequencies, but also exhibit a better temperature stability and a lower acoustoelectric coupling.

Applications of the APM device involve the study of the interaction of proteins with the modified device surface and the detection of double-stranded DNA. APM immunosensors have been developed after coupling IgG to the amino-silanized (Andle *et al.* 1993; Dahint *et al.* 1994; Bender *et al.* 1997; Dahint *et al.* 1999), dextran and poly(etherurethane)-activated device surface (Renken *et al.* 1996). Dahint *et al.* (1999) emphasized the significance of the operating frequency in developing an immunosensor that can differentiate between specific and non-specific binding. It was shown that, by varying the frequency of operation, the sensor's response toward nonspecific protein adsorption was reduced as the frequency increased, while the specific response remained almost unchanged (Figure 7.11). This was explained on the basis that non-specifically adsorbed proteins are less firmly bound to the surface than specifically-bound ones resulting in reduced coupling to the acoustic wave field. The above effect is pronounced at higher frequencies indicating that, by using multi-frequency sensors, it is possible to distinguish between different states of adsorbed proteins. The APM sensor has also been used for the study of the kinetics of the adsorption of proteins at artificial surfaces (Ros Seigel *et al.* 1997). In this paper the adsorption of fibrinogen on a hydrophobic surface was reported and explained based on the Langmuir model while the real-time signal was used for deriving information on the kinetics of the adsorption process.

Finally, the APM device has been used for the detection of the hybridization between complementary single-stranded DNA molecules (Andle *et al.* 1992). Amino-modified single-stranded DNA was covalently bound to the silanized device surface and subsequently used for the *in situ* detection of the binding of its complementary DNA strand. In another work, viral DNA a few hundred nucleotides long was chemically denatured and the resulting single-stranded molecules were bound to the activated device surface (Andle *et al.* 1995). The four-times higher sensitivity observed during the DNA binding, compared to what theory predicts, was attributed to the morphology of the bound molecules and their influence on local viscosity. Surface-attached viral DNA extends a few hundred nucleotides in the liquid causing an increase in the wave's penetration depth through extra coupling to the DNA chains. This is in contrast with surface bound protein films which are much thinner and, thus, a smaller discrepancy is found between experimental results and theoretical predictions (Figure 7.12).

Love wave sensors

Love wave sensors comprise a waveguide geometry where a shear-operating SAW device, such as an SSBW device, is overlayed by a layer of a dielectric material (see Figure 7.2). During the last few years, there has been an increased interest in using the Love wave device for biosensing applications. This is mainly due to the fact that this device exhibits a higher surface sensitivity than other SH-sensors as a result of the high energy confinement of the wave on the sensing surface (Gizeli *et al.* 1992a; Kovacs and Venema 1992). Also, polymers have been found to be equally good or even superior waveguide layers than silica ones, making the construction of a Love

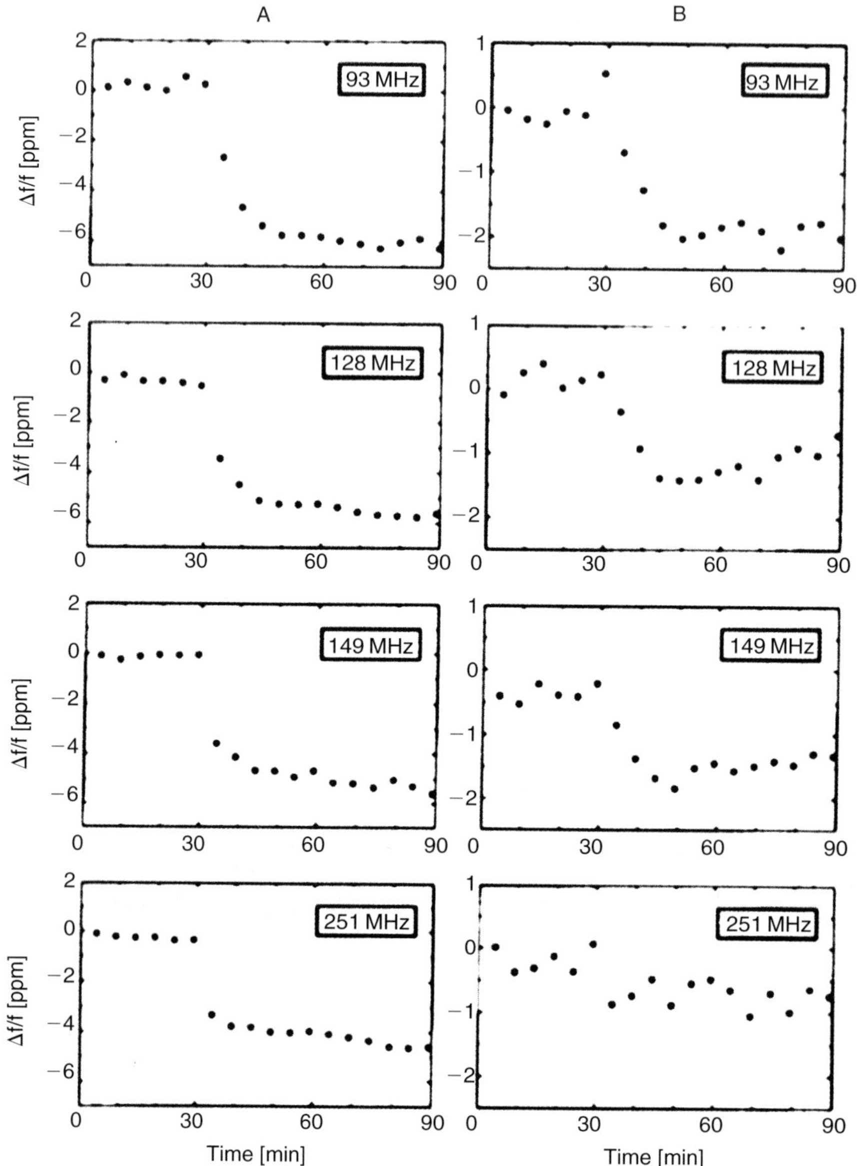

Figure 7.11 Detecting of specific (A) and unspecific (B) binding of IgG as a function of the normalised frequency ($\Delta f/f$). The detection of non-specific binding decreases significantly as frequencies increases (Dahint *et al.* 1999)

wave device very simple. Finally, the application of a waveguide layer on the surface of the device has the additional advantage that it isolates electrically the transducers and prevents energy leaking from the input to the output IDT through the liquid sample.

THEORY

The mass sensitivity of the Love wave sensor has been described theoretically by several investigators (Gizeli *et al.* 1992b; Kovacs and Venema 1992; Kovacs *et al.* 1994; Wang *et al.* 1994; Jakoby and Vellekoop 1997; McMullan *et al.* 2000). The configuration of the Love wave sensor consists of a semi-infinite substrate, overlayed by a thin solid layer which acts as the waveguiding layer. Perturbation theory is used to show that a thin layer of mass (density ρ_3, shear velocity V_{s3}, thickness ϵ) deposited on the device surface will change the frequency according to:

$$\frac{\Delta f}{f_0} = - \frac{\rho_3 \epsilon}{4U_a} \left[\left(1 - \frac{V_{S3}^2}{V_0^2} \right) |u_x|^2 \right]_{y=0} \tag{7.11}$$

where f_o is the unperturbed frequency and Δf the difference between the perturbed and unperturbed frequencies, U_a the average stored energy in the waveguide, and $(u_x)_{y=0}$ the vibration velocity of the mass loaded on the surface (Wang *et al.* 1994). In the above analysis it is assumed that the thickness of the deposited mass layer is much smaller than the wavelength and, thus, does not have the characteristics of a waveguide layer.

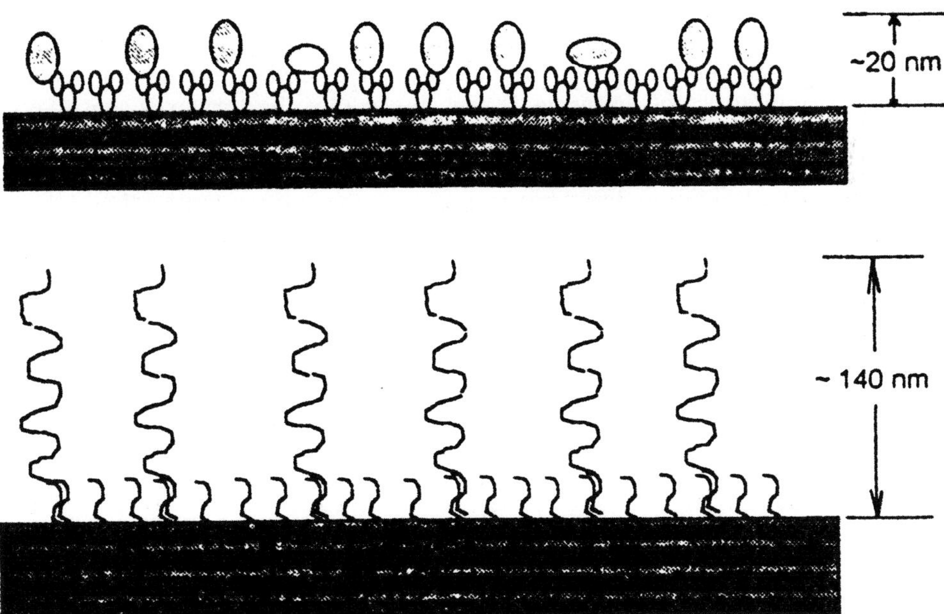

Figure 7.12 Comparison of the device interface when a 20 and 140 nm layer of protein and nucleic acid, respectively, is deposited on the activated device surface. The globular shaped proteins are distinctively different from the long, thin and stiff nucleic acids (Andle *et al.* 1995)

The effect of a viscous liquid applied on the device surface has also been mathematically analyzed for a shear horizontal wave travelling on the device/viscous liquid interface (Josse and Shana 1988; 1989). This analysis is more general and applies to a number of SH waves, including the Love wave.

APPLICATIONS

During the last few years a number of publications have appeared on the optimization of the Love wave device for biosensing applications. The design considerations for this device include the type of material used for the waveguide layer and its thickness. Two materials have been mainly applied: elastic layers of amorphous silica (SiO_2) (Du *et al.* 1996) and inelastic ones of polymethylmethacrylate (PMMA) (Gizeli *et al.* 1992b) and photoresist (Rasmusson and Gizeli 2001; McHale *et al.* 2002). In general, it has been shown that if the shear acoustic velocity in the overlayer is low enough, as is true for PMMA, good coupling of the acoustic energy and high mass sensitivity can be achieved with thin overlayers. In contrast, if the shear acoustic velocity in the overlayer is relatively high, as for SiO_2, thick overlayers are required in order to achieve the same sensitivity. The combination of the above materials in order to construct a multilayer waveguide structure has been also shown to be a good alternative for developing a Love wave biosensor (Harding and Du 1997; Du and Harding 1998).

The Love wave sensor has been used for the study of the interaction of a vesicle suspension with a solid surface in order to form a membrane-type interface as a function of the surface wetting properties and topography (Melzak *et al.* 2001; Melzak and Gizeli 2002). A supported lipid monolayer, bilayer and vesicle layer was formed on a hydrophobic, hydrophilic and relatively hydrophilic surface, respectively. By recording the change in the velocity (measured as phase) and energy (measured as amplitude) of the wave, it was shown that the former is sensitive to both mass and viscoelastic changes while the latter is only sensitive to viscoelastic changes. This is clearly shown in Figure 7.13 where amplitude does not respond to the deposition of a lipid monolayer but responds to the formation of a lipid bilayer and vesicle layer. A supported lipid monolayer deposited directly on a hydrophobic surface would act as an ideal elastic mass since there would be no water molecules trapped between the lipid and the surface and there would be no energy loss through viscous coupling. This is not the case with the supported lipid bilayer and vesicle layer where the water molecules around the lipid will form a lossy interface layer giving rise to viscoelastic coupling.

The Love wave sensor has been also applied to the study of immunological applications. The silanized device surface was used to covalently attach IgG and subsequently detect anti-IgG in the concentration range of 6.6×10^{-12} to 0.8×10^{-7} M (Harding *et al.* 1997). Figure 7.14 shows the relationship between frequency change and IgG concentration as this was measured experimentally and predicted, theoretically based on the assumption that frequency is directly related to deposited mass. The higher frequency change observed during mass deposition in solution than in air is attributed to the presence of trapped water and, subsequently, viscous coupling. It is mentioned that the proportion of the contribution of the antibody mass loading and viscous coupling to frequency shift varies with the antibody concentration and the amount of the surface-covered analyte.

Another approach for the immobilization of the antibody involves the formation of a specific multilayer on the device surface (Gizeli *et al.* 1997). A biotinylated supported lipid layer was attached on the device surface followed by the addition of streptavidin and subsequently, biotinylated IgG. The above surface was used to detect the analyte, i.e. anti-IgG in the concentration range of 3×10^{-8} to 10^{-6}M. By using the anti-IgG binding isotherm and the time-resolved measurements of antibody binding, both the binding and rate constants of the reaction were determined.

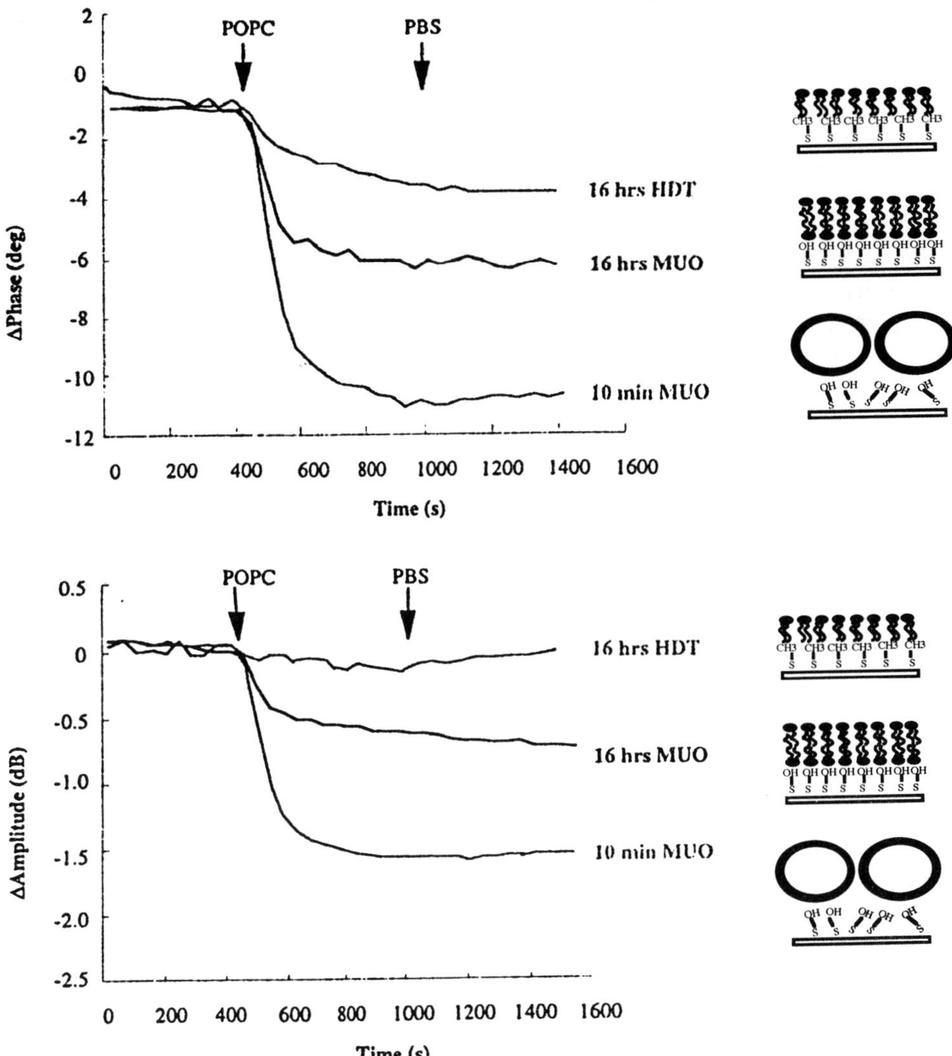

Figure 7.13 Phase and amplitude change as a function of time during the application of a vesicle solution of POPC on three gold coated Love wave devices. The surfaces used were modified with hexadecanethiol (HDT) for 16 h, mercaptoundecanol (MUO) for 16 h and MUO for 10 min, in order to give a hydrophobic, hydrophilic and relatively hydrophilic surface, respectively (Melzak *et al.* 2001)

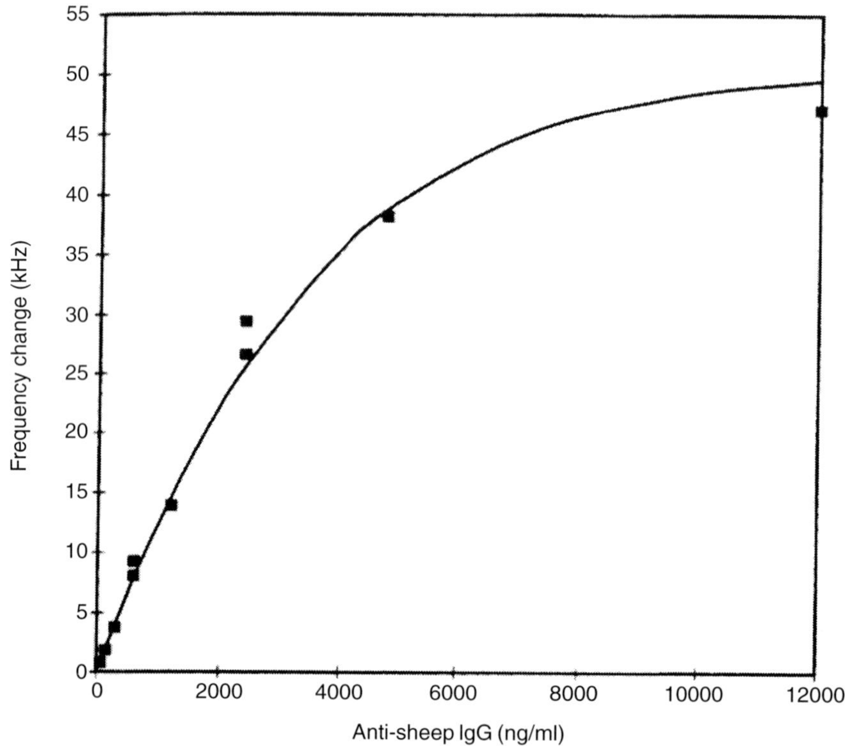

Figure 7.14 Binding isotherm of sheep anti-IgG to the IgG-immobilized device surface plotted as frequency change against the antibody concentration. The sensor used was a silica-coated Love wave device. The solid line was predicted theoretically (Harding *et al.* 1997)

Comparative measurements using surface plasmon resonance allowed the response of the acoustic immunosensor to be quantitatively correlated with mass binding to the surface.

Comparison of acoustic sensors

The sensitivity S_m provides a general quantitative method for comparing the effect of mass loading on different devices. Sensitivity is defined as the ratio of the fractional change of the frequency of the acoustic wave Δf as a result of mass loading to the deposited mass per unit area ($h\rho$) divided by the operating frequency f_o:

$$S_m = \frac{1}{f_0} \lim_{h \to 0} \frac{\Delta f}{[h\rho]} \tag{7.12}$$

Table 7.1 summarizes the experimental values of the mass sensitivity of some of the acoustic devices, measured during mass deposition in air. Based on this table we can

Table 7.1 Comparison of the measured mass sensitivities in air for the thickness shear resonator (TSM), acoustic plate mode (APM) and Love wave devices. The limiting factor for further improvement is also included for each system

Sensors	*TSM[a]*	*APM[a]*	*Love wave* *SiO$_2$[b]*	*PMMA[c]*	*SiO$_2$/PMMA[d]*
• f_o (MHz)	6	97	124	110	124
• Sensitivity (cm^{2}g^{-1})	14	9.5 (0th mode) 19.4 (higher mode)	380	430	519
• Limiting factor	device thickness	device thickness	overlayer acoustic properties and thickness		

Notes
a Ballantine *et al.* 1997
b Du *et al.* 1996
c Gizeli *et al.* 1992
d Harding and Du 1997

conclude that the sensitivity of the TSM-resonator is the lowest and is limited by the operating frequency of the device which, in turn, is limited by the thickness of the quartz plate. The APM and Love wave devices both operate at a higher frequency. Of these two devices, the APM has the lowest sensitivity; further increase of its sensitivity is limited by the distribution of the energy between the two surfaces and, consequently, the thickness of the plate. The Love wave sensor is limited by the acoustic properties and thickness of the overlayer. Optimization of these two parameters can result in the confinement of the energy at the sensing surface and in mass sensitivities which can be up to fifty times higher than the sensitivity of the fundamental APM.

The interesting question for biosensing applications is which device exhibits the highest sensitivity in the presence of liquid and, consequently, is the most promising for studying biomolecular interactions. Unfortunately, based on existing data, it is impossible to quantify mass sensitivity in the presence of liquid for the following reasons:

1   The effect of the electric coupling of the liquid to the piezoelectric substrate varies for the different devices and is quite significant when the high frequency SH–SAW devices are used. This effect is even more pronounced when strong piezoelectric substrates such as $LiNbO_3$ are used for SAW devices and in some cases the acoustoelectric response can be greater than the mass loading response. This interaction can be largely suppressed by applying a thick metal layer on the device surface between the transducers. An additional interference occurs when the liquid sample is localized between the IDTs and energy leaks from one IDT to the other through coupling to the dielectric or conducting liquid placed in the middle. This effect can be also minimized in the Love wave device by applying high dielectric overlayers which can largely isolate the two transducers. However, in all cases it is difficult to quantify any remaining electric interaction in terms of what fraction of the frequency change is due to the electric effect rather than mass and viscoelastic loading.

2   The standard way to quote sensitivity is by relating frequency change to the bulk concentration of the analyte in the solution. This information does not take

into account the number of receptor molecules attached to the device surface and the affinity of the analyte for the receptor so cannot be used to compare different biosensors and their sensitivities. The only valid way to compare different acoustic biosensors would be if frequency change was related to the surface mass density of the deposited molecules for the same biomolecular interaction. Unfortunately, currently this information is not available.

However, according to theoretical predictions and on the basis that electric interactions are completely eliminated, the high frequency SH–SAW devices are expected to respond with higher sensitivity to the binding of low concentrations of smaller analytes.

Conclusions

Acoustic wave devices have been shown to be capable of detecting and characterizing biological binding events in real time with high levels of sensitivity and selectivity and without the need of additional labels. In contrast to optical techniques such as surface plasmon resonance and ellipsometry, shear acoustic waves can be used to provide more information on biological interactions than just the mass of the bound analyte. The measurement of the energy loss occurring at the solid/liquid interface can be a powerful method to study changes of the tertiary structure of proteins or elasticity of membranes. In addition, from the frequency measurement it is possible to derive information on the stoichiometry of the binding, the "on" and "off" kinetic constants and the overall affinity of the soluble analyte for the immobilized receptor. Further advances in the design of high frequency acoustic wave devices operating in the GHz range will offer higher sensitivities. The next step in the development of powerful acoustic wave sensors for biomolecular interactions will be to better define the interactions between the acoustic wave and the properties of the solid/liquid interface and also quantify them by using model biological systems.

Acknowledgements

I would like to thank Dr. K. Melzak and Dr. F. Bender for reading the manuscript and making useful comments and Mr. J. Lapinski and Mr. K. Kabilan for their assistance with the reproduction of the figures; my sister, Katerina, is also acknowledged for her support during the preparation of this manuscript.

References

Andle, J.C., J.F. Vetelino and M.W. Lade (1992) An acoustic plate mode biosensor, *Sens. Act. B* 8, 191–198.

Andle, J.C., J. Weaver, D.J. McAllister, F. Josse and J.F. Vetelino (1993) Improved acoustic-plate-mode biosensor, *Sens. & Act. B* 13–14, 437–442.

Andle, J.C., J. Weaver, J.F. Vetelino and D.J. McAllister (1995) Selective acoustic plate mode DNA sensor, *Sens. & Act. B* 24–25, 129–133.

Ballantine, D.S., R.M. White, S.J. Martin, A.J. Ricco, E.T. Zellers, G.C. Frye and H. Wohltjen (1997) *Acoustic Wave Sensors*. Academic Press.

Bandey, H.L., S.J. Martin, R.W. Cernosek and R. Hillman (1999) Modeling the responses of thickness-shear mode resonators under various loading conditions, *Anal. Chem.* 71(11), 2205–2214.

Barnes, C., C. D'Silva, J.P. Jones and T.J. Lewis (1991) A concanavalin A-coated piezoelectric crystal biosensor, *Sens. & Act. B* 3, 295–304.

Bastiaans, G.J. (1988) Piezoelectric transducers. In: *Chemical Sensors*, T.E. Edmonds (ed.), Blackie and Son, London, pp. 295–319.

Ben-Dov, I., I. Willner and E. Zisman (1997) Piezoelectric immunosensor for urine specimens of chlamydia trachomatis employing quartz crystal microbalance microgravimetric analyses, *Anal. Chem.* 69, 3506–3512.

Bender, F., R. Dahint, F. Josse, A. Ricco and S. Martin (1999) Characteristics of acoustic plate modes on rotated Y-cuts of quartz utilised for biosensing applications, *Anal. Chem.* 71, 5064–5068.

Bender, F., F. Meimeth, R. Dahint, M. Grunze and F. Josse (1997) Mechanisms of interaction in acoustic plate mode immunosensors, *Sens. Act. B* 40, 105–110.

Browning, T.I. and M.F. Lewis (1977) New family of bulk-acoustic-wave devices employing interdigital transducers, *Electr. Lett.* 13(5), 128–130.

Cady, W.G. (1924) *Proc. Inst. Radio Eng.* 12, 805.

Cady, W.G. (1964) *Piezoelectricity, an Introduction to the Theory and Applications of Electromechanical Phenomena in Crystals.* Dover, New York.

Carey, W.P. and B.P. Kowalski (1986) Chemical piezoelectric sensor and sensor array characterisation, *Anal. Chem.* 58, 3077–3084.

Cavic, B.A., G.L. Hayward and M. Thompson (1999) Acoustic waves and the study of biochemical macromolecules and cells at the sensor-liquid interface, *Analyst* 124, 1405–1420.

Dahint, R., F. Bender and F. Morhard (1999) Operation of acoustic plate mode immunosensors in complex biological media, *Anal. Chem.* 71, 3150–3156.

Dahint, R., M. Grunze, F. Josse and J. Renken (1994) Acoustic plate mode sensor for immunochemical reactions, *Anal. Chem.* 66, 2888–2892.

Davis, K.A. and T.R. Leary (1989) Continuous liquid-phase piezoelectric biosensor for kinetic immunoassays, *Anal. Chem.* 61(11), 1227–1230.

Dieulesaint, E. and D. Royer (1980) Acoustic surface waves. In: *Elastic Waves in Solids, Applications to Signal Processing*, J. Wiley, New York, pp. 65–190.

Du, J. and G.L. Harding (1998) A multilayer structure for Love-mode acoustic sensors, *Sens. & Act. A* 65, 152–159.

Du, J., G.L. Harding, J.A. Ogilvy, P.R. Dencher and M. Lake (1996) A study of Love-wave acoustic sensors, *Sens. & Act. A* 56, 211–219.

Dubrovsky, T., A. Tronin, S. Dubrovskaya, S. Vakula and N. Nicolini (1995) Immunological activity of IgG Langmuir films oriented by protein A sublayer, *Sens. & Act. B* 23, 1–7.

Ebersole, R.C. and M.D. Ward (1988) Amplified mass immunoabsorbent assay with a quartz crystal microbalance, *J. Am. Chem. Soc.* 110(26), 8623–8628.

Furtado, L.M., H. Su, M. Thompson, D.P. Mack, G.L. Hayward (1999) Interactions of HIV-1 Tar RNA with tat-derived peptides discriminated by on-line acoustic wave detector, *Anal. Chem.* 71, 1167–1175.

Furusawa, H., Y. Ebara and Y. Okahata (1999) *In situ* monitoring of poly(L-glutamine)-bound dsDNA selection on a quartz-crystal microbalance, *Chem. Lett.* 823–824.

Furusawa, H., A. Watanabe, Y. Ebara and Y. Okahata (1998) *In vitro* selection of DNA on a quartz-crystal microbalance, *Nucl. Acis Symp. Ser.* 39, 79–80.

Ghafouri, S. and M. Thompson (1999) Interfacial properties of biotin conjugate-avidin complexes studied by acoustic wave sensor, *Langmuir* 15, 564–572.

Gizeli, E., M. Liley, C.R. Lowe and H. Vogel (1997) Antibody binding to a functionalised supported lipid layer; a direct immunosensor, *Anal. Chem.* 69, 4808–4813.

Gizeli, E., A.C. Stevenson, N.J. Goddard and C.R. Lowe (1992a) A novel Love-plate acoustic sensor utilizing polymer overlayers, *IEEE Trans. Ultras. Ferroel. Frequ. Contr.* 39(5), 657–659.

Gizeli, E., A.C. Stevenson, N.J. Goddard and C.R. Lowe (1992b) A Love plate biosensor utilizing a polymer layer, *Sens. & Act. B* 6(1–3), 131–137.

Gulyaev, Y.V. (1998) Review of shear surface acoustic waves in solids, *IEEE Trans. Ultras. Ferroel. Frequ. Contr.* 45, 935–937.

Ha, T.H., C.H. Kim, J.S. Park and K. Kim (2000) Interaction of indolicidin with model lipid bilayer: quartz crystal microbalance and atomic force microscopy study, *Langmuir* 16, 871–875.

Harding, G.L. and J. Du (1997) Design and properties of quartz-based Love wave acoustic sensors incorporating silicon dioxide and PMMA guiding layers, *Smart Mat. Struct.* 6, 716–720.

Harding, G.L., J. Du, P.R. Dencher, D. Barnett and E. Howe (1997) Love wave acoustic immunosensor operating in liquid, *Sens. & Act. A* 61, 279–286.

Hauck, H., C. Kößlinger, S. Drost and H. Wolf (1998) Biosensor system for the determination of ochratoxin A, *Biosensors* 1, 515–518.

Ijiro, K., H. Ringsdorf, E. Birch-Hirschfeld, S. Hoffmann, U. Schilken and M. Strube (1998) Protein-DNA double and triple layers: interaction of biotinylated DNA fragments with solid supported streptavidin layers, *Langmuir* 14, 2796–2800.

Janshoff, A., H. Galla and C. Steinem (2000) Piezoelectric mass-sensing devices as biosensors – an alternative to optical biosensors? *Angew. Chem. Int. Ed.* 39, 4004–4032.

Jakoby, B. and M.J. Vellekoop (1997) Properties of Love waves: applications in sensors, *Smart Mater. Struct.* 6, 668–679.

Josse, F., R. Dahint, J. Schumacher, M. Grunze, J. Andle and J. Vetelino (1996) On the sensitivity of acoustic-plate-mode sensors, *Sens. & Act. A* 53, 243–248.

Josse, F. and Z.A. Shana (1988) Analysis of shear horizontal surface waves at the boundary between a piezoelectric crystal and a viscous fluid medium, *J. Acoust. Soc. Am.* 84(3), 978–984.

Josse, F. and Z.A. Shana (1989) Effect of liquid relaxation time on SH surface waves liquid sensors, *J. Acoust. Soc. Am.* 85(4), 1556–1559.

Kanazawa, K.K. and J.G. Gordon (1985) The oscillation frequency of a quartz resonator in contact with a liquid, *Anal. Chim. Acta.* 157, 99–105.

King, W.H. (1964) Piezoelectric sorption detector, *Anal. Chem.* 36, 1735–1739.

König, B. and M. Grätzel (1993) Detection of human T-lymphocytes with a piezoelectric immunosensor, *Anal. Chim. Acta.* 281, 13–28.

Kößlinger, C., S. Drost, F. Aberl, H. Wolf, S. Koch and P. Woias (1992) A quartz crystal biosensor for measurements in liquids, *Bios. & Bioel.* 7, 397–404.

Kovacs, G., M.J. Vellekoop, G.W. Lubking and A. Venema (1994) A Love wave sensor for (bio)chemical sensing in liquids, *Sens. & Act. A* 43(1–3), 38–43.

Kovacs, G. and A. Venema (1992) Theoretical comparison of sensitivities of acoustic shear wave mode for (bio)chemical sensing in liquids, *Appl. Phys. Lett.* 61(6), 639–641.

Lewis, M.F. (1982) High frequency acoustic plate mode device employing interdigital transducers, *Electronic Letters* 17, 819–821.

Love, A.E.H. (1911) *Some Problems of Geodynamics.* Cambridge University Press, Cambridge.

McHale, G., M.I. Newton, F. Martin, E. Gizeli and K. Melzak (2001) Resonant conditions for Love wave guiding layer thickness, *App. Phys. Lett.* 79(21), 3542–3543.

McMullan, C., H. Mehta, E. Gizeli and C.R. Lowe (2000) Modelling of the mass sensitivity of the Love wave device in the presence of a viscous liquid, *J. Phys. D: Appl. Phys.* 33, 1–7.

Martin, S.J., V.E. Granstaff and G.C. Frye (1991) Characterisation of a quartz crystal microbalance with simultaneous mass and liquid loading, *Anal. Chem.* 63, 2272–2281.

Martin, S.J., A.J. Ricco and G.C. Frye (1989) Characterization of SH acoustic plate mode liquid sensors, *Sens. Act.* 20, 253–268.

Melzak, K., and E. Gizeli (2002) A Silicate gel for promoting deposition of lipid bilayers, *J. Coll. Int. Sc.* 246, 21–28.

Melzak, K., E. Ralph and E. Gizeli (2001) Effect of the surface hydrophilicity on the formation of a membrane-type interface; study using an acoustic wave device, Langmuir 17(5), 1599–1603.

Morgan, D.P. (1985) *Surface Wave Devices for Signal Processing.* Elsevier, New York.

Muramatsu, H., J.M. Dicks, E. Tamiya and I. Karube (1987) Piezoelectric crystal biosensor modified with protein A for determination of immunoglobulins, *Anal. Chem.* 59, 2760–2763.

Muratsugu, M., F. Ohta, Y. Miya, T. Hosokawa, S. Kurosawa, N. Kamo and H. Ikeda (1993) Quartz crystal microbalance for the detection of microgram quantities of human serum albumin: relationship between the frequency change and the mass of protein adsorbed, *Anal. Chem.* 65, 2933–2937.

Niikura, K., H. Matsuno and Y. Okahata (1998) Direct monitoring of DNA polymerase reactions on a quartz-crystal microbalance, *J. Am. Chem. Soc.* 120(3), 8537–8538.

Niikura, K., H. Matsuno and Y. Okahata (1999) Binding behavior of lysine-containing helical peptides to DNA duplexes immobilised on a 27 MHz quartz-crystal microbalance, *Chem. Eur. J.* 5(5), 1609–1616.

Okahata, Y. and H. Ebato (1989a) Application of a quartz-crystal microbalance for detection of phase transitions in liquid crystals and lipid multilayers, *Anal. Chem.* 61, 2185–2188.

Okahata, Y., M. Kawase, K. Niikura, F. Ohtake, H. Furusawa and Y. Ebara (1998b) Kinetic measurements of DNA hybridisation on an oligonucleotide-immobilised 27 MHz quartz-crystal microbalance, *Anal. Chem.* 70, 1288–1296.

Okahata, Y., Y. Matsunobu, K. Ijiro, M. Mukae and K. Makino (1992) Hybridisation of nucleic acids immobilised on a quartz crystal microbalance, *J. Amer. Chem. Soc.* 114, 8299–8300.

Okahata, Y., K. Niikura, M. Sawada and T. Morii (1998a) Kinetic studies of sequence-specific binding of GCN4-bZIP peptides to DNA strands immobilised on a 27-MHz quartz-crystal microbalance, *Bioch.* 37, 5666–5672.

Pierce, G.W. (1923) *Proc. Am. Acad. Art. Sci.* 59, 81.

Raman Suri, C., P.K. Jain and G.C. Mishra (1995) Development of piezoelectric crystal based microgravimetric immunoassay for determination of insulin concentration, *J. Biotech.* 39, 27–34.

Raman Suri, C., M. Raje and G.C. Mishra (1994) Determination of immunoglobulin M concentration by piezoelectric crystal immunobiosensor coated with protamine, *Biosens. & Bioel.* 9, 535–542.

Rasmusson, A. and E. Gizeli (2001) Comparison of poly(methylmethacrylate) and Novolac wave guide coatings for an acoustic biosensor, *J. App. Phys.* 90(12), 5911–5914.

Lord Rayleigh (1888) On waves propagated along the plane surface of an elastic solid, *Proc. London Math. Soc.* 17, 4–11.

Renken, J., R. Dahint, M. Grunze and F. Josse (1996) Multifrequency evaluation of different immunosorbents on acoustic plate mode sensors, *Anal. Chem.* 68, 176–182.

Ricket, J., A. Brecht and W. Goepel (1997) QCM operation in liquids: constant sensitivity during formation of extended protein multilayers by affinity, *Anal. Chem.* 69, 1441–1448.

Rodahl, M. and B. Kasemo (1996) *Rev. Sc. Instr.* 67, 3238–3241.

Ros Seigel, R., P. Harder, R. Dahint, M. Grunze, F. Josse, M. Mrksich and M. Whitesides (1997) Characteristics of acoustic plate modes on rotated Y-cuts of quartz utilised for biosensing applications, *Anal. Chem.* 69, 3321–3328.

Sakai, G., T. Saiki, T. Uda, N. Miura and N. Yamazoe (1995) Selective and repeatable detection of human serum albumin by using piezoelectric immunosensor, *Sens. & Act. B* 24–25, 134–137.

Sato, T., T. Serizawa, F. Ohtake, M. Nakamura, T. Terabayashi, Y. Kawanishi and

Y. Okahata (1998) Quantitative measurements of the interaction between monoganglio-side monolayers and wheat germ agglutinin (WGA) by a quartz crystal microbalance, *Biochim. Biophys. Acta.* 12, 82–92.

Sato, T., T. Serizawa and Y. Okahata (1996) Binding of influenza A virus to monosialogan-glioside (GM3) reconstituted in glucosylceramide and sphingomyelin membranes, *Biochim. Bioph. Acta.* 3, 14–20.

Sauerbrey, G. (1959) *Z. Physik*, 155, 206–222.

Snellings, S., J. Fuller and D. Paul (2001) Response of a thickness-shear-mode acoustic wave sensor to the adsorption of lipoprotein particles, *Langmuir* 17(8), 2521–2527.

Steinem, C., A. Janshoff, J. Wegener, W.-P. Urlich, W. Willenbrink, M. Sieber and H-J. Galla (1997) Impedance and shear wave resonance analysis of ligand–receptor interactions at functionalised surfaces and of cell monolayers, *Bios. Bioel.* 12, 787–808.

Stoneley, R. (1924) Elastic waves at the surface of separation of two solids, *Proceedings of the Royal Society* A106, 416–428.

Su, H., S. Chong and M. Thompson (1996) Interfacial hybridisation of RNA homopolymers studied by liquid phase acoustic network analysis, *Langmuir* 12, 2247–2255.

Su, H. and M. Thompson (1995) Kinetics of interfacial nucleic acid hybridisation studied by acoustic network analysis, *Bios. & Bioel.* 10, 329–340.

Thompson, M., G.K. Dhaliwal, C.L. Arthur and G.S. Calabrese (1987) The potential of the bulk acoustic wave device as a liquid-phase immunosensor, *IEEE Trans. Ultras., Ferroel., Frequ. Contr. UFFC* 34(2), 127–135.

Uberall, H. (1972) Surface waves in acoustics. In: *Physical acoustics*, W.P. Mason and R.N. Thurston (eds), Academic Press, New York, pp. 1–59.

Wang, J., P.E. Nielsen, M. Jiang, X. Chai, J.R. Fernandes, D.H. Grant, M. Ozsoz, A. Beglieter and M. Mowat (1997) Mismatch-sensitive hybridisation detection by peptide nucleic acid immobilised on a quartz crystal microbalance, *Anal. Chem.* 69, 5200–5202.

Wang, Z., J.D.N. Cheeke and C.K. Jen (1994) Sensitivity analysis for Love mode acoustic gravimetric sensors, *Appl. Phys. Lett.* 64(22), 2940–2942.

Ward, M.D. and B.A. Buttry (1990) *In situ* interfacial mass detection with piezoelectric trans-ducers, *Science* 249, 1000–1007.

White, R.M. (1970) Surface elastic waves, *Proc. of IEEE* 58(8), 1238–1277.

White, R.M. and F.W. Voltmer (1965) Direct piezoelectric coupling to surface elastic waves, *Appl. Phys. Lett.* 7, 314–316.

Yang, M., F.L. Chung and M. Thompson (1993) Acoustic network analysis as a novel tech-nique for studying protein adsorption and denaturation at surfaces, *Anal. Chem.* 65, 3713–3716.

Yang, M. and M. Thompson (1993) Multiple chemical information from the thickness shear mode acoustic wave sensor in the liquid phase, *Anal. Chem.* 65, 1158–1168.

Yun, K., E. Kobatake, T. Haruyama, M. Laukkanen, K. Keinänen and M. Aizawa (1998) Use of a quartz crystal microbalance to monitor immunoliposome–antigen interaction, *Anal. Chem.* 70, 260–264.

8 Immunoassays using enzymatic amplification electrodes

*Frieder W. Scheller, Christian G. Bauer,
Alexander Makower, Ulla Wollenberger,
Axel Warsinke and Frank F. Bier*

Introduction

Immunoassay has become an important analytical technique in the last decade. Coupling immunoassays with enzymatic analysis was one of the breakthroughs in immunoanalysis, resulting in enzyme linked immunosorbed assay (ELISA), which is applicable for many of the analytes. The coupling is usually done by conjugation of the antibody or the analyte with a suitable enzyme, the label. Alkaline phosphatase (AP), horse radish peroxidase (HPR) and β-galactosidase (GAL) have proven to be suitable because of stability, purity and availability of a spectrum of substrates, the product of which can be detected easily by photometric, fluorimetric or chemiluminometric means.

Immunoassays with electrochemical detection have gained growing attention recently (Green *et al.* 1991; Ho *et al.* 1995; Niwa 1995; Bier *et al.* 1996; La Gal La Salle *et al.* 1995), because it is possible to measure current precisely even in colored and turbid samples. Electroanalytical techniques are fairly sensitive and currents as low as 10^{-10} A can be recorded with commercial devices. Thus under optimum conditions – with high enzyme loading under fast mass transport in thin layers and efficient external mass transfer – enzyme electrodes can measure substrates down to 10^{-6} mol/l with acceptable precision. For the measurement of substrates in the nanomolar range, an increase of sensitivity of the enzyme electrode is required. One way to solve this problem is the continuous regeneration of the analyte in cyclic reactions.

Coupling of immunoassays with enzymatic recycling electrodes

Enzymatic substrate regeneration is a tool to enhance the sensitivity of enzyme electrodes both for substrate analysis and immunoassays.

The combination of immunoreactions and electrode based substrate recycling connects very specific recognition of an analyte with highly sensitive detection. Most important for this field of application is the sensitivity, which permits the detection of a label at very low concentration. For enzyme immunoassays in particular, several goals have to be achieved:

i   product is indicated very sensitively,
ii   enzyme has high turn over number,
iii   enzyme substrate produce only a low background response,
iv   substrate and product are stable,
v   close pH optimum of marker enzyme and detector.

Since the sensor is used to quantify the change of one particular substance the broad substrate specificity of the enzymes used is not critical.

Therefore sensors using acceptor dependent dehydrogenases in combination with phenol oxidases (Ghindilis *et al.* 1995b; Bauer *et al.* 1996; Bier *et al.* 1996b) and the lactate oxidase/lactate dehydrogenase couple (Makower *et al.* 1994) have been used. The enzymes measured are alkaline phosphatase and β-galactosidase (Scheller *et al.* 1995).

In the "simplest" approach for the determination of antigens the recycling electrode can be used in combination with a conventional enzyme linked immunosorbent assay. In both sandwich-immunoassays, and competitive immunoassays the electrode traces the enzyme marker, which generates the substrate for the recycling electrode from a non-detectable precursor (Figure 8.1).

For hapten determination the new dimension of sensitivity of enzyme electrodes permits a novel approach to competitive immunoassays where a substrate of the recycling electrode itself is used as a marker of a hapten. The competition between the labeled hapten and the analyte hapten for binding to the respective antibody is indicated with the sensor. Since only the free (unbound) labeled hapten is measurable the whole assay can be performed in a homogeneous format. The indication of a displaced immobilized antibody-bound labeled hapten in a flow system is a further step towards automatic immunoassays.

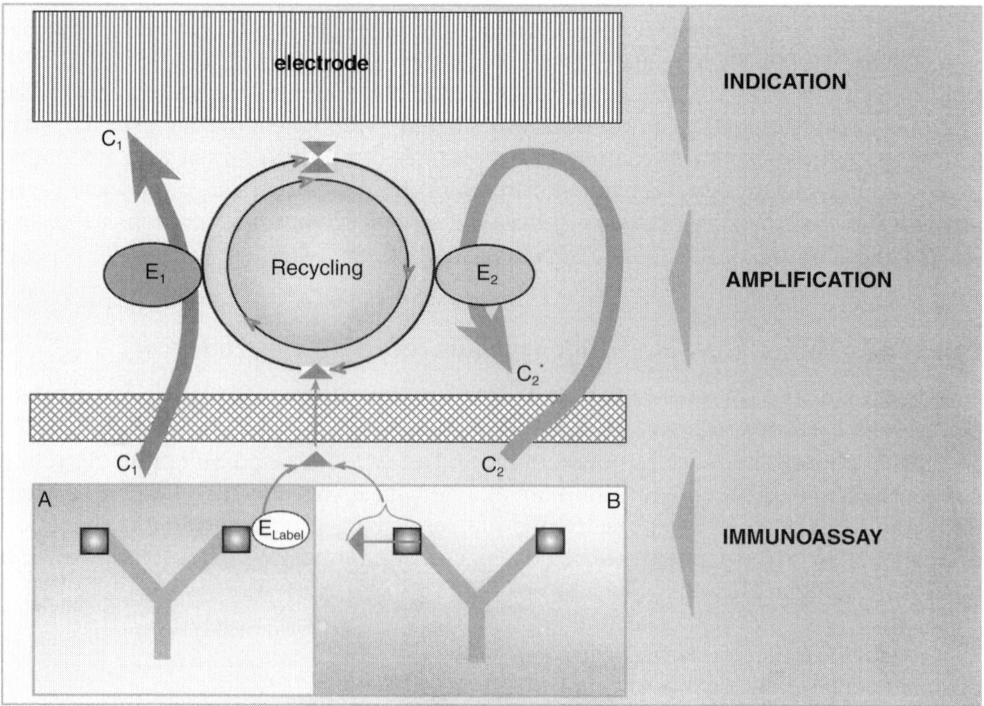

Figure 8.1 Scheme of coupling amplification and immunoassays. A – Enzyme immunoassay; B – Redoxlabel immunoassay

Signal amplification by recycling

The combination of the electrochemical detection principle and the recycling of the analyte can be performed in a number of ways (Figure 8.2).

In the electrochemical (Niwa *et al.* 1991; Wollenberger *et al.* 1994) and bioelectro-catalytic (Scheller *et al.* 1987; Ortega *et al.* 1993) approach, the target analyte is recy-cled between electrodes or electrode and redox centre of the enzyme and is therefore necessarily limited to reversible redox species. In contrast, bi-enzymatic substrate recycling (Schubert *et al.* 1985; Wollenberger *et al.* 1993) may be per-formed between enzymes, where one enzyme forms the substrate of the other, while a co-reactant is measured directly or in an additional analytical step.

Bi-enzymatic substrate recycling

In the bi-enzymatic approach, the sensitivity enhancement is provided by shuttling the analyte between enzymes acting in cyclic series of reactions accompanied by co-substrate consumption and accumulation of by-products.

The target analyte can be one of the substrates or coenzymes of the participating enzymes. Assuming a sufficiently high activity of one enzyme in the presence of its co-substrate and an analyte at a concentration far below its Michaelis constant, an amplification is achieved by turning on the second enzyme by addition of its co-substrate. By measuring the concentration change of one of these co-reactants directly or in an additional analytical step, the recycling system is used as a biochem-ical amplifier for the analyte, which can be each of the substrates of the participating enzymes.

This principle of signal amplification has been first successfully applied to systems with soluble enzymes for the indication of pyridine di-nucleotides (Lowry and Passonneau 1972). For soluble enzymes at steady state the overall cycling constant,

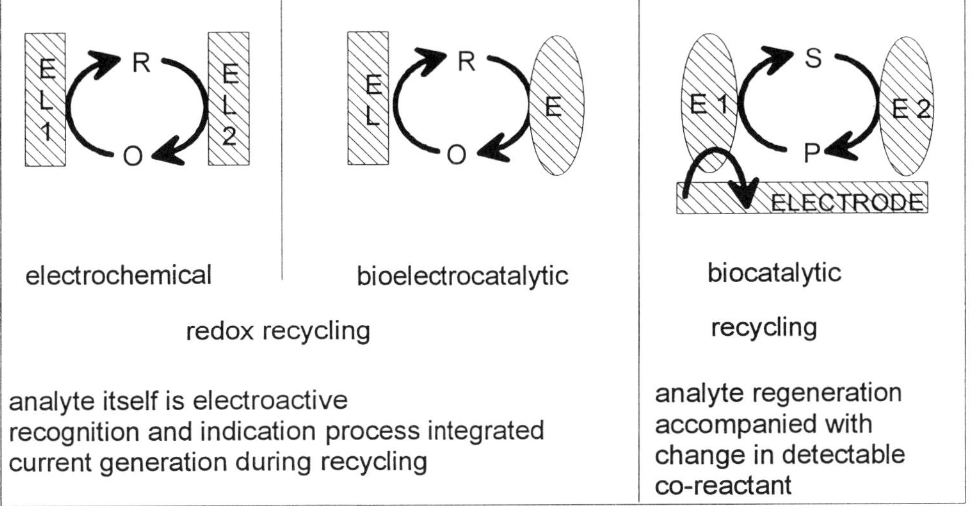

Figure 8.2 Principles of analyte recycling electrodes (with permission from *Frontiers in Biosensorics II. Practical Applications*, 1997, Basel: Birkhäuser Verlag, p. 46)

k, which represents the amplification, is only dependent on the apparent first order rate constants, k_i, of the enzymes used:

$$k = k_1 k_2 / (k_1 + k_2) \tag{8.1}$$

For enzymes in solution in some cases rates of 20,000 per hour can be obtained.

For enzyme electrodes, additionally, diffusion has to be considered. Under steady state conditions (Kulys *et al.* 1986) the amplification factor, *G*, can be expressed as:

$$G = k_1 k_2 L^2 / (k_1 + k_2) D \tag{8.2}$$

where *L* is the membrane thickness, and *D* the diffusion coefficient. Obviously, at high activities of both enzymes immobilized into the enzyme layer with high characteristic diffusion time (L^2/D), the possible amplification is very large, too. The influence of the enzyme loading on the sensor's sensitivity was exemplified for the analyte phenol using the tyrosinase/glucose dehydrogenase electrode (Figure 8.3).

When one molecule of product is formed per substrate molecule the total concentration of intermediate substrates remains constant and the concentrations of the co-reactants increase or decrease linearly with time. Then the number of cycles in which the substrate is turned over in a given time is a function of the substrate concentration.

The concept of linear enzymatic signal amplification has been realized by coupling dehydrogenases with oxidases or transaminases, phosphorylases and phophatases, or by coupling kinases with each other (Table 8.1). When oxidases are

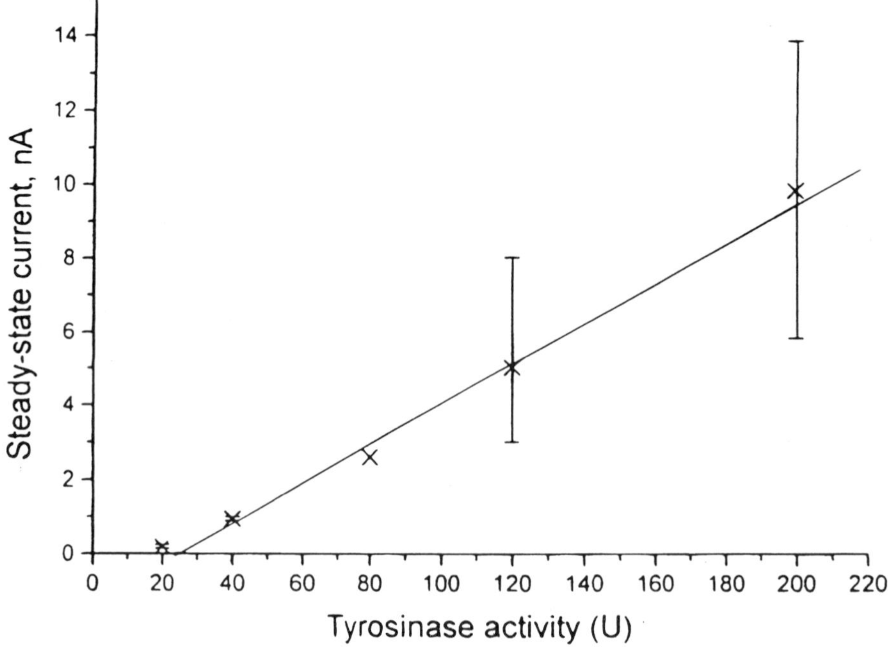

Figure 8.3 Dependence of the GDH/tyrosinase electrode response on the tyrosinase activity for 1 μM tyrosine at 30 U GDH/cm² coimmobilized in PVA (with permission from *J. Chem. Tech. Biotech.* (1996), 65, p. 41)

Table 8.1 Enzymatic substrate recycling electrodes

Analyte	Enzyme couple	Transducer	Amplification	Comment	Reference
ADP	myokinase/pyruvate kinase	oxygen electrode with pyruvate oxidase	800	exponential cycle	Pfeiffer *et al.* 1994
ADP/ATP	hexokinase/pyruvate kinase	oxygen electrode with lactate DH/lactate monooxygenase modified carbon electrode with glucose-6-phosphate DH	220 1200	with intermediate accumulation	Wollenberger *et al.* 1987b Yang *et al.* 1991
Adrenaline *p*-aminophenol	oligosaccharide DH/laccase	oxygen electrode	3000	for enzyme immunoassay	Bier *et al.* 1996a
Adrenaline phenol derivatives	(PQQ)glucose DH/tyrosinase	oxygen electrode	100–1000	determination of catecholamines phenols in water alkaline phosphatase in ELISA[1]	Makower *et al.* 1995
Adrenaline, *p*-aminophenol	(PQQ)glucose DH/laccase	oxygen electrode potentiometric redox electrode antimony pH electrode pH-FET	5000 10,000 200 1000	also ferrocenes, for redox label and enzyme[2] immunoassay potentiometric bioelectrocatalytic detection	Ghindilis *et al.* 1995b Ghindilis *et al.* 1995a Ghindilis *et al.* 1995c Eremenko *et al.* 1995b
Benzoquinone/ hydroquinone	fructose DH/laccase	oxygen electrode	700		Wollenberger unpublished
Benzoquinone /hydroquinone	cytochrome b2/laccase	oxygen electrode	500		Scheller *et al.* 1992

continued

Table 8.1 continued

Analyte	Enzyme couple	Transducer	Amplification	Comment	Reference
Ethanol	alcohol oxidase/ alcohol DH	oxygen electrode			Hopkins 1985
Glucose	glucose oxidase/ glucose DH	oxygen electrode	10		Schubert *et al.* 1985
Glutamate	glutamate DH/alanine aminotransferase	modified carbon electrode	15		Schubert *et al.* 1986
Glutamate	glutamate oxidase/ glutamate DH	oxygen electrode	20		Wollenberger *et al.* 1989
Glutamate	glutamate oxidase/ alanine aminotransferase	hydrogen peroxide electrode	500		Yao *et al.* 1989
L-Leucine	leucine DH/amino acid oxidase	oxygen electrode	40		Scheller *et al.* 1990
Lactate/pyruvate	cytochrome b2/ lactate DH	Pt-electrode glassy carbon electrode	10		Schubert *et al.* 1985 Vidziunaite and Kulys 1985
Lactate/pyruvate	lactate oxidase/ lactate DH	oxygen electrode	48,000 4100 250	used for enzyme immunoassay[3] sialinic acid determination[4]	Scheller *et al.* 1992 Wollenberger *et al.* 1987a Mizutani *et al.* 1985
		Pt-ring electrode	10^9	rotating dual reactor electrode	Raba and Mottola 1994
Malate/oxalacetate	malate DH/lactate monooxygenase	oxygen electrode	3		Scheller *et al.* 1988
NADH/NAD$^+$	peroxidase/glucose DH	oxygen electrode	60		Schubert *et al.* 1985

NADH/NAD[+]	diaphorase/glycerol DH	glassy carbon electrode	800–1200		Tang and Johannson 1995
NADH/NAD[+]	NADH oxidase/alcohol dehydrogenase	oxygen electrode	100	enzyme pair used in conjunction with H_2O_2 detector for enzyme immuno assays[5]	Mizutani *et al.* 1993
NADPH/NADP+	NADPH oxidase/g lucose-6-phosphate DH	oxygen electrode	50		*ibid.*
Peptides (containing tyrosine)	(PQQ)glucose DH/ tyrosinase	oxygen electrode			Eremenko *et al.* 1996
Phosphate	nucleoside phosphorylase/ alkaline phosphatase	oxygen electrode with xanthine oxidase	20		Wollenberger *et al.* 1992
Phosphate	maltose phosphorylase/ acid phosphatase	hydrogen peroxide electrode with glucose oxidase	15		Conrath *et al.* 1995

Notes
1 Bauer *et al.* 1996
2 Scheller *et al.* 1995
3 Makower *et al.* 1994
4 Pfeiffer *et al.* 1994
5 Athey and McNeil 1994

involved, electrode detectable species are included in the reaction scheme. Therefore, the change of co-reactant concentration can be measured directly at the electrode onto which the recycling enzyme pair is immobilized. As can be seen in Table 8.1, in most cases oxygen consumption has been followed, but also potential and pH changes have been indicated. The recycling systems based on kinases and phosphorylases/phosphatase have been combined with indicator enzymes in order to form an electrochemically detectable species.

A tremendous signal amplification is expected if, in the cycling reaction, more than one analyte molecule is regenerated. Here the total amount of intermediates and by-products is increasing exponentially with time. Theoretical considerations have shown that the concentration of any of the cycling intermediates or byproducts at any given time is a linear function of the initial substrate concentration (Kopp and Miech 1972). An example illustrating this principle is the ADP/ATP cycling system using myokinase/pyruvate kinase (Pfeiffer *et al.* 1994). With an optimized sensor configuration, an increase of sensitivity for ADP by a factor of 800 was obtained, resulting in a measuring range between 50 nmol/l and 2 µmol/l.

For those sensors where high activity of both enzymes could be applied an amplification by several orders of magnitude has been realized. When dealing with extremely high amplification one has to bear in mind, however, that the sensor signal becomes highly susceptible to minute amounts of contaminants affecting the enzyme reactions.

So far, the highest amplification was obtained for pyruvate and lactate determination by using the lactate oxidase/lactate dehydrogenase pair. The oxygen consumption in the gelatin membrane bearing both enzymes was enhanced by a factor of up to 4100 (Wollenberger *et al.* 1987a). This value is reasonable using the characteristic diffusion time of 90 s from the enzyme loading test in Equation 8.2. With this sensor lactate concentrations as low as 1 nmol/l could be determined with acceptable precision. When the immobilization was performed with polyurethane, which permits higher enzyme loading up to 48,000-fold enhancement for the lactate response was achieved (Scheller *et al.* 1992). However, the reproducibility and stability of this large amplification is poor. The amplification of the lactate response decreases with progressive enzyme inactivation during sensor operation, while the un-amplified lactate signal (using only lactate oxidase) remains stable (Figure 8.4). This behavior is due to the fact that the enzyme excess maintains diffusion control of the sensor for the simple process. Amplification can be tuned by limiting the co-substrate. Thus, employing the recycling sensor 1–2 orders above the possible detection limit, a more stable response is obtained, allowing for the combination with a displacement enzyme immunoassay for the determination of cocaine (Makower *et al.* 1994).

Sensors with high sensitivity can be also constructed on the basis of phenol oxidizing enzymes, which are currently available with high catalytic activity. For example phenol oxidases, i.e. laccase and tyrosinase, which oxidize a wide range of substances including catecholamines, phenols, and redox dyes by dissolved oxygen (Peter and Wollenberger 1995), have been used in combination with heme and pyrrolo-quinoline quinone (PQQ) containing (NAD(P)H independent) dehydrogenases. The highest sensitivity was obtained for aminophenol (subnanomolar concentration) when quinoprotein glucose dehydrogenase from *Acinetobacter calcoaceticus* and laccase from *Coriolus hirsutus* were co-entrapped (Ghindilis *et al.* 1995b). Because of its group specificity, the electrode can also be used for the detec-

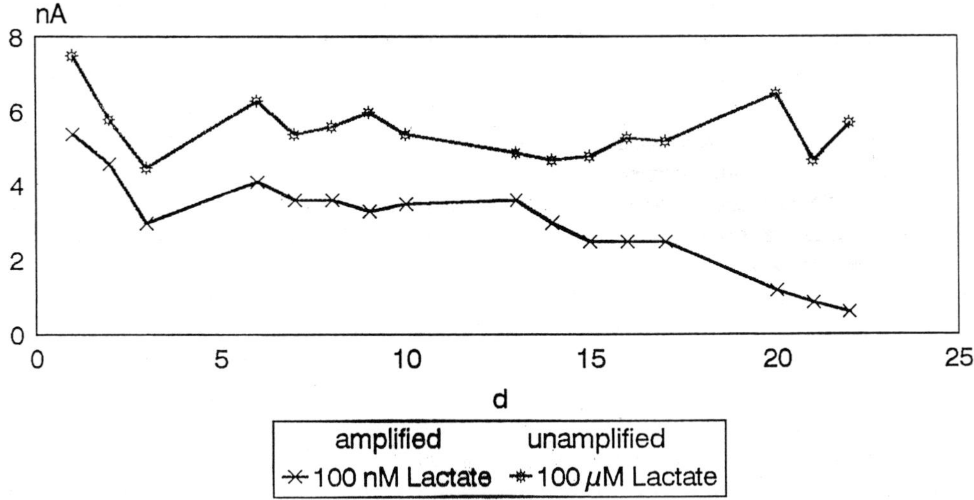

Figure 8.4 Working stability of the unamplified and amplified signal for lactate of the LOD/LDH recycling electrode

tion of low amounts of redox dyes (Ghindilis *et al.* 1995a), which is of particular interest for the development of immunoassays (see below, page 217).

Similar sensitivity is obtained when mushroom tyrosinase substitutes laccase. The change in specificity makes this arrangement suitable for sensitive determination of phenols in aqueous samples (Makower *et al.* 1995). Other acceptor dependent dehydrogenases such as fructose dehydrogenase and oligosaccharide:acceptor oxidase (oligosaccharide dehydrogenase) have been employed in place of the quinoprotein glucose dehydrogenase. When using the latter enzyme a rather broad spectrum of substances, such as aminophenols, diamines, and catecholamines is accessible to sensitive detection, but with different sensitivities as compared with the laccase–glucose dehydrogenase couple (Bier *et al.* 1996a).

Recycling systems are not necessarily limited to reactions in which electrochemically active compounds are involved. In these other cases the recycling enzyme pair is combined with an indicator enzyme (or sequence) transforming one of the cycle co-reactants (mostly a product) into a measurable species. Owing to their usually favorable equilibrium constants, kinase reactions are well applicable to such recycling experiments (Wollenberger *et al.* 1987b; Yang *et al.* 1991).

Mono-enzymatic cycles, where one enzyme is catalyzing forward and backward reaction when the respective reagents are present (Schubert *et al.* 1999a) might facilitate the design of novel, simple enzymatic amplification schemes. On the other hand, the multiplication of the amplification by a second recycling enzyme pair in a double amplification system may yield a further increase in sensitivity. This has already been accomplished with pyruvate kinase/hexokinase-lactate dehydrogenase/lactate oxidase.

Further progress is expected from engineering appropriate recognition and signal generating systems, such as hybrid enzymes generated either by chemical "site-to-site" linkage of interacting enzymes or by gene fusion. In this way the biochemical

signal may be effectively channeled, which may lead to high sensitivity and specificity.

Bioelectrocatalytic substrate recycling

One enzyme of the pair may be substituted by a redox electrode. Then the enzymatic reaction is followed by an electrochemical regeneration (or vice versa) resulting in a catalytic current, which is the measured signal. As has been mentioned
above, this type of substrate recycling is limited to reversible redox species.

Already, in 1984, the enhancement by laccase of the sensor response for nor-
adrenaline, hydroquinone, and *p*-phenylene diamine had been reported (Wasa *et al.*
1984). Since then a number of other enzymes have been investigated, which belong
in the great majority of cases to the family of copper oxidases, but flavoenzymes,
heme or PQQ containing enzymes have also been used (Table 8.2).

Effective bioelectrocatalytic recycling requires (i) high catalytic activity, (ii) sufficient (co)substrate of the enzyme is present (in order not to limit the reaction), and
(iii) the analyte is stable in both redox states. To avoid electrochemical interferants
the potential for the regeneration is low.

Vital for a rapid heterogeneous electron transfer is the close contact of enzyme
and electrode material. Therefore surface immobilization using adsorption, covalent
binding, and entrapment of the redox enzyme, and bulk modification procedures
have been established, the latter appearing to be the most effective way. Improvement results from the additional integration of modifier and promoter, which binds
the protein to the electrode surface and, while not itself taking part in the electron
transfer process, encourages electron transfer with the protein to proceed (Gorton
1995). Recent developments profit from modern fabrication technologies by combining enzymes with screen printed electrodes (Wang and Chen 1995b) for remote
sensors and microfabricated interdigitated gold electrodes for gas phase application
(Dennison *et al.* 1995).

The substances measured so far with the bioelectrocatalytic approach are redox
mediators, electroactive metabolites, phenolics and drugs. Marker enzymes liberating those compounds from electroinactive precursors (in the potential region investigated) may be quantified, too.

The highest sensitivity may be obtained when both the electrode and enzyme
reaction are very fast and the intermediate substances (oxidized and reduced form
of the analyte) are stable. As can be seen from Table 8.2, in most reported cases bioelectrocatalysis has been accomplished with carbonaceous electrode material. This
material efficiently facilitates fast redox conversion of quinones and related compounds. For example, the quinoprotein glucose dehydrogenase, tyrosinase and
laccase preparations are of very high enzyme activity. Therefore the combination of
these enzymes with carbonaceous electrodes results in ultra-sensitive sensors with
detection limits even below 1 nmol/l. At the same time the regeneration prevents
fouling of the electrode surface by polymeric products and therefore induces
stability. Remarkably short response times (a few seconds) are achieved when the
enzyme is covalently bond to an activated self-assembled monolayer on a gold
surface (Jin *et al.* 1995). Furthermore, this binding procedure prolonged considerably the working stability of quinoprotein glucose dehydrogenase modified sensors
as compared to bulk modified (Wollenberger *et al.* 1997) and polymer membrane

(Eremenko *et al.* 1995a) detectors. With intermittent PQQ incubations the electrodes could be used for more than four weeks.

Substituting the electrode by a reducing agent may also gain sensitivity due to a regeneration process (Macholan 1990; Uchiyama *et al.* 1993). For example, the addition of ascorbic acid to a phenol oxidase modified electrode improves the sensitivity by a factor of 300 (Uchiyama *et al.* 1993). The rate of the chemical regeneration, i.e. recycling efficiency, is lower than the electrode reaction and therefore the sensors are not as sensitive as the bioelectrocatalytic systems.

Sensitive detection of marker enzymes

Most of electrochemical immunoassays apply alkaline phosphatase (AP) as label (Kronkvist *et al.* 1997; Niwa *et al.* 1995; Duan and Meyerhoff 1994; La Gal La Salle *et al.* 1995; Ho *et al.* 1995; Bier *et al.* 1996b).

For measurement of AP, a standard method has been established. AP dephosphorylates *p*-aminophenyl posphate (PAPP) enzymatically to produce *p*-aminophenol (PAP), which is measured amperometrically at low potential with a glassy carbon electrode. This method has been suggested by Kulys *et al.* 1980, and its use in electrochemical immunoassay has been pioneered by Heineman and Halsall (Tang *et al.* 1988).

THE PRINCIPLE OF ALKALINE PHOSPHATASE MEASUREMENT

To detect minute amounts of AP, a bienzyme-substrate recycling biosensor based on tyrosinase and quinoprotein glucose dehydrogenase was operated in a flow injection analysis (FIA) system (Bauer *et al.* 1996). In this case phenyl phosphate was the substrate of AP. The stability and purity of the substrate phenyl phosphate, the stability of the product phenol, and the stability of the sensor response were examined.

The principle of AP-measurement consists of the combination of the enzymatic dephosphorylation of phenyl phosphate with a biosensor for the amplified measurement of phenol (Figure 8.5). Membrane-entrapped tyrosinase consumes oxygen while it oxidizes phenol via catechol to o-quinone. GDH reconverts o-quinone back to catechol (substrate cycling). The oxygen consumption in the presence of phenol is indicated by the current decrease at a Clark type electrode behind the enzyme membrane.

The lower detection limit (>3 SD) for phenol was 10 nM. The electrode response was linear between 20 and 1000 nM phenol. Signals after injections were obtained as a transient decrease in current that was 30 percent of the steady state response measured after some minutes of continuous input of the same sample. Since the flow system diluted the samples 1:1 to shift the buffer composition for the AP reaction to the conditions required by the biosensor.

Phenol detection in this concentration range and below has been reported with carbon electrode based analyte recycling systems. Ortega *et al.* (1993) reported 3 nM detection limit for phenol at a tyrosinase-modified graphite electrode. Kotte *et al.* (1995) reported 0.25 nM with a thick film carbon electrode modified with tyrosinase and a mediator. But for AP detection, not only high sensitivity is important, but also very good selectivity of the sensor for the enzyme product in the presence of a great

Table 8.2 Bioelectrocatalytic sensors for selected phenolic compounds[1]

Enzyme	Electrode material	Substrate[1]	Detection limit, nmol/l[1]	Comment	Reference
Cytochrome b$_2$	glassy carbon		500	gelatin membrane, +lactate	Scheller and Schubert 1989
Cellobiose dehydrogenase	graphite	hydroquinone/catechol	2 3.5	Adsorbed, +cellobiose	Lindgren *et al.* 1999
Diaphorase	glassy carbon	*p*-aminophenol	0.5	aP determination; diaphorase in solution, +NADH	Yamaguchi *et al.* 1992
Fructose dehydrogenase	glassy carbon			+ fructose	Ikeda *et al.* 1991
Glucose dehydrogenase. PQQ	glassy carbon	*p*-aminophenol	0.5–20	PVA, +glucose	Eremenko *et al.* 1995a
	carbon paste	*p*-aminophenol	0.5	bulk modification with PEI promoter, +glucose	Wollenberger *et al.* 1997
	carbon ink	dopamine	2	screen printed electrodes, +glucose	Lisdat *et al.* 1998
	gold	*p*-aminophenol	5.0	covalent multi-layer assembly, +glucose	Jin *et al.* 1995
Glucose oxidase	glassy carbon	hydroquinone/catechol	1.0	photocrosslinked PVA-stilbazolium, +glucose	Mizutani *et al.* 1991
	edge-plain pyrolytic graphite		100	PVA/glutardialdehyde + glucose	Moore *et al.* 1994
	graphite	acetaminophen		+ glucose	Ikeda, 1984
Oligosaccharide dehydrogenase	carbon paste	*p*-aminophenol	5	+ glucose	
	carbon ink	*p*-aminophenol	50	photopolymer, +glucose	
Peroxidase HRP	glassy carbon	2-amino-4-chlorophenol	20	adsorption, halogenated phenols + H$_2$O$_2$,	Ruzgas *et al.* 1995
	carbon paste			bulk modification. with lactitol	

Enzyme	Electrode	Substrate[1]	Detection limit	Remarks	Reference
Tobacco peroxidase	graphite	o-aminophenol	10	aromatic amines	Munteanu *et al.* 1998
Laccase	reticulated vitreous carbon	hydroquinone *p*-phenylene diamine	70	glutardialdehyde	Wasa *et al.* 1984
	pyrographite	hydroquinone		gelatin	Scheller *et al.* 1987
	glassy carbon	hydroquinone	600	AQ, measurements in organic solvents	Wang *et al.* 1993
	carbon paste	*p*-aminophenol	2		author
	carbon ink	*p*-aminophenol		photopolymer, screen printed	Lisdat *et al.* 1998
	carbon paste	catechol	10	bulk modification	author
	carbon based electrodes	catechol		coimmobilized tyrosinase	Yaropolov *et al.* 1995
Tyrosinase	carbon paste, glassy carbon, graphite	catechol/phenol	10/13	fixed on surface with dialysis membrane	Skladal 1991
	glassy carbon	phenol	100	AQ, organics Accumulation	Wang *et al.* 1993 Wang and Chen 1995a
	carbon paste	phenol	3	bulk modification with PEI promoter	Ortega *et al.* 1993
	carbon paste	phenol	10	bulk modification	author
	graphite	catechol	2	adsorption, carbodiimide	Ortega *et al.* 1994
	carbon paste	catechol/phenol	40/1000	graphite epoxy	Önnerfjord *et al.* 1995
	carbon paste	phenol	10	graphite modified with PTFE (gas permeable electrode)	Kaisheva *et al.* 1995
	carbon ink	catechol	100	thick-film el.	Wang and Chen 1995b
	carbon	phenol	1000	adsorbed, measurement in chloroform	Hall *et al.* 1988
	gold	phenol		in glycerol gas phase sensor	Dennison *et al.* 1995

Notes

1 Substrate which was measured with the highest sensitivity

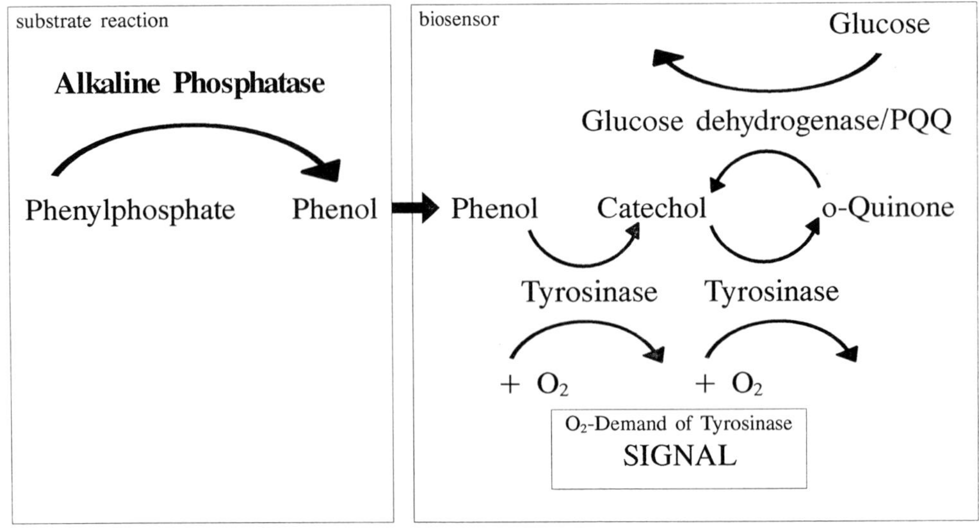

Figure 8.5 Schematic representation of the principle of AP measurement (with permission from *Anal. Chem.* (1996), 68, p. 2455)

excess of substrate is essential. A glassy carbon electrode covered with a tyrosinase/PVA membrane showed a low selectivity for phenol. The signal for phenyl phosphate was 1.6 percent of the signal for the same phenol concentration. Therefore this electrode-based approach would lead to a phenol detection system limited by responses to the AP-substrate phenyl phosphate.

The K_m of AP was determined with $36 \pm 3.9\,\mu$M phenyl phosphate. The turnover number k_{cat} was $47{,}500 \pm 1400\,$min^{-1}. The K_m is in the same range as for p-nitrophenol ($82 \pm 5\,\mu$M) and PAPP ($56 \pm 5\,\mu$M).

The sensor response to phenol and the stability of the substrate phenyl phosphate have been studied. 100 nM phenol and 0.1 mM phenyl phosphate were injected repeatedly over a period of 120 min, alternating every 5 min between the two samples. Phenol and phenyl phosphate were incubated in buffer at room temperature during the experiment.

The plot of electrode response vs time (Figure 8.6) indicates stability of the phenol signal (RSD = 1.77 percent, $n = 11$, 100 nM phenol). Phenol produced by the enzymatic reaction does not decompose during the time of measurement. This is an advantage compared to PAP.

The response to phenyl phosphate was nearly four orders of magnitude smaller than for an equal concentration of phenol and indicated an eighty-fold improvement in selectivity compared to recycling between glassy carbon electrode and tyrosinase. With a bienzyme analyte recycling electrode, a highly amplified response is obtained only if the analyte is converted rapidly by both enzymes. GDH is essential for the phenol-selectivity in the presence of phenyl phosphate, since electrochemical reactions of the substrate at the electrode surface can be avoided this way.

Phenyl phosphate showed good stability. The spontaneous decomposition was 0.1 percent in 2–3 h and is neglectable. No problems with phenol signals deviating from an earlier calibration curve have been found operating the FIA with the same substrate buffer for up to 10 h.

The purity and stability of reagents PAP and PAPP combined with selectivity of the sensor response have been the key problems in systems previously investigated for the measurement of AP based on GDH-modified glassy carbon electrodes or laccase-containing-amplification systems.

MEASUREMENT OF ALKALINE PHOSPHATASE

AP is measured by incubating for a certain time in substrate buffer taken from the buffer reservoir of the FIA, and then injected into the FIA-biosensor system. A signal–time plot from FIA for repeated injections of 100 fM AP in buffer (with 1 mg/ml BSA) shows a linear accumulation of product for at least 1 h.

A calibration curve for AP measured after an external 57.5 min incubation gave the lower detection limit of 320 zmol/100 μl AP. The calibration curve was linear up to 1000 fM.

The lower detection limit for AP can also be calculated from the data presented. The lower detection limit of phenol, the measured turnover number, the K_m and the substrate concentration give an AP detection limit per minute of 290 fM. This is comparable to the measured detection limits 92 fM min^{-1} and 184 fM min^{-1} (calibration curve for AP).

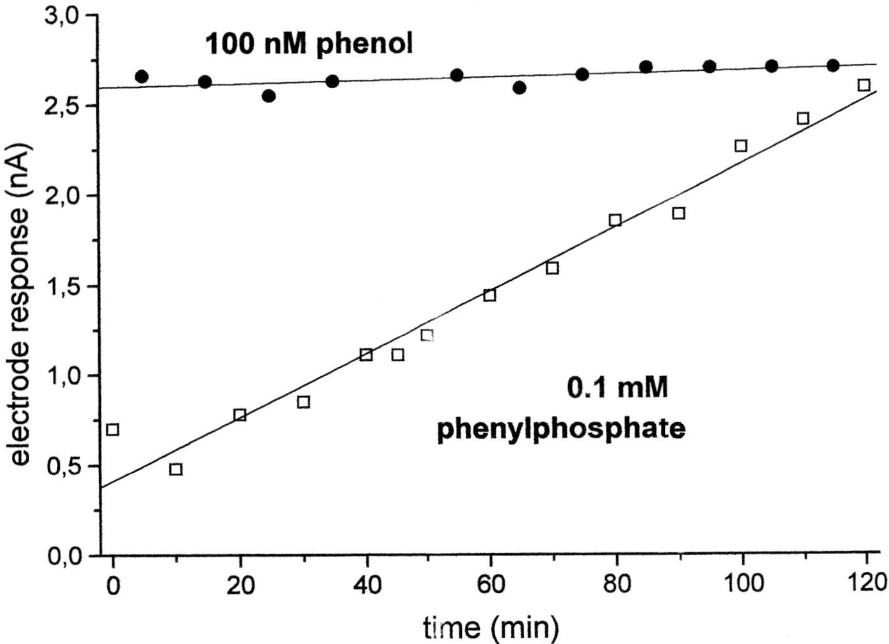

Figure 8.6 Stability of the response of the GDH/tyrosinase sensor to phenol and to a 1000-fold higher phenyl phosphate concentration (with permission from *Anal. Chem.* (1996), 68, p. 256)

Other highly sensitive electrochemical measurements of AP published recently show detection limits of $250\,fM\,min^{-1}$ (recalculated from 10 min incubation at 37 °C) for the recycling of hydroquinone and benzoquinone between glucose oxidase in solution and a carbon paste electrode at 350 mV vs Ag/AgCl (Della Ciana *et al.* 1995). The system requires degassed solutions. It has not been coupled to an immunoassay.

A calculated detection limit of $110\,fM\,min^{-1}$ AP (recalculated from 5 min incubation) has been reported by Ho *et al.* (1995). It is based on kinetic measurement of hydrogen peroxide released from a phosphate ester of an indolyl-phosphate derivative. The hydrogen peroxide has been measured by a peroxidase-modified carbon electrode at 0 mV vs Ag/AgCl. This measurement has been applied to an immunoassay performed directly on the electrode surface.

La Gal La Salle *et al.* (1995) detected $420\,fM\,min^{-1}$ AP (recalculated from a 15 min incubation at 37 °C). They used a ferrocene derivative that liberates a charged product by dephosphorylation. In a negatively charged membrane this product is accumulated while the substrate is excluded. Then they detect the ferrocene by square wave voltammetry. This system has also been applied to immunoassay detection.

The final evaluation of the bienzyme electrode involved its application as a detector for a competitive enzyme-linked immunosorbent assay or 2,4-D. The competitive immunoassay for 2,4-D uses an immobilized anti-2,4-D antibody in a microtiterplate and competition between 2,4-D and 2,4-D-AP conjugate (see below, page 223).

Immunoassays using enzyme recycling electrodes

Enzyme labeled immunoassays

During the last years an increasing number of electrochemical enzyme-immunoassays with alkaline phosphatase as label have been described.

Substrate cycling is a means well-suited to achieve high sensitivity for detection of marker enzymes. Here the product of the enzyme label is not only measured once, like PAP, but is reconverted to be measured again. There are analyte recycling systems known with electrochemical as well as optical detection. The sensitive measurement is not only dependent on a high amplification but also on stability and purity of the reagents.

The formation of NAD+ from NADP+ (Athey and McNeil 1994), pyruvate from phosphoenolpyruvate (Makower *et al.* 1994), phenol from phenol phosphate (Bauer *et al.* 1996) and aminophenol from aminophenylphosphate (Ghindilis *et al.* 1995a; Bier *et al.* 1996b) have been followed using enzymatic substrate recycling and electrodes (see Tables 8.1 and 8.2). Besides, β-galactosidase can also be employed; here an aminophenylated galactoside is hydrolyzed.

Determination of goat IgG and human thyroid stimulating hormone (hTSH) has been performed in a sandwich type immunoassay using an oligosaccharide dehydrogenase/laccase electrode (Bier *et al.* 1996b). In a first investigation, alkaline phosphatase has been used for the model compound IgG, and the liberation of *p*-aminophenol from *p*-aminophenyl phosphate was followed. This reaction, however, is known to suffer from drawbacks related to the limited stability of both *p*-aminophenyl phosphate and *p*-aminophenol in alkaline solution (Tang *et al.* 1988).

Therefore a large blank signal, which in some cases exceeds the dynamic measuring range of the electrode, is obtained and the incubation time is limited. Furthermore, the use of alkaline phosphatase requires a change of the pH between the immunoassay and the electrode reaction.

Therefore, β-galactosidase label was used. Since the optimum pH of β-galactosidase is close to that of the bi-enzyme sensor (pH 6.5) the whole assay could be performed under the same conditions. The amount of bound β-galactosidase has been traced after incubation with 1 mmol/l *p*-aminophenyl-β-D-galactoside (p-apgal) for 30 min. The sensitivity of the total assay was comparable to that of the photometric test. For the determination of hTSH, the sandwich-type assay has been performed using biotinylated tracer antibody and streptavidin-β-galactosidase conjugate. The measuring range extends from 0.0005 ng/ml to 20 ng/ml with a detection limit of 0.3 pg/ml (Figure 8.7).

Bauer *et al.* (1996) used a tyrosinase and quinoprotein glucose dehydrogenase covered oxygen electrode as fast readout of a competitive immunoassay for the herbicide 2,4-D involving an alkaline phosphatase label where phenyl phosphate is used as substrate for alkaline phosphatase and the liberated phenol is indicated. When phenyl phosphate is a permanent additive of the buffer, the measurement can be performed free of substrate blanks. The competitive immunoassay uses an immobilized antibody against 2,4-D in a microtiter plate. After competition between 2,4-D-alkaline phosphatase conjugates and 2,4-D for 1 h and a washing step, substrate buffer has been added and transferred to a biosensor equipped FIA after a further 2 min. This means a reduction of the incubation time from overnight, which is the time required for the standard optical method with *p*-nitrophenyl phosphate. The working range is ca. 0.1–10 μg/l. The results are in good agreement with the optical reference method (Figure 8.8).

Cocaine has been determined using conjugate displacement and a laccase/quinoprotein glucose dehydrogenase modified Clark-type oxygen electrode (Scheller *et al.* 1995). The displacement assay has been performed in two ways. Monoclonal antibodies against cocaine have been immobilized on a microtiter plate and cellulose beads (AvidGel AX) in a microcolumn saturated with alkaline phosphatase–cocaine conjugate. The displacement of the conjugate by cocaine is followed by indicating the enzyme in the supernatant (off-line) in the well or in the eluent of the column. Whereas a long incubation time permits 10 nmol/l–10 μmol/l cocaine to be detected, with the flow system the detection limit is about 200 nmol/l (100 pmole).

Flow immunoassays

Fast analysis by an enzyme-labeled immunoassay requires not only an excellent immunoassay, but also an ultrasensitive detection method for the enzyme label, to keep the substrate incubation time short. A good means to achieve this ultrasensitivity is signal amplification by substrate recycling (Tables 8.1 and 8.2).

A fast and sensitive screening method for the repetitive analysis of cocaine was developed. This method, amplified flow immunoassay (AFIA, Bauer *et al.* 1998), consisted of three steps: first the immunorecognition, where cocaine binds to the AP-labeled antibody (pAb-AP). Second, excess pAb-AP is removed by a cocaine-modified affinity column. Third, the cocaine-pAb-AP complex is detected in the column effluent. This AP-detection includes the enzymatic hydrolysis of phenyl

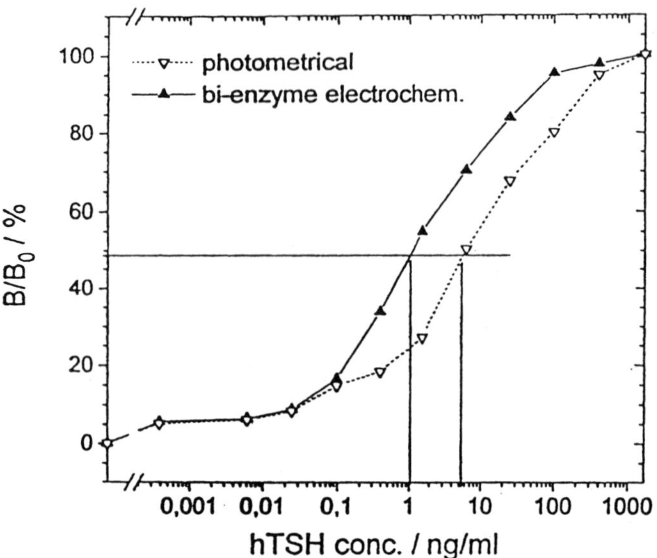

Figure 8.7 Immunoassay for the determination of human thyroid stimulating hormone (hTSH) using β-galactosidase as label and bienzyme amplified electrochemical detection (filled triangles) and comparison with photometrical determination (open triangles). The label substrate, ρ-apgal, was reduced to $10\,\mu\text{M}$. The c_{50}-value is shifted by a factor of 6 using the bienzyme electrode (with permission from *Anal. Chim. Acta.* (1996), 328, p. 31)

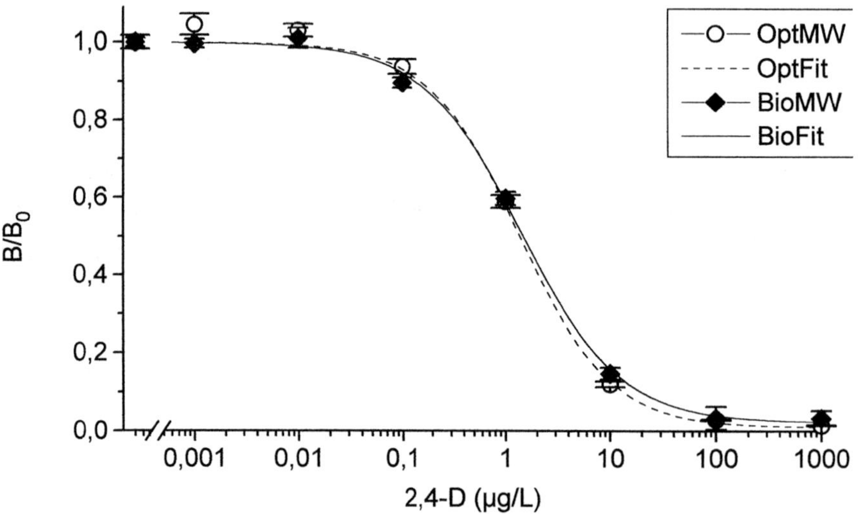

Figure 8.8 2,4-D concentration dependence measured by biosensor FIA (2 min incubation with phenyl phosphate) and optically (overnight incubation with ρ-nitrophenyl phosphate). The normalized mean (OptMW, BioMW) and the standard deviation of four-fold 2.4-D determinations for both methods are plotted. The data are fitted to a four-parameter function (OptFit, BioFit) (with permission from *Anal. Chem.* (1996), 68, p. 2457)

phosphate to phenol and the amplified detection of the generated phenol. The phenol-detector is a Clark-type oxygen electrode covered by an enzyme layer containing tyrosinase and quinoprotein glucose dehydrogenase.

MEASUREMENT OF COCAINE

The development of signal with the time for immuno-recognition was studied first. The time between stop and injection of the mixture of sample and reagent was taken as the time of immuno-recognition. The signal of AFIA increased with time. The reaction reaches equilibrium after 2 min. The immuno-recognition time has been set to 30 s in the following experiments (80 percent of the equilibrium signal).

Cocaine could be quantified between 0.38 and 3.2 nM (Figure 8.9). The lower detection limit 0.38 nM was calculated from the nonlinear fit (>3 SD, fourfold injections). The cocaine-calibration curve of AFIA is a typical IEMA calibration curve. The signal increased with increasing cocaine concentration. And the increase was sigmoidal as expected for bivalent binders. The standard deviations of linear regressions for each cocaine concentration were between 3.4 and 1.6 percent ($n = 8$).

The response time of AFIA was 75 s after injection of cocaine. This includes calculated contact times of 30 s for the immuno-recognition, 5 s for the affinity

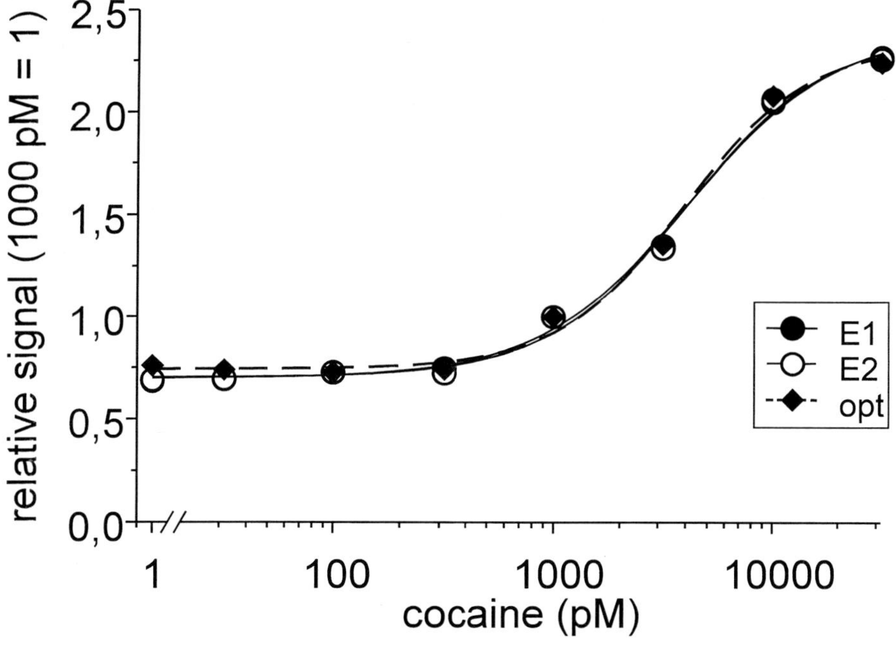

Figure 8.9 Comparison of optical and amperometrical detection of enzyme label alkaline phosphatase in a cocaine immunoassay. The results and nonlinear fit functions of two biosensors (X over open circles, straight lines, phenylphosphate) and an optical reference (solid diamonds, dashed line, ρ-nitrophenyphosphate) were normalized to baseline. The cocaine immunoassay was measured using conjugate pAb-AP (BK 227 1, 1:1000) (with permission from *Anal. Chem.* (1998), 70, p. 4628)

separation, 2 s for the substrate reaction and 38 s for the transport to the detector and the development of its maximal signal. This fast response time of the immunoassay was made possible (i) by a high affinity antiserum that allowed binding of the sample in less than two minutes, (ii) by a high speed and high capacity perfusion chromatography for removing the free antibodies, (iii) by an on-line detector for the enzyme label that responded after a few seconds of substrate incubation time, (iv) by an injection technique that used transport time for immuno-recognition and mixed sample and reagents before dilution by transport.

The performance of AFIA can be compared to a manually operated cocaine-immunoassay with off-line detection based on the same reagents (Eremenko *et al.* 1998). A preincubated mixture of cocaine and pAb-AP was injected there and the detection limit of 0.5 nM cocaine in this mixture was similar to that found here. The response time to cocaine was 1 hr there and decreased to 75 s for AFIA. This fifty-fold acceleration was achieved avoiding dilution by on-line detection and integrating the multistep manual protocol into one automated procedure with reduced times for immuno-recognition and label detection.

The optical detection of the enzyme label by *p*-nitrophenyl phosphate required incubation overnight. Both detection principles gave superimposable results.

The signal for repetitive cocaine measurements decreased 6.2 ± 0.2 percent per hour for all cocaine concentrations. The same detector could be used for continuous measurements over two days. The decrease in signal for the label AP, however, does not change the detection limit for cocaine.

AFIA could perform more than 200 consecutive cocaine assays without regeneration of the affinity column. Regeneration was not necessary, because the 50 μl affinity column contains thirteen-million-fold molar excess over the amount of pAb-AP injected per cocaine-assay. This is in good agreement with earlier findings that a column could be used for more than 2000 injections over six month without regeneration.

AFIA is forty times more sensitive than the displacement flow immunosensor for cocaine (Ogert *et al.* 1992). Both methods detect cocaine in similar times. But the lifetime of the displacement column is limited to 32 hours or 5–15 cocaine injections. AFIA is the most sensitive and one of the fastest immunoassays for cocaine.

Measurement of real samples

Two groups of samples were examined: samples of different origin that were not spiked with cocaine and cocaine-spiked samples of street dust. This cocaine-contaminated sample was measured as a 10,000-fold dilution and the result corresponds to approximately 3.4 μg collected cocaine.

The results for cocaine-spiked samples of street-dust were calculated from a line in a calibration-plot connecting the mean of the lower and higher cocaine standard in the adjacent calibration curves. The cocaine concentrations spiked could be quantitated reproducible (the coefficient of variation was 9.5 ± 2.5 percent, $n = 3$), but with a recovery of only 49 ± 3 percent for the various concentrations. The low recovery may be due to the fact that the method of sample preparation was developed for higher cocaine concentrations. Nevertheless AFIA could discriminate a suitcase surface touched with cocaine-contaminated hands from the non-contaminated suitcase.

Redox label immunoassay

The major drawback of immunoassays, and also most of the proposed immunosensors, is the necessity of multiple washing steps and prolonged incubation times of the label enzyme making immunoassays time consuming. Immunosensors ideally link an immunoassay directly to the transducer, which produces a signal from the binding event without any further processing and washing. However, most approaches to immunosensor development still need addition of reagents or incorporate washing steps. It is therefore a subject of current investigations to facilitate reagentless measurements by homogenous or pseudo-homogenous assay formats.

The concept of redox-label overcomes some of these limitations. This concept was first introduced by Green (1987) who proposed the use of ferrocene as a label in combination with glucose oxidase modified electrodes for transduction. A redox-label is a bi-functional molecule, which is well recognized by an antibody as a hapten on one hand, and on the other hand it is also electrochemically active at an electrode or the substrate of a redox-enzyme or enzyme system. These two functionalities of the molecule can be realized in one structure, i.e. the hapten is redox-active by itself, or the hapten has to be conjugated to a redox-active molecule forming a heterofunctional molecule with both functionalities in separate parts of the molecule.

The bi-enzyme cycle consisting of laccase or tyrosinase (TYR) as oxidases and oligosaccharide dehydrogenase (ODH) or NADPH independent glucose dehydrogenase (GDH) as reductases allows for the highly sensitive determination of phenolic compounds and ferrocene. The lower limit of detection approaches the concentration region of conventional immunoassay. Thus no signal amplification by an enzyme label is required but the concentration of a redox labeled hapten can be evaluated by these sensors.

The ability to detect nanomolar concentration of ferrocene derivatives with a quinoprotein glucose dehydrogenase/laccase sensor allows for the use of a ferrocene tracer in an immunoassay for low molecular weight antigens. When an antibody is bound to the ferrocene-labeled hapten the redox label can not penetrate the dialysis membrane and is therefore not measurable. The basis of the immunoassay which can be performed with this redox label is a competition between analyte hapten and ferrocene-hapten conjugate for the antibodies and a signal generation proportional to the analyte concentration.

A model immunoassay was developed for antibodies against a ferrocene benzoic acid isothiocyanate conjugate (fer-benz) (Ghindilis *et al.* 1995b). The conjugate is a good substrate for laccase and easily regenerated by quinoprotein glucose dehydrogenase; the lower detection limit is 0.5 nmol/l. After an external preincubation of fer-benz with antibody for 15 min the solution was transferred to the measuring cell. The response reflected the concentration of unbound conjugate, which is inversely proportional to the antibody concentration. The effect of antibody on the electrode response was examined without preincubation as well. In this case the antibody solution was injected directly into the batch and the response recorded for 10 nmol/l conjugate. Antibody without conjugate did not cause any response.

The results indicate that ferrocene derivatives can be used to label haptens, and the combination with a glucose dehydrogenase/laccase sensor is promising for the development of homogeneous immunoassays in the nanomolar concentration range.

Eremenko *et al.* (1997) found that the cycling system also reacts on peptides containing tyrosine. This stimulated Bier *et al.* (1997) to investigate the redox-label approach using tyrosine as redox-active functionality.

In this study a heterofunctional molecule with two distinguished functionalities: 2,4-dichlorophenoxyacetic acid-tyrosine (2,4-D-Tyr) is applied. 2,4-D is the hapten recognized by a monoclonal antibody, and tyrosine is the substrate of the bi-enzyme electrode.

The immunosensor was constructed by sandwiching the antibody loaded membrane and the bi-enzyme-membrane between a dialysis membrane and a polyethylene support in front of a Pt vs Ag/AgCl electrode. A schematic picture of the system is given in Figure 8.10.

The stirred cell contains two electrodes, one loaded with antibody and bi-enzyme membrane, the other only with bi-enzyme membranes as a reference.

The immunoassay on top of the electrode has the format of a displacement assay. The electrode is loaded by the redox-label, which is captured by the immobilized antibody; the analyte, 2,4,-D, displaces the redox-label, which is now registered by the bi-enzyme cycling system via oxygen consumption. Both functionalities of the redox-label have been investigated independently.

2,4-D-Tyr was treated as an analyte in a competitive immunoassay using β-GAL as enzyme label. Anti-2,4-D-antibodies were immobilized in a microtiter plate and 2,4-D or 2,4-D-Tyr compete against 2,4-D-β-GAL for the antibody binding sites. The typical sigmoidal curve for the competitive assay is shifted due to the lower affinity of the antibody to the redox-label as compared to 2,4-D (Figure 8.11). The comparison of the Ic_{50} values gives 10 percent of the affinity of the antibody to the redox-label compared to the analyte. This is a necessary condition for a displacement type immunoassay.

The ODH/tyrosinase electrode responds best to phenol and several catecholamines. It also responds to tyrosine and tyrosine containing peptides (Eremenko *et al.* 1997) as well as to 2,4-D-Tyr. Calibration curves for tyrosine and 2,4-D-Tyr are given in Figure 8.12, no response was observed for 2.4-D alone. Also no inhibitory effect of 2,4-D to one of the enzymes of the bi-enzyme cycle was determined. The lower limit of detection for tyrosine was at 50 nM, for 2,4-D-Tyr it was found at 80 nM.

The antibody layer of the immunoelectrode was incubated for at least 2 h at room temperature with 1 mM 2,4-D-Tyr. After extensive washing with buffer (without glucose) the base-line was stable indicating no measurable loss of the redox-label.

The immunoelectrode responds to the addition of 2,4-D down to the sub μM range. Figure 8.13 shows an original recording of the response on the addition of 2,4-D. The reference electrode which was covered by a bi-enzyme membrane alone gives no response at all on the addition of 2,4-D. For the immunoelectrode a calculated detection limit of 25 nM (S/R = 3) was determined.

Depending on the absolute amount of analyte, i.e. either after several measurements of low concentrations or one measurement of higher concentration, reloading becomes necessary. It was possible to reload the membrane with loss of sensitivity of 30 percent as is also shown in Figure 8.4.

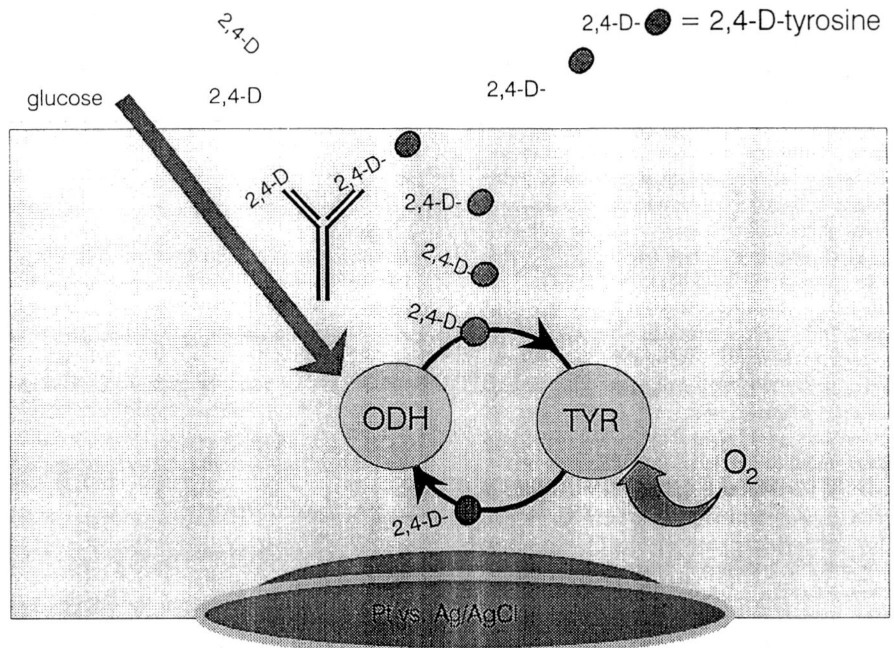

Figure 8.10 Scheme of the redox labeled immunosensor for 2,4 D (with permission from *Anal. Chim. Acta.* (1997), 344, p. 121)

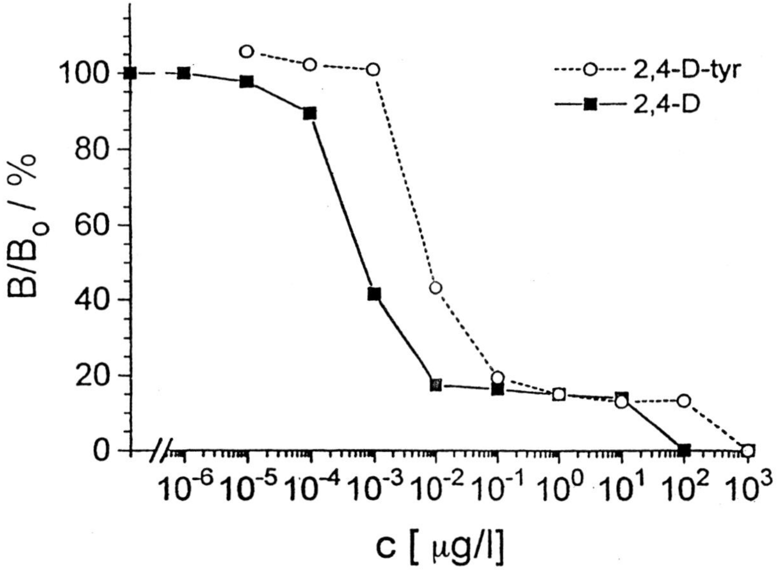

Figure 8.11 Competitive immunoassay for 2.4-D and 2,4-D-tyrosine using 2,4-D-β-GAL as enzyme label, determination was performed photometrically using nitrophenyl-β-D-galactopyranoside as substrate, measured at 405 nm (with permission from *Anal. Chim. Acta.* (1997), 344, p. 122)

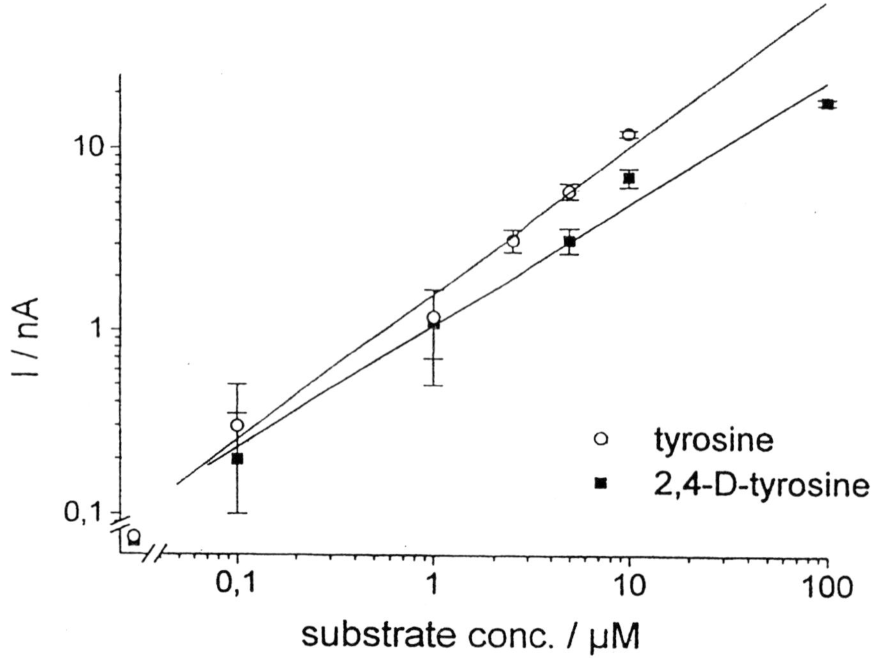

Figure 8.12 Calibration curve of the ODH/tyrosinase cycling electrode for tyrosine and 2.4-D-tyrosine (with permission from *Anal. Chim. Acta.* (1997), 344, p. 122)

Conclusions

The literature provides examples of fast-flow injection immunoassays (FIIA) for haptens. To illustrate the performance of the electrochemical immunoassays discussed above they are compared to a variety of detection principles that allow us to quantify a hapten in 15 min or less (Figure 8.14). These assays evaluate changes in fluorescence (Ogert *et al.* 1992; Reeves *et al.* 1994; Palmer *et al.* 1995; Schmalzing *et al.* 1995; Oosterkamp *et al.* 1996; Palmer *et al.* 1996; Narang *et al.* 1997), fluorescence polarization (Gaikwad *et al.* 1993; Matveeva, *et al.* 1997), surface plasmon resonance (Sakai *et al.* 1997; Bauer *et al.* 1998), and current (Dzantiev 1996; Bier *et al.* 1997; Kronkvist *et al.* 1997; Wilmer 1997; Bauer *et al.* 1998).

They are fast compared to microtiterplate immunoassays. Most of the FIIA use competitive (Reeves *et al.* 1994; Palmer and Miller, 1995; Osipov *et al.* 1996; Chiem and Harrison 1997; Wilmer *et al.* 1997), displacement (Ogert *et al.* 1992; Yu *et al.* 1996; Kronkvist *et al.* 1997; Narang *et al.* 1997) or noncompetitive formats.

Competitive FIIA requires at least 5 min total assay time for analyte concentrations of less than 10 nM, for subnanomolar concentrations about 10 min are needed (Osipov *et al.* 1996; Wilmer *et al.* 1997). Competitive assays (triangles) are most sensitive when the analyte-antibody reaction is allowed to proceed to equilibrium. There are a number of sensitive assays that allow fast sampling rates (Kaneki *et al.* 1994; Schmalzing *et al.* 1995; Locascio-Brown *et al.* 1996; Chiem and Harrison 1997; Kronkvist *et al.* 1997), but much of the signal response time is spent waiting for the

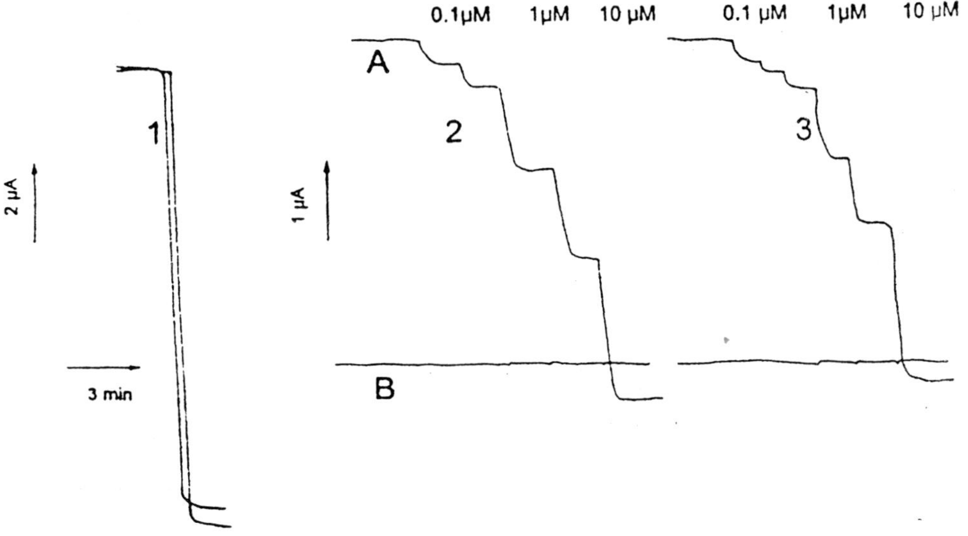

Figure 8.13 Electrode response of the immunoelectrode (A) and a reference electrode without immobilized antibodies (B) to 100 nM phenol (1) and various concentrations of 2,4-D after the first (2) and second loading (3) and 2,4-D-tyrosine (with permission from *Anal. Chim. Acta.* (1997), 344, p. 122)

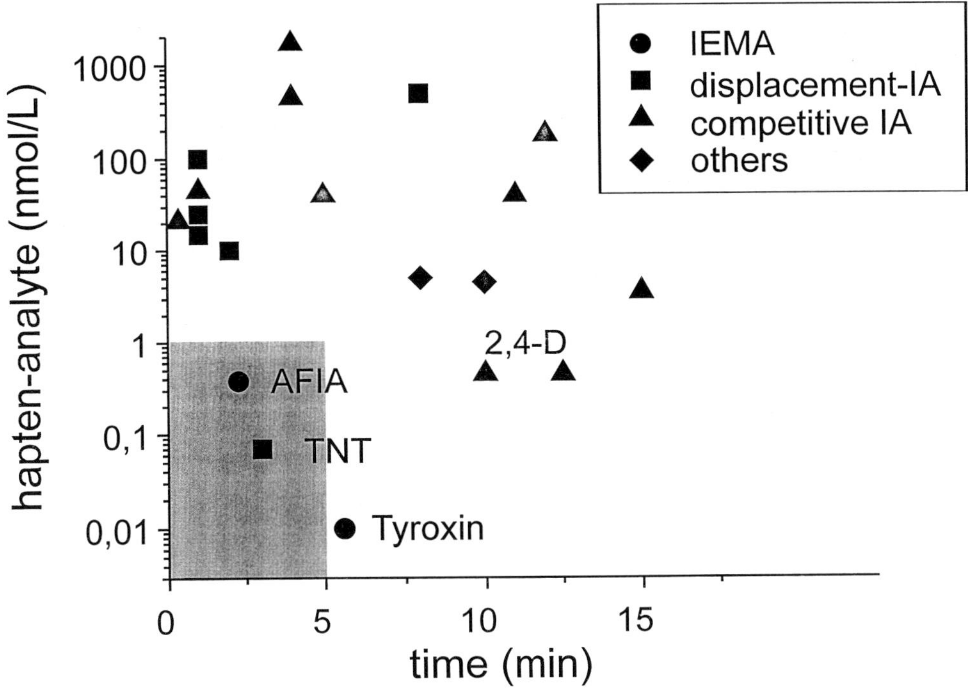

Figure 8.14 Lower limit of detection and measuring time for fast flow immunoassays

binding of analyte by a limited amount of antibodies. An additional drawback of a typical competitive assay is the time-consuming task to regenerate the column between each measurement. The harsh regeneration conditions can decrease the binding capacity and hence the lifetime of the column (Nilsson *et al.* 1991; Nilsson *et al.* 1992). The regeneration of the column can be avoided by using postcolumn detection of the nonbound fraction of the label, this also accelerates (4 min analysis time) and stabilizes the assay (Nilsson *et al.* 1991; Nilsson *et al.* 1992; Palmer *et al.* 1995; Wilmer *et al.* 1997). Shorter signal response times of 1 min or less are only achieved detecting competitive binding in homogeneous solution (Gaikwad *et al.* 1993; Matveeva, *et al.* 1997).

Displacement assays (squares) are inherently faster than competitive assays. They can be as sensitive as competitive assays when antibodies with medium affinities are employed. This, in turn, severely limits the lifetime of the column (Ogert *et al.* 1992; Yu *et al.* 1996). The solution to resaturate the antibody-column with labeled hapten after each injection doubles the assay time (Kronkvist *et al.* 1997). Displacement assays as well as competitive or sandwich assays obviously show markedly increased sensitivity (0.07 nM TNT) when performed in capillaries (De Frutos *et al.* 1993; Kaneki *et al.* 1994; Narang *et al.* 1997).

Noncompetitive immunoassays (circles), which use an excess of labeled antibodies vs analyte, are more sensitive than competitive or displacement assays, because, theoretically, one molecule of label is provided for every molecule of analyte. Noncompetitive immunoassays rely exclusively on fast association reactions and, thus, achieve binding between analyte and antibodies faster than displacement or competitive assays where dissociation contributes to signal generation. Prior to label detection these noncompetitive immunoenzymometric assays (IEMA) use an affinity column with immobilized analyte to capture the excess labeled antibodies. The analyte-complex of the labeled antibodies passes the column. This affinity purification takes several minutes using conventional chromatography (Gunaratna and Wilson 1993). Perfusion chromatography used by Bauer *et al.* (1998) allows separations with ten-fold speed, maintaining high capacity and high resolution. Receptors or antibodies have also been used in noncompetitive assays as reagents in selective postcolumn detection in liquid chromatography (Oosterkamp *et al.* 1996).

In summary, noncompetitive immunoassays, like AFIA, are the immunoassay type of choice, when sensitivity and speed and multiple or continuous measurements are important. Enzymatic amplification is a very competent tool for speeding up and simplifying the whole procedure. The most sensitive of these assays achieve detection limits that are lower than for techniques based on masspectrometry.

Acknowledgement

Financial support from the European Community (ENV4-CT97–0476), the Deutsche Forschungsgemeinschaft (INK 16 B1-1), and the BMBF, Germany, is gratefully acknowledged.

References

Athey, D. and McNeil, C. (1994) Amplified electrochemical immunoassay for thyrotropin using thermophilic β-NADH oxidase, *J. Immunol. Meth.* 176, 153–162.

Bauer, C.G., Eremenko, A.V., Ehrentreich-Förster, E., Bier, F.F., Makower, A., Halsall, H.B., Heineman, W.R. and Scheller, F.W. (1996) Zeptomole-detecting biosensor for alkaline phosphatase in an electrochemical immunoassay for 2,4-dichlorophenoxyacetic acid, *Anal. Chem.* 68, 2453–2458.

Bauer, C.G., Eremenko, A.V., Kühn, A., Kürzinger, K., Makower, A. and Scheller, F.W. (1998) Automated amplified flow immunoassay for cocaine, *Anal. Chem.* 70, 4624–4630.

Bier, F.F., Ehrentreich-Förster, E., Bauer, Ch. and Scheller, F.W. (1996a) High sensitive competitive immunodetection of 2,4-dichlorophenoxyacetic acid using enzymatic amplification with electrochemical detection, *Fresenius J. Anal. Chem.* 354, 861–865.

Bier, F.F., Ehrentreich-Förster, E., Dolling, R., Eremenko, A.V. and Scheller, F.W. (1997) A redox-label immunosensor on basis of a bienzymeelectrode, *Anal. Chim. Acta* 344, 119–124.

Bier, F., Förster, E., Makower, A. and Scheller, F. (1996b) An enzymatic amplification cycle for highly sensitive immunoassay, *Anal. Chim. Acta.* 328, 27–32.

Burstedt, E. Nistor, C., Schagerlöf, U. and Emneus, J. (2000) An enzyme flow immunoassay that uses beta-galactosidase as the label and a cellobiose dehydrogenase biosensor as the label detector, *Anal. Chem.* 72, 4271–4177.

Chen, F.C., Lin, N.N., Kuo, J.S., Cheng, L.J., Chang, F.M. and Chia, L.G. (1994) Simultaneous measurement of plasma serontonin, catecholamines, and their metabolites by an in vitro microdialysis–microbore HPLC and a dual amperometric detector, *Electroanal.* 6, 871–877.

Chiem, N. and Harrison, J. (1997) Microchip-based capillary electrophoresis for immunoassays: analysis of monoclonal antibodies and theophyllin, *Anal. Chem.* 69, 373–378.

Ciolkowski, E.L., Maness, K.M., Cahill, P.S., Wightman, R.M., Evans, D.H., Fosset, B. and Amatore, C. (1994) Disproportionation during electrooxidation of catecholamines at carbon-fiber microelectrodes, *Anal. Chem.* 66, 3611–3617.

Conrath, N., Gründig, B., Hüwel, S. and Cammann, K. (1995) A novel enzyme sensor for the determination of inorganic phosphate, *Anal. Chim. Acta.* 309, 47–52.

De Frutos, M., Paliwal, S.K. and Regnier, F.E. (1993) Liquid chromatography based enzyme-amplified immunological assays in fused-silica capillaries at the zeptomole level, *Anal. Chem.* 65, 2159–2163.

Della Ciana, L., Bernacca, G., Bordin, F., Fenu, S. and Garetto, F. (1995) Highly sensitive amperometric measurement of alkaline phosphatase activity with glucose oxidase amplification, *J. Electroanal. Chem.* 381, 129–135.

Dennison, M.J., Hall, J.M. and Turner, A.P.F. (1995) Gas-phase microbiosensor for monitoring phenol vapor at ppb levels, *Anal. Chem.* 67, 3922–3927.

Duan, C. and Meyerhoff, M.E. (1994) Separation-free sandwich enzyme immunoassays using microporous gold electrodes and self-assembled monolayer/immobilized capture antibodies, *Anal. Chem.* 66,1369–1377.

Dzantiev, B., Yulaev, M.F., Sitdikov, R.A., Dimitrieva, N.M., Moreva, I.Y. and Zherdev, A.V. (1996) Electrochemical immunosensors for determination of the pesticides 2,4-dichlorophenoxyacetic and 2,4,5-trichlorophenoxyacetic acids, *Biosensors Bioelectronics* 1/2, 179–185.

Eremenko, A.F., Makower, A., Bauer, C., Kurochkin and Scheller, F.W. (1997) A bienzyme electrode for tyrosine containing peptides determination, *Electroanal.* 4.

Eremenko, A.F., Makower, A. and Scheller, F.W. (1995b) Measurement of nanomolar diphenols by substrate recycling coupled to a pH-sensitive electrode, *Fresenius J. Anal. Chem.* 351, 729–731.

Eremenko, A.F., Makower, A., Wen, J., Rüger, P. and Scheller, F.W. (1995a) Biosensor based on an enzyme modified electrode for highly-sensitive measurement of polyphenols, *Biosens. Bioelectr.* 10, 717–722.

Eremenko, A.V., Bauer, C.G., Makower, A., Kanne, B., Baumgarten, H. and Scheller, F.W. (1998) The development of a non-competitive immunoenzymometric assay of cocaine, *Anal. Chim. Acta.* 358, 5–13.

Ewing, A.G., Strein, T.G. and Lau, Y.Y. (1992) Analytical chemistry in microenvironments: single nerve cells, *Acc. Chem. Res.* 25, 440–447.

Gaikwad, A., Gomezhens, A. and Perezbendito, D. (1993) Estimation of opiate, urine by stopped flow fluoroimmunoassay, *Fresenius J. Anal. Chem.* 347, 450–453.

Ghindilis, A.L., Makower, A., Bauer, C.G., Bier, F.F. and Scheller, F.W. (1995a) Picomolar determination of *p*-aminophenol and catecholamines based on recycling enzyme amplification, *Anal. Chim. Acta.* 304, 25–31.

Ghindilis, A.L., Makower, A. and Scheller, F.W. (1995b) Nanomolar determination of ferrocene derivatives using a recycling enzyme electrode. Development of a redox label immunoassay, *Anal. Lett.* 28, 1–11.

Ghindilis, A.L., Makower, A. and Scheller, F.W. (1995c) A laccase-glucose dehydrogenase recycling-enzyme electrode based on potentiometric mediatorless electrocatalytic detection, *Anal. Meth. Instr.* 2, 129–132.

Gorton, L. (1995) Carbon paste electrodes modified with enzymes, tissues, and cells, *Electroanal.* 7, 23–45.

Green, M.J. (1987) Electrochemical immunoassays, *Phil. Trans. R. Soc. Lond. B* 316, 135–142.

Green, M.J., Barrance, D.J. and Hilditch, P.I. (1991) Electrochemical immunoassay. In: Price, C.P. and Newman, L.J. (eds) *Principles and Practice of Immunoassay*, Macmillan, pp. 482–514.

Gunaratna, P.C. and Wilson, G.S. (1993) Noncompetitive flow injection immunoassay for a hapten, α-(difluoromethyl)ornithine, *Anal. Chem.* 65, 152–157.

Hall, G.F., Best, D.A. and Turner, A.P.F. (1988) Amperometric enzyme electrode for the determination of phenols in chloroform, *Enzyme Microb. Technol.* 10, 543–546.

Ho, W.O., Athey, D. and McNeil, C. (1995) Amperometric detection of alkaline phosphatase activity at a horseradish peroxidase enzyme electrode based on activated carbon: potential application to electrochemical immunoassay, *Biosensors Bioelectronics.* 10, 683–691.

Hopkins, T.R. (1985) A multipurpose enzyme sensor based on alcohol oxidase, *Int. Biotech. Lab.* 3, 20–25.

Ikeda, T., Katasho, I., Kamei, M. and Senda, M. (1984) Electrocatalysis with glucose oxidase immobilized graphite electrode, *J. Electroanal. Chem.* 48, 1969–1979.

Ikeda, T., Matsushita, F. and Senda, M. (1991) Amperometric fructose sensor based on direct bioelectrocatalysis, *Biosens. Bioelectr.* 6, 299–304.

Jankowski, J.A., Tracht, S. and Sweedler, J.V. (1995) Assaying single cells with capillary electrophoresis, *TRAC* 14, 170–176.

Jin, W., Bier, F., Wollenberger, U. and Scheller, F.W. (1995) Construction and characterization of a multi-layer enzyme electrode: covalent binding of quinoprotein glucose dehydrogenase onto gold electrodes, *Biosens. Bioelectr.* 10, 823–829.

Kaisheva, A., Iliev, I., Kazareva, R., Christov, S., Petkova, J., Wollenberger, U. and Scheller, F.W. (1996) Enzyme/gas-diffusion electrodes for determination of phenol, *Sensors and Actuators* 33, 39–43.

Kaneki, N., Xu, Y., Kumari, A., Halsall, H.B., Heineman, W.R. and Kissinger P.T. (1994) Electrochemical enzyme immunoassay using sequential saturation technique in a 20-μl capillary: digoxin as a model analyte, *Anal. Chim. Acta.* 287, 253–258.

Kopp, L.E. and Miech, R.P. (1972) Nonlinear enzymatic cycling systems: the exponential cycling system, *J. Biol. Chem.* 247, 3558–3563.

Kotte, H., Gründig, B., Vorlop, K.-D., Strehlitz, B. and Stottmeister, U. (1995) Methylphenazonium-modified enzyme sensor based on polymer thick films for nanomolar detection of phenols, *Anal. Chem.* 67, 65–70.

Kronkvist, K., Lövgren, U., Edholm, L.E., Johansson, G. (1993) A new approach to competitive flow ELISA for steroids yielding and increase in sensitivity and sample throughput, PhD thesis, University of Lund, Sweden.

Kronkvist, K., Lövgren, U., Svenson, J., Edholm, L.E., Johansson G. (1997) Competitive flow injection enzyme immunoassay for steroids using a post-column reaction technique, *J. Immunological Methods* 200, 145–153.

Kulys, J.J., Razumas, V. and Malinauskas, A. (1980) Electrochemical oxidation of catechol and *p*-aminophenol esters in the presence of hydrolases, *Bioelectrochem. Bioenerg.* 7, 11–24.

Kulys, J.J., Sorochinskii, V.V. and Vidziunaite, R.A. (1986) Transient response of bienzyme electrodes, *Biosensors* 2, 135–146.

La Gal La Salle, A., Limoges, B. and Degrand, C. (1995) Enzyme immunoassays with an electrochemical detection method using alkaline phosphatase and a perfluorosulfonated ionomer-modified electrode. Application to phenytoin assays, *Anal. Chem.* 67, 1245–1253.

Lisdat, F., Ho, W.O., Wollenberger, U., Scheller, F.W., Richter, T. and Bilitewski, U. (1998) Recycling systems based on screen-printed electrodes, *Electroanalysis* 10, 803.

Liu, A.H. and Wang, E.K. (1994) Amperometric detection of catecholamines with liquid chromatography at a polypyrrole-phosphomolybdic anion-modified electrode, *Anal. Chim. Acta.* 296, 171–180.

Locascio-Brown, L., Martynova, L., Christensen, R.G. and Horvai, G. (1996) Flow immunoassay using solid-phase entrapment, *Anal. Chem.* 68, 1665–1670.

Lowry, O.H. and Passonneau, J.V. (1972) In: *A Flexible System of Enzymatic Analysis*, New York: Academic Press, pp. 129–145.

Macholan, L. (1990) Phenol-sensitive enzyme electrode with substrate recycling for quantification of certain inhibitory aromatic acids and thio compounds, *Coll. Czech. Chem. Commun.* 55, 2152–2159.

Makower, A., Bauer, C.G., Ghindilis, A. and Scheller, F.W. (1994) Sensitive detection of cocaine by combining conjugate displacement and an enzyme substrate recycling sensor. In: Turner, T. (ed.), *Biosensors 94, Proceedings of the 3rd World Congress on Biosensors*, Oxford: Elsevier Advanced Technology, Poster 3.27.

Makower, A., Eremenko, A.V., Streffer, K., Wollenberger, U. and Scheller, F.W. (1995) Tyrosinase-glucose dehydrogenase substrate recycling biosensor. Highly sensitive measurement of phenolic compounds, *J. Chem. Tech. Biotech.* 65, 39–44.

Matveeva, E.G., Aguilarcaballos, M.P., Eremin, S.A., Gomezhens, A. and Perezbendito, D. (1997) Use of stopped-flow fluoroimmunoassay in pesticide determination, *Analyst* 122, 863–866.

Matysik, F.M., Nagy, G. and Pungor, E. (1992) Analytical distinction between different catechols by means of reverse differential-pulse voltammetry, *Anal. Chim. Acta.* 264, 177–184.

Mizutani, F., Yabuki, S. and Asai, M. (1991) Highly-sensitive measurement of hydroquinone with an enzyme electrode, *Biosens. Bioelectr.* 6, 305–310.

Mizutani, F., Yabuki, S. and Katsura, T. (1993) Amperometric enzyme electrode with the use of dehydrogenase and NAD(P)H oxidase, *Sensors Actuators B* 13–14, 574–575.

Mizutani, F., Yamanaka, T., Tanabe, Y. and Tsuda, K. (1985) An enzyme electrode for L-lactate with a chemically amplified response, *Anal. Chim. Acta.* 177, 153–166.

Moore, T.J., Nam, G.G., Pipes and Coury, Jr., L.A. (1994) Chemically amplified voltammetric enzyme electrodes for oxidizable pharmaceuticals, *Anal. Chem.* 66, 3158–3163.

Munteanu, F.D., Lindgren, A., Emneus, J., Gorton, L., Ruzgas, T., Csoregi, E., Ciucu, A., van Huystee, R.B., Gazaryan, I.G. and Lagrimini, L.M. (1998) Bioelectrochemical monitoring of phenols and aromatic amines in flow injection using novel plant peroxidases, *Anal. Chem.* 70. 2596–2600.

Narang, U., Gauger, P.R. and Ligler, F.S. (1997) Capillary-based displacement flow immunosensor, *Anal. Chem.* 69, 1961–1964.

Nilsson, M., Håkansson, H. and Mattiasson, B. (1991) Flow-injection ELISA for process monitoring and control, *Anal. Chim. Acta.* 249, 163–168.

Nilsson, M., Håkansson, H. and Mattiasson, B. (1992) Evaluation of direct versus competitive assays in flow injection systems, *Process Control and Quality* 4, 37–45.

Niwa, O. (1995) Electroanalysis with interdigitated array microelectrodes, *Electroanal.* 7, 606–613.

Niwa, O., Morita, M. and Tabei, H. (1991) Highly sensitive and selective voltammetric detection of dopamine with vertically separated interdigitated array electrodes, *Electroanal.* 3, 163–168.

Niwa, O., Xu, Y., Halsall, H.B. and Heinemann, W.R. (1993) Small-volume voltammetric detection of 4-aminophenol with interdigitated array electrodes and its application to electrochemical enzyme immunoassay, *Anal. Chem.* 65, 1559–1563.

Ogert, R.A., Kusterbeck, A.W., Wemhoff, G.A., Burke, R. and Ligler, F.S. (1992) Detection of cocaine using flow immunosensors, *Anal. Letters* 25, 1999–2019.

Önnerfjord, P., Emneus, J., Marko-Varga, G. and Gorton, L. (1995) Tyrosinase graphite-epoxy based composite electrodes for detection of phenols, *Biosens. Bioelectr.* 10, 607–619.

Oosterkamp, A.J., Villaverde-Herraiz, M.T., Irth, H., Tjaden, U.R. and van der Greef, J. (1996) Reversed-phase liquid chromatography coupled on-line to receptor affinity detection based on the human estrogen receptor, *Anal. Chem.* 68, 1201–1206.

Ortega, F., Dominguez, E., Burestedt, E., Emneus, J., Gorton, L. and Marko-Varga, G. (1994) Phenol oxidase-based biosensors as selective detection units in column liquid chromatography for the determination of phenolic compounds, *J. Chromatogr.* 675, 65–78.

Ortega, F., Dominguez, E., Jonsson-Pettersson, G. and Gorton, L. (1993) Amperometric biosensor for the determination of phenolic compounds using a tyrosinase graphite electrode in a flow injection system, *J. Biotech.* 31, 289–300.

Osipov, A.P., Zaitseva, N.V. and Egorov, A.M. (1996) Chemiluminescent immunoenzyme biosensor with a thin-layer flow-through cell. Application for study of a real-time bimolecular antigen–antibody interaction, *Biosens. Bioelectr.* 11, 881–887.

Palmer, D.A., Evans, M., French, M. and Miller, J.N. (1996) Rapid fluorescence flow injection immunoassay using a novel perfusion chromatographic material, *Analyst* 119, 943–947.

Palmer, D.A. and Miller, J.N. (1995) Thiophilic gels: applications in flow-injection immunoassays for macromolecules and haptens, *Anal. Chim. Acta.* 303, 223–230.

Peter, M.G. and Wollenberger, U. (1997) Phenol-oxidizing enzymes: mechanisms and applications in biosensors. In: Scheller, F.W., Schubert, F. and Fedrowitz, J. (eds), *Frontier in Biosensorics I. Fundamental Aspects,* Basel: Birkhäuser Verlag, pp. 63–82.

Pfeiffer, D., Scheller, F.W., McNeil, C. and Schulmeister, T. (1994) Cascade like exponential substrate amplification in enzyme electrodes, *Biosens. Bioelectr.* 10, 169–180.

Pfeiffer, D., Scheller, F.W., Wollenberger, U., Makower, A., Bier, F.F., Szeponik, J., Klimes, N., Gajovic, N., Ghindilis, A. and Scholz, C. (in press) Enzyme sensors for the detection of subnano- and milimolar concentrations. Development and application of biosensors, *Analyst.*

Raba, J. and Mottola, H.A. (1994) On-line enzymatic amplification by substrate cycling in a dual bioreactor with rotation and amperometric detection, *Anal. Biochem.* 220, 297–302.

Razumas, V.J., Kulys, J.J. and Malinauskas, A.A. (1980) Kinetic amperometric determination of hydrolase activity, *Anal. Chim. Acta.* 117, 387–390.

Reeves, S.G., Rule, G.S., Roberts, M.A., Edwards, A.J. and Durst, R.A. (1994) Flow-injection liposome immunoanalysis (FILIA) for alachlor, *Talanta* 41, 1747–1753.

Rosen, I. and Rishpon, J. (1989) Alkaline phosphatase as a label for a heterogeneous immunoelectrochemical sensor, *J. Electroanal. Chem.* 258, 27–39.

Ruzgas, T., Emneus, J., Gorton, L. and Marko-Varga, G. (1995) The development of a peroxidase biosensor for monitoring phenol and related aromatic compounds, *Anal. Chim. Acta.* 311, 245–253.

Sakai, G., Ogata, K., Miura, N. and Yamazoe, N. (1997) Highly sensitive detection of morphine by using immunosensor based on surface-plasmonresonance, Paper presented at Transducers '97, the International Conference on Solid-State Sensors and Actuators, Chicago, June 16–19.

Scheller, F.W., Makower, A., Ghindilis, A.L., Bier, F.F., Förster, E., Wollenberger, U., Bauer, C.G., Micheel, B., Pfeiffer, D., Szeponik, J., Michael, N. and Kaden, H. (1995) Enzyme sensors for subnanomolar concentrations. In: Rogers, K.R., Mulchandani, A. and Zhou, W. (eds), *ACS Symposium Series 613, Biosensor and Chemical Sensor Technology*, Washington DC: American Chemical Society, pp. 70–81.

Scheller, F.W., Pfeiffer, D., Hintsche, R., Dransfeld, I. and Wollenberger, U. (1990) Analytical aspects of internal signal processing in biosensors, *Ann. NY Acad. Sci.* 613, 68–78.

Scheller, F.W. and Schubert, F. (1989) *Biosensoren*, 1st edn, Berlin: Akademie Verlag.

Scheller, F.W., Schubert, F., Pfeiffer, D., Wollenberger, U., Renneberg, R., Hintsche, R., Kühn, M. (1992) Fifteen years of biosensor research in Berlin-Buch, *GBF Monographs* 17, 3–10.

Scheller, F.W., Schubert, F., Weigelt, D., Mohr, P. and Wollenberger, U. (1988) Molecular recognition and signal processing in biosensors, *Makromolek. Chem.* 17, 429–439.

Scheller, F.W., Wollenberger, U., Schubert, F., Pfeiffer, D. and Bogdanovskaya, V.A. (1987) Amplification and switching by enzymes in biosensors, *GBF Monographs* 10, 39–49.

Schmalzing, D., Nashabeh, W., Yao, X.W., Mhatre, R., Regnier, F.E., Afeyan, N.B. and Fuchs, M. (1995) Capillary electrophoresis-based immunoassay for cortisol in serum, *Anal. Chem.* 67, 606–612.

Schubert, F., Kirstein, D., Scheller, F., Appelqvist, R., Gorton, L. and Johansson, G. (1986) Enzyme electrodes for L-glutamate using chemical redox mediators and enzymatic substrate amplification, *Anal. Lett.* 19, 1273–1288.

Schubert, F., Kirstein, D., Schröder, K.L. and Scheller, F.W. (1985) Enzyme electrodes with substrate and coenzyme amplification, *Anal. Chim. Acta.* 169, 391–396.

Schubert, F., Scheller, F.W. and Krasteva, N. (1990a) Lactate-dehydrogenase-based biosensors for glyoxylate and NADH determination: a novel principle of analyte recycling, *Electroanal.* 2, 347–351.

Schubert, F.W., Wollenberger, U., Scheller, F. and Müller, H.G. (1990b) Artificially coupled enzyme reactions with immobilized enzymes: biological analogs and technical consequences. In: Wise, D. (ed.), *Bioinstrumentation and Biosensors*, New York: Marcel Decker, pp. 19–39.

Skladal, P. (1991) Mushroom tyrosinase-modified carbon paste electrode as amperometric biosensor for phenols, *Coll. Czech. Chem. Comm.* 56, 1427–1433.

Tang, H.T., Lunte, C.F., Halsall, H.B. and Heineman, W.R. (1988) *p*-Aminophenyl phosphate: an improved substrate for electrochemical enzyme immunoassay, *Anal. Chim. Acta.* 214, 187–195.

Tang, X. and Johansson, G. (1995) Enzyme electrode for amplification of NAD+/NADH using glycerol dehydrogenase and diaphorase with amperometric detection, *Anal. Lett.* 28, 2595–2606.

Uchiyama, S., Hasebe, Y., Shimizu, H. and Ishihara, H. (1993) Enzyme-based catechol sensor based on the cyclic reaction between catechol and 1,2-benzoquinone, using L-ascorbate and tyrosinase, *Anal. Chim. Acta.* 276, 341–345.

Vidziunaite, R.A. and Kulys, J.J. (1985) Kinetic regularities of cyclic substrate conversion in enzyme membranes, (russ.) *Liet. TSR Mokslu Akad. darbai Ser.* C2, 84–91.

Wang, J., Lin, Y. and Chen, Q. (1993) Organic-phase biosensors based on entrapment of enzymes within poly(ester-sulfonic acid) coatings, *Electroanal.* 5, 23–28.

Wang, J. and Chen, Q. (1995a) Highly sensitive biosensing of phenolic compounds using bioaccumulation/chronoamperometry at a tyrosinase electrode, *Electroanal.*, 7, 746–749.

Wang, J. and Chen, Q. (1995b) Microfabricated phenol biosensors based on screen printing of tyrosinase containing carbon ink, *Anal. Lett.* 28, 1131–1142.

Wang, J., Lin, Y., Eremenko, A.V., Ghindilis, A.L. and Kurochkin, I.N. (1993) A laccase electrode for organic-phase enzymatic assays, *Anal. Lett.* 26, 197–207.

Wasa, T., Akimoto, K., Yao, T. and Murao, S. (1984) Development of laccase membrane electrode by using carbon electrode impregnated with epoxy resin and its response characteristics, *Nippon Kagaku Koishi* 9, 1398–1403.

Wightman, R.M., Finnegan, J.M. and Pihel, K. (1995) Monitoring catecholamines at single cells, *TRAC* 14, 154–158.

Wilmer, M., Trau, D., Renneberg, R. and Spener, F. (1997) Amperometric immunosensor for the detection of 2,4-dichlorophenoxyacetic acid (2,4-D) in water, *Anal. Letters* 30, 515–525.

Wollenberger, U., Paeschke, M. and Hintsche, R. (1994) Interdigitated array electrodes for the determination of enzyme activities, *Analyst* 119, 1245–1249.

Wollenberger, U., Neumann, B. and Scheller, F.W. (1997) Quinoprotein glucose dehydrogenase modified carbon paste electrodes for the detection of phenolic compounds, *Electroanalysis* 9, 366–371.

Wollenberger, U., Scheller, F.W., Pavlova, M., Müller, H.G., Risinger, L. and Gorton, L. (1989) Glutamate oxidase based biosensors, *GBF Monographs* 13, 33–36.

Wollenberger, U., Schubert, F. and Scheller, F.W. (1992) Biosensor for sensitive phosphate detection, *Sensors and Actuators B* 7, 412–415.

Wollenberger, U., Schubert, F., Pfeiffer, D. and Scheller, F.W. (1993) Enhancing biosensor performance using multienzyme systems, *TIBtech* 11, 255–262.

Wollenberger, U., Schubert, F., Scheller, F.W., Danielsson, B. and Mosbach, K. (1987a) Coupled reactions with immobilized enzymes in biosensors, *Studia Biophys.* 119, 167–170.

Wollenberger, U., Schubert, F., Scheller, F.W., Danielsson, B. and Mosbach, K. (1987b) A biosensor for ADP with internal substrate amplification, *Anal. Lett.* 20, 657–668.

Yamaguchi, S., Ozawa, S., Ikeda, T. and Senda, M. (1992) Sensitive amperometry of 4-aminophenol based on catalytic current involving enzymatic recycling with diaphorase and its application to alkaline phosphatase assay, *Anal. Sci.* 8, 87–88.

Yang, X., Pfeiffer, D., Johansson, G. and Scheller, F.W. (1991) Enzyme electrodes for ADP/ATP with enhanced sensitivity due to chemical amplification and intermediate accumulation, *Electroanal.* 3, 659–663.

Yao, T., Yamamoto, H. and Wasa, T. (1989) Chemical amplified enzyme electrodes by substrate recycling, *Proceedings ISE-Meeting, Kyoto*, Tokyo: Kodanasha, pp. 1014–1015.

Yaropolov, A.I., Kharybin, A.N., Emneus, J., Marko-Varga, G. and Gorton, L. (1995) Flow-injection analysis of phenols at a graphite electrode modified with co-immobilized laccase and tyrosinase, *Anal. Chim. Acta.* 308, 137–144.

Yu, H., Kusterbeck, A.W., Hale, M.J., Ligler, F.S. and Whelan, J.P. (1996) Use of the USDT flow immunosensor for quantitation of benzoylecgonine in urine, *Biosens. Bioelectr.* 11, 725–734.

Part IV
Applications

9 Surface plasmon resonance

Development and use of BIACORE instruments for biomolecular interaction analysis

Bengt Ivarsson and Magnus Malmqvist

Introduction

After a long period of genomic research, an increased interest in proteins is a logical step in biological research. Technology development is one of the most important factors behind the rapid increase of knowledge about molecular biology, protein structure and interactions in networks and pathways. Important goals for new bioanalytical technology include setting up experiments to answer new questions and increased laboratory productivity. High throughput screening in pharmaceutical industry for receptor ligands and miniaturised technologies for analysis of DNA hybridisation are two examples of how advances in technology can increase laboratory productivity.

For characterisation of protein interactions both structure and activity have to be analysed. The functional properties of a binding site and corresponding ligand in terms of reaction kinetics and affinity under standardised conditions is the biophysical way to describe molecular interactions. In order to compare results, high accuracy and precision measurements are needed not least in thermodynamic calculations of entropy and enthalpy contributions to the binding.

During a long period of time determination of structure has been possible with X-ray crystallography and more lately by NMR. However, ligand design and calculation of functional properties based on structure have been very difficult, despite major investments of effort and resources. Today, detailed functional analysis can be the basis for better understanding of how structural change actually causes a modified function of a molecule. From the early determinations of function in binding/no binding assays we are now using biosensor technology for a characterisation of kinetic properties with better than 10 per cent precision for dissociation rate constants (Rauffer-Bruyère 1997). The need for carefully determined kinetic constants and active concentrations is obvious for the analysis and description of functional complexes. Future characterisation and modelling of molecular networks also require precision measurements of functional properties.

The biosensor system concept for detailed characterisation of biomolecular interactions was first introduced on the market in 1990 (Jönsson *et al.* 1991). This technology was approached through the use of surface sensitive optical techniques for studies of protein adsorption on solid surfaces. Ellipsometry is a classical optical technique for real-time label-free adsorption studies, and has been used for characterisation of protein adsorption on modified surfaces and covalent immobilisation of proteins (Jönsson *et al.* 1985), as well as for biospecific adsorption in flow systems

(Jönsson 1985). Whereas the probing light beam in external reflection ellipsometry passes through the sample solution, the light beam in surface plasmon resonance (SPR) measurement does not pass through the sample. The light is instead used to create an evanescent wave penetrating about a wavelength into the sample solution. The SPR technique was first used by Liedberg *et al.* (1983) for detection of protein adsorption and antibody binding to adsorbed antigens.

The use of SPR-refractometry for surface optical detection of biomolecular interactions has several direct advantages in analysis and characterisation of biomolecular interactions. Label-free and real-time detection of mass changes within a surface confined volume are two important factors in this context. By immobilising one of the components of a biospecific pair on a surface that acts both as an affinity binding surface and the transducer of the changed mass into a sensor signal, the adsorption characteristics can be evaluated continuously with time, Figure 9.1. Controlled liquid flow over the sensor surface is a third parameter that gives the system important properties suitable for kinetic and affinity measurements as well as for mass-transport controlled concentration determinations. Automation of the system enables accuracy and precision in all steps of the analysis, and increases both ease of use and reproducibility. Figure 9.2 gives an overview of the methods that have been developed based on this biosensor technique.

These physical properties have to be combined with controlled chemistry of the sensor surfaces in order to avoid other adsorption phenomena which can obscure the studied reaction. The surface should provide chemical handles for immobilisation of biomolecules as well as a controlled chemical environment is the milieu for the studied reaction.

Characterisation of biomolecular interactions in kinetic or affinity terms requires a known active concentration of the interacting component. Conversely, with known or standardised molecular binding properties the concentration can be determined.

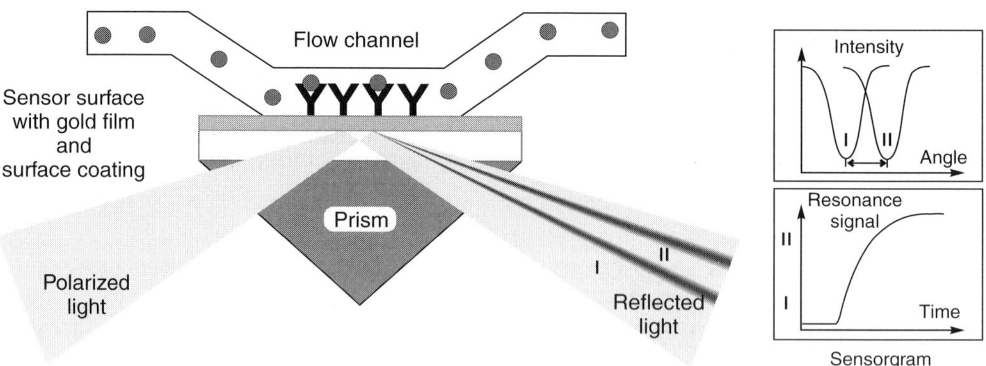

Figure 9.1a BIACORE system. The incident *p*-polarised light is focused into a wedge-shaped beam providing simultaneously a continuous interval of light wave-vector k_x at sensor spots along the focal line. This range covers the working range for the plasmon wave-vector k_{sp} during biomolecular interaction analysis. An increased sample concentration in the surface coating of the sensor chip causes a corresponding increase in refractive index which shifts the SPR-angle monitored as a change in the detector position for the reflected intensity dip (from I to II). By monitoring the SPR-angle as a function of time for each sensor spot, the kinetic events in the surface concentration are displayed in a sensorgram

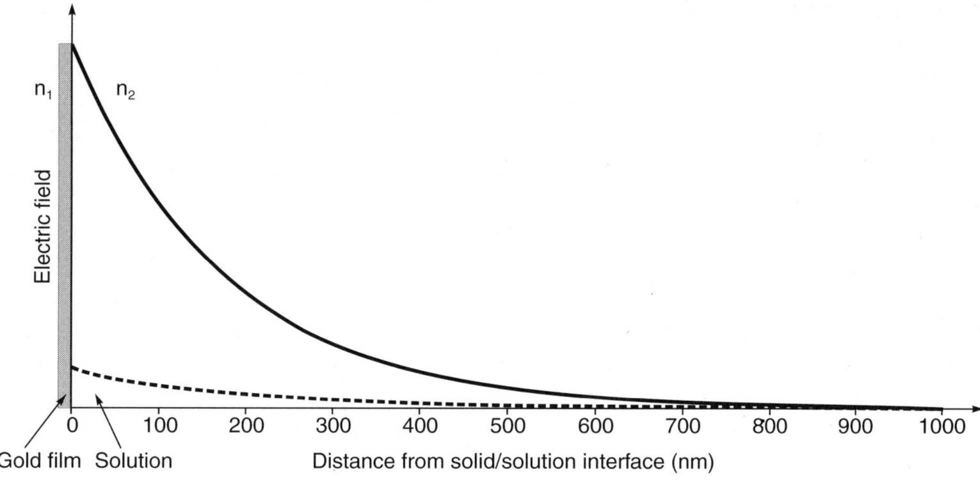

Figure 9.1b Relative evanescent electric field amplitude (E) versus distance from solid/solution interface into sample. Continuous line for SPR-evanescent wave (gold film), dashed line for non-absorbing TIR (no gold film)

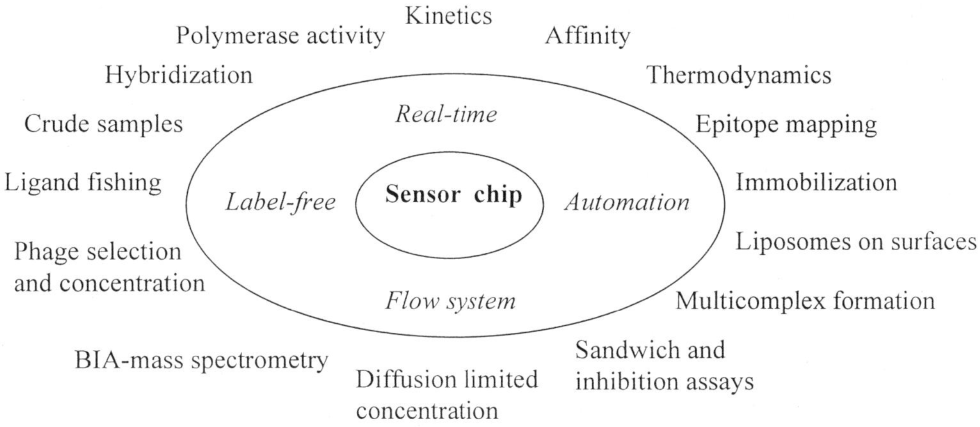

Figure 9.2 This figure illustrates how the centre for the technology is the adsorption behaviour at the sensor chip surface. Around the molecular interaction automation, liquid handling and surface plasmon resonance creates the basis for a whole series of methods used with BIACORE instruments

Early biosensor technology was very much driven by demands from clinical concentration analysis and easy testing of clinical samples. The basic idea of using label-free surface optical detection was that very cheap analytical testing could be performed compared with labelled reagents. To use this new technology for clinical samples in very complex mixtures just changed the position of the problem from analysis of the label to analysis of all interactions taking place in complex mixtures. Using label-free technology set much higher goals for non-wanted binding to surfaces and molecular binding sites than traditional labelled technologies.

Application demands

A general goal for the function and performance in routine operation for the first launched SPR-sensor product was to enable new applications and functions such as qualitative as well as quantitative label-free specific determination of at least one biomolecule in a sample in real-time either simultaneously or sequentially.

Concentration measurement

An application should have an analysis time <15 minutes at precision $= 5\%$ CV dose, and typical ranges of interests were:

High molecular weight analytes
    direct assay 10^{-5}–10^{-9} M
    sandwich assay 10^{-3}–10^{-11} M

Low molecular weight analytes
    inhibition assay 10^{-3}–10^{-9} M

- qualitative as well as quantitative relative structural information of macromolecules through the detection of interactions between surface-exposed structural elements on the macromolecule and various ligands;
- kinetic information about molecular interactions.

Typical ranges of interests are listed below.

Affinity measurements

    K_A 10^4–10^{11} M^{-1}

Kinetic measurements

    k_a 10^3–10^6 M^{-1}s^{-1}
    k_d 10^{-5}–10^{-1} s^{-1}

- the use of multiple flow-cells enable specific functionalisation of at least one sensing surface and non-functionalisation of at least one reference surface on the sensor chip.

Precision of immobilised protein

Between flow-cells on one chip   typically $<5\%$ CV
Between sensor chips   typically $<10\%$ CV

- regeneration or changing of the specific functionalisation of at least one sensing surface on the sensor unit;
- simultaneous multiple flow-cell analysis;
- programmable application methods and computer controlled sample liquid preparation and sample flow path across the individual sensor surfaces.

In order to shed light on the possibilities and limitations of bioanalytical applications employing SPR-technology, we will here first give a short description of the technology behind optical configurations known for SPR-detection, followed by an analysis of the application driven system technology required to support the SPR-optics. Finally a number of applications are shown for a commercial SPR-based instrument.

SPR-refractometer instrumental configurations

Refractometric biomolecular detection based on surface plasmon resonance may be classified according to the optical construction of the SPR-detector into three optical configurations, discrete elements optics, miniature optics, and integrated optics, each providing its characteristic SPR-optical function and performance. The function of the SPR-detector for real-time analysis is to measure the time-dependent change in angle or wavelength for the reflected light intensity spectrum, or change in intensity at constant angle or wavelength, at SPR in relation to a biomolecular interaction taking part within the evanescent wave of the plasmon supporting sensor surface. For a comprehensive review see Garland (1996).

Depending on the intended applicability and performance for the analytical instrumental system or probe, the SPR-detector is integrated with other appropriate modules.

In the case of an automatic analytical instrument, various modules are required for the operator-friendly, controllable, and efficient biomolecular interaction analysis. These modules provide functions such as:

- handling of a sensor component, so-called sensor chip, carrying plasmon supporting metal film with receptors or ligands which interact selectively with one or more biomolecules;
- controlled reaction time and flow-rate for liquid transport of sample reactants to and from the sensor surface and rinsing;
- controlled temperature during the analysis;
- controlled treatment of measured SPR-angle, SPR-wavelength or reflectance versus time data, models for its interpretation into surface concentration-time data, and kinetic constants for the sample interactions. In the high-performance SPR-analytical instruments of today, the SPR-detector consists of well proven discrete optical and optoelectronic elements.

In the case of a remote, portable, or in-process line instrument, where the SPR-sensor surface may be manually or automatically brought to the sample solution, or mounted in a pipe-line, the optimum SPR-detector takes the form of either a compact assemblage of discrete miniature optical elements, or an integrated design like an optical fibre or wave-guide. This provides a less costly and less complex instrument; however, the necessary control of sample transport and temperature requires either external supporting modules, or complex integrated liquid handling and reference sensors, in order to enable the accuracy and a sensitivity provided by the analytical system instrument.

Photon excited plasmons

As was previously described in this book, SPR is a result of energy and momentum of light being transformed into electronic charge density waves in the surface of a solid which has a free electron-like structure. The momentum transfer from the photon to the plasmon requires that the momentum, or wave-vector, of the photon is increased to equal that of the plasmon.

The works of Turbadar, Otto, Kretschmann and Raether introduced the use of an attenuated total reflection (ATR) prism or lens, for coupling of the photon energy and momentum into surface plasmon excitation, for a review see, for example, Welford (1991). The principle is that light is totally internally reflected (TIR) at a prism surface, whereby the refractive index of the prism increases the wave-vector of the evanescent wave, and this wave-vector is during detection, by providing the required wavelength and angle of incidence, tuned to match the wave-vector of the plasmon wave during its change caused by the changed refractive index within the electric field of the plasmon evanescent wave (see Figure 9.3).

In the Otto-configuration, this ATR-prism surface is brought into such a close proximity to a plasmon supporting metal film deposited on a separate glass slide that the ATR-evanescent wave possessing the increased wave-vector penetrates the air gap and excites plasmons at the metal/air interface (see Reference 2 in Welford 1991). In the Kretschmann–Raether configuration (Figures 9.1 and 9.3) the plasmon supporting metal is deposited directly onto the ATR-prism surface, where the ATR-evanescent wave penetrates the metal film and excites plasmons at the metal/surrounding interface (References 1 and 3 in Welford 1991). Both configurations have been used in numerous publications for analysis of metals, semiconductors and organic layers deposited onto the plasmon supporting metal.

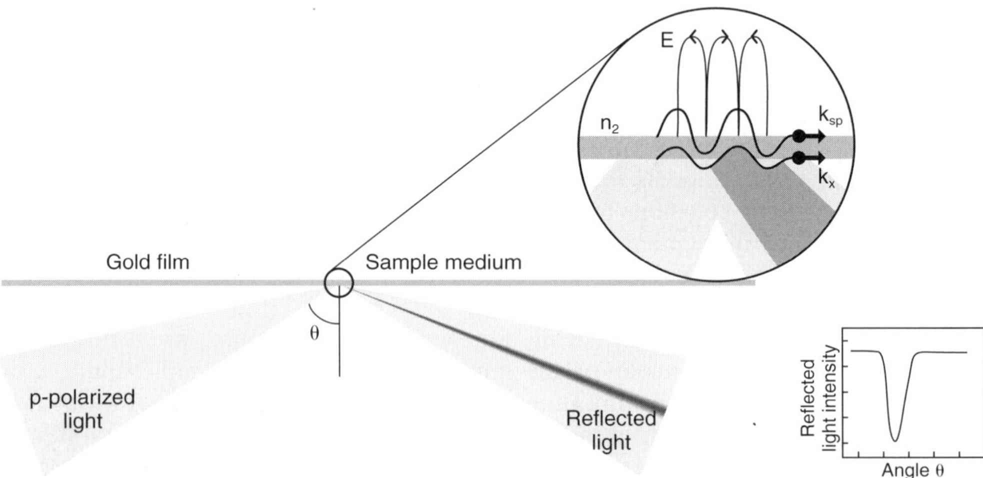

Figure 9.3 SPR excited by *p*-polarised totally internally reflected light at a glass metal film interface when wave-vector k_x of light equates the plasmon wave-vector k_{sp}, the surface plasmon enhancing the evanescent field amplitude, *E*. In BIACORE systems which use a sensor chip, this interface takes the form of an exchangeable gold-coated glass slide. SPR is observed as a dip in the reflected light intensity at a specific angle of reflection

Moveable and stationary optics

In most laboratory SPR-instruments, the wave-vector match is obtained by scanning either the angle of incidence of a collimated light beam incident onto the ATR-spot by pivoting the mechanical axes carrying the optical components and/or rotating the coupling prism/lens (so-called goniometer mechanism scanning), or by monochromatic scanning of the wavelength. Computer controlled stepping motors driving the scanning mechanisms and computer processing of reflected p-polarised light intensity versus angle data affords accurate data acquisition.

A drawback of the high-inertia mechanical angle-scanning methods is the relative low rate of reflectance spectrum measurement due to the recording of the reflectivity, point-by-point, as the incident angle is varied, gradually building the ATR-spectrum, resulting in an insufficient time resolution in the SPR-angle determination. As a consequence, their use for studying of the rapid changes in reflectance spectrum characteristic for dynamic behaviour of molecular adsorption and desorption at a surface or interface is limited.

Moreover, such bulk mechanical constructions are less suitable for practical commercial instruments; they are too complex and cumbersome.

The speed of goniometer scanning of a collimated beam can be improved by use of a focusing optics and one photo-detector element together with a low-inertia rotational mirror scanning as has been described by Oda and Fukui (1986). Here, the time required to measure one ATR-spectrum is fundamentally limited by rotation speed of the mirror.

Following a suggestion of Kretschmann (1978), the above mentioned drawbacks of moveable optical systems are avoided by the use of a focused incident beam, so-called focused light ATR. Since this optical system is stationary, and the reflected light cone simultaneously may contain the whole p-polarised reflectance versus angle of incidence spectrum, a set of fixed photo-detectors at different angles, as suggested by Kretschmann, could be used for detection of a rapid variation of reflectivity with time. This detection principle enables a high accuracy determination of the SPR-angle at high time resolution which is fundamentally limited by the fast electronic read out from a photo-detector array. Additionally, this focused beam solves the problem of so-called beam walk, i.e., that the position of the ATR-spot oscillates due to refraction of the collimated beam in a plane sided coupling prism, during a goniometer mechanism scanning.

Linear multi-site SPR refractometry

By use of a cylindrical focusing lens, simultaneous detection of ATR- and SPR-spectra at separate surface areas along a line of focused light of identical incident angular range at a hemicylindrical ATR-lens, was shown by Benner *et al.* (1979)

The usefulness of surface plasmons for a sensitive monitor of coating thickness in the order of 2–50 nanometres was described by Swalen *et al.* (1980). Following the principle of cylindrically focused light ATR, simultaneous multi-site SPR-detection was demonstrated in this work for a series of steps of fatty acid salt films.

In addition, Swalen therein described a brilliant optical technique for simultaneous detection of SPR-angle and SPR-wavelength by simultaneous presence of a wavelength spectra and angle spectra, in which the beam to be focused first passes a dispersion-prism, then a cylindrical focusing lens which creates a wedge-shaped beam, whereby the position of the rays along the focused line corresponds to the wavelength. This principle of wavelength resolved focused light ATR creates an image of reflectance versus both wavelength and angle wherein the SPR is observed as a line of intensity minima. Since the penetration depth into the sample by the plasmon evanescent wave is proportional to the wavelength, this principle can provide thickness-structural information in the detected mass change.

Plasmon resonance reflectance curve and time resolved analysis

The first experimental work describing the potential of SPR-technique for monitoring biomolecular interaction directly with time without labelling was published by Liedberg *et al.* (1983). By choosing a fixed angle of incidence halfway down the reflectance minimum, and recording the altered reflected intensity at that angle, changes in refractive index at the SPR-sensor surface of about 10^{-5} were easily detected. Moreover, Liedberg *et al.* suggested the application of SPR for monitoring enzyme activity, hormone-receptor and antibody interactions, as was demonstrated by recording the reflectivity versus time for antigen–antibody interactions and varying the antibody concentration.

The drawback of measuring reflectivity at constant angle, or wavelength, is that this steep part of the ATR-spectrum covers a limited refractive index range, since the shift in the SPR reflectivity curve makes the fixed angle (or wavelength) to coincide with either the more stable higher reflectivity, or even pass the zero reflectivity, depending on the adsorption–desorption character of the studied interaction. Further on, alterations in incident light intensity appears as reflectivity disturbances which has to be corrected for by a simultaneously monitored reference beam in order to obtain a required accuracy and detection limit for this SPR-sensor principle.

The application driven requirements given above, together with the need to eliminate the prior-art technologies' limitations in detection rate, sensitivity, and dynamic range, and moreover the prior-art's dependence on highly skilled operators for sample and liquid handling in order to obtain accurate results, all together called for a new SPR-detection concept. These requirements were first achieved with the BIACORE SPR-technology, by which the advantages of a stationary optics SPR-detector were integrated with ease-of-use components. With exchangeable sensing surfaces, accurate ease-of-use sample handling and SPR-data reading together with interpretation and presentation of the biomolecular interaction in real-time, SPR-technology-based analytical system could be produced. This will now be described in further detail.

Illumination unit

The market driven need for a simultaneous test on a plurality of specific interactions is enabled by the used wedge-shaped focused light ATR (Figure 9.4a). To this end, the line shaped ATR sensing surface takes the form of a series of sensitised areas, each comprising a different reactant. In the BIACORE instrument, the light source is a near-infrared light-emitting diode of peak intensity wavelength 760 nm, for which the penetration depth of the SPR-evanescent wave into the sample is about

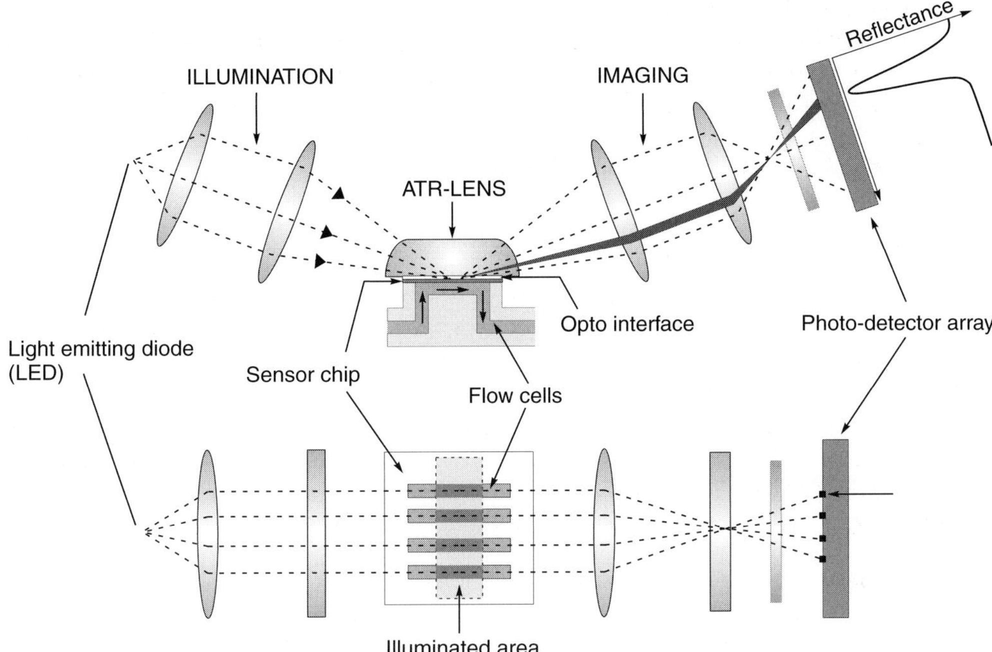

Figure 9.4a BIACORE optical system. Upper view shows light focused into a wedge-shaped beam the focal line of which lies in the plane of the gold film of the sensor chip. Incident and reflected wedge-shaped beams cover an angle of incidence (reflectance) range corresponding to a suitable sample refractive index range. The optical module presses an elastic opto-interface and the sensor chip against the fixed liquid handling cartridge. Light rays are coupled between the ATR-lens and the sensor chip via the opto-interface as the gold film side of the chip forms a wall of a flow-cell at the flow-cell plateau of the liquid handling cartridge. In this plane, the image system projects rays onto the columns of a two-dimensional photo-detector array whereon the rays' position corresponds to the rays' angle of incidence. The SPR-angle is derived from the measured light intensity along a detector column (SPR-reflectance curve).

Lower view shows the focal line crossing four flow-cells where the reflected light from each flow-cell is imaged on each its own detector column, providing a simultaneous multitude of SPR-detected flow-cells

300 nm (distance over which the evanescent wave electric field strength decays to e^{-1} ($\approx$37 per cent) of its value at the metal surface).

Since surface plasmons are excited within the illuminated focused line area on the sensing surface, the size and position of this SPR-sensor area in relation to the sample cell geometry can be controlled in order to optimise the interaction of the SPR-evanescent wave with the immobilised reactants. The size of the SPR-sensor area must be carefully chosen. A small illuminated area in relation to the chemical sensor layer may reduce the effects of inevitable variations in a commercial product-line metal film and molecular coating. However, the fact that the surface concentration of bound sample molecules can also be non-uniform across the sensitive layer and dependent on mass transport conditions calls for an

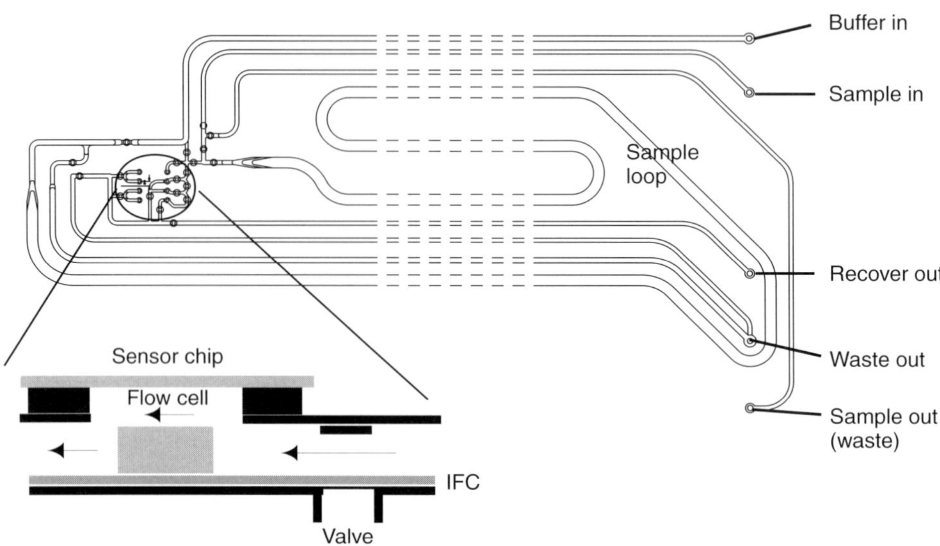

Figure 9.4b Diagram of the integrated μ-fluidic cartridge (IFC) as seen from above. The liquid handling cartridge consists of a series of channels and pneumatic valves encased in a plastic housing, and serve to automatically control delivery of thermostatted liquid (sample or buffer) to the sensor chip surface. Samples are injected from an auto-sampler into the IFC, which connects directly with the detector flow-cells

illuminated area large enough to provide a good average value. Thus, too small a sensing area will be very sensitive to local variations between different sensor chips and sample transfer to this small area, resulting in a non-competitively low accuracy in the SPR response, while too large a sensing area will lead to a low sensitivity, larger sample volume and a distributed surface concentration across the sensor area.

Imaging unit

Multi-sensor area monitoring requires that the site-specific interaction can be identified by the mapping of each sensor area to at least one detector column, while the corresponding interaction is quantified by the position of the SPR-reflectance curve along this detector column. In the BIACORE instrument an anamorphic imaging objective projects light rays reflected from each separate sensor area onto separate photo-detector columns in a detector array, while projecting rays at each angle of incidence onto separate positions along the columns within the array (Figure 9.4a).

The stationary integrated optics and photo-detector array minimises the risk for mechanical vibrations disturbing the SPR-angle measurement. In BIACORE 2000 the measurement accuracy is still further increased by establishing an optical and electronic mean value procedure for the SPR-angle determination. Since each point within the sensor surface is probed by rays at the same angle interval in the plane of incidence, and rays of identical angle of reflectance are projected onto the same photo-detector row in the corresponding column, an optical averaging of the distributed refractive index within the plasmon evanescent wave is inherent in the created SPR-reflectance curve.

Additionally, since the anamorphic imaging system also produces a reduced image of each sensor area on the photo-detector column, this real imaging contributes to the average SPR-reflectance curve for the sensing area. By measuring the shift in this averaged SPR-reflectance curve caused by a changed amount of bound biomolecules, a high degree of accuracy of the SPR-response curve required in a commercial-type analytical instrument is obtained together with practicable reproducibility and control of:

i   variation in the optical properties within the SPR supporting metal film, metal coating and metal substrate;
ii   flow profile in the sample liquid (diffusion layer thickness transversely of the flow-cell);
iii   homogeneity of ligand density and ligand accessibility within the matrix on the sensing surface.

Photo-detector-array unit

The illumination unit, coupling lens or prism, sensor chip, imaging unit, and photo-detector are all mounted in the stationary configuration in a robust mechanical housing.

Stationary optics and focused light ATR enables the sample rate in the SPR-angle versus time data to be limited only by the rate for the data acquisition from the photo-detector array and the data processing in the calculation of the SPR reflectivity curve versus time. This provides the high time resolution of the sensorgram required for an accurate determination of the kinetic constants for association and dissociation characteristic for the biomolecular interaction.

The inevitable variation in light intensity-voltage response between the detector elements in a photo-detector array may be normalised by collecting a number of intensity curves at various different set light source intensities and total internal reflection at the sensor chip. In addition, the noise in the detected SPR-reflectance curve is reduced by averaging the voltage response over time for each detector element.

SPR-detector calibration and control software

In order to provide a quantitative measurement, the measured shift in SPR-angle, defined as the SPR-angle response, must be calibrated against known changes in the refractive index sensed by the SPR-evanescent wave. SPR causes a dip in the reflected light intensity distribution versus length along the photo-detector column. A curve-fitting algorithm determines the position of the reflectance dip in terms of photo-detector element number, and from this position and a known angle of incidence interval projected on the photo-detector column, the SPR-angle is calculated. Due to the high sensitivity of the SPR-response for a changed refractive index, a small practical unit for refractive index change was defined, $1\,RU = 10^{-6}\,RIU$ (Refractive Index Unit). Hence, in each BIACORE instrument the SPR-angle response is calibrated against standard refractive index solutions, whereby a scale factor (angle shift°)/(refractive index shift) in unit °/RU is calculated. It is found that

1 RU in the solution causes a SPR-angle response, of approximately 0.0001°. However, the SPR-response to 1 RU depends on, above all, the choice of SPR-supporting metal, and the optical characteristics of the metal coating, substrate glass, and wavelength, respectively.

By use of this calibration scale factor (°/RU), the SPR-angle response versus time course for a biomolecular interaction is monitored in RU/sec, which is defined as a sensorgram (Figure 9.5). Thus, the SPR-angle response to the refractive index change caused by interaction of biomolecules is monitored as the corresponding refractive index change of the bulk of a liquid phase sample. Stenberg *et al.* have quantified the correlation between absolute surface concentration of protein on the bio-specific active surface and the measured SPR-angle shift, defined as the SPR-response, by use of radio-labelled proteins (Stenberg *et al.* 1991) A linear correlation between the SPR-angle shift and the surface concentration was shown to be valid up to a surface concentration of $50\,\mathrm{ng\,mm^{-2}}$, wherein 1000 RU corresponds to approximately $1\,\mathrm{ng\,mm^{-2}}$. No significant difference was found between the proteins studied (chymotrypsinogen, transferrin and anti-transferrin monoclonal antibody).

For an automated routine SPR-bioanalytical instrument all module functions during analysis are controlled and displayed from a control software running on a desktop computer. The control software includes an interface to evaluation routines for interpreting sensorgram data. BIA evaluation is a stand-alone software package for presentation and evaluation of: kinetic constants from sensorgram data using numerical integration and global fitting methods; pre-defined and user-defined

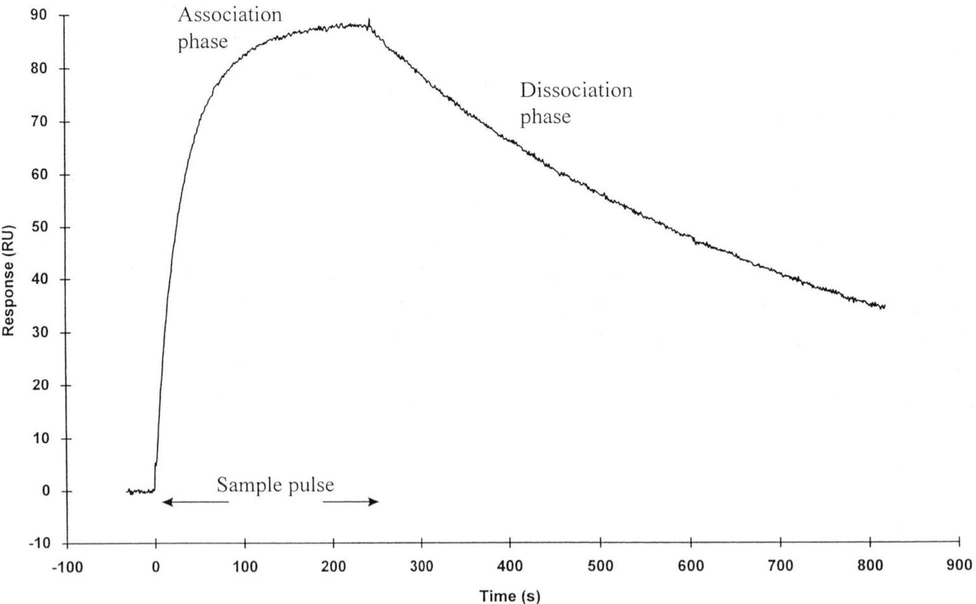

Figure 9.5 The sensorgram shows the detector response versus time and is the basis for all molecular analysis. The sample is injected and during the association phase adsorption takes place. After the sample pulse has passed, the molecular complex on the sensor chip surface dissociate and molecules leave with the flow system. The analysis of kinetic properties is normally done with a series of concentrations and global fit of kinetic constants to experimental data in an interaction model

molecular interaction models; simulation of sensorgrams from specified models and kinetic parameters; determination of affinity constants for both surface and solution interactions; analysis of concentration measurements from either binding level or rate data. Simultaneous model fitting to a number of sensorgrams for a given application improves the robustness of the fitting procedure. The global fitting method wherein one or more variables (e.g. association or dissociation constants) are in addition constrained to have the same value for each sensorgram, contributes greatly to the stability of the fitting procedure for complex models, and also reflects many physical interaction systems more faithfully (a set of sensorgrams for the same interaction at different analyte concentrations should a priori have identical rate constants).

In BIACORE 2000 a microprocessor performs the basic processing of the SPR signal into SPR-response and sensorgram. The SPR angle is measured at a frequency of 160 Hz, which data is averaged by the microprocessor in the instrument and transmitted to a control software at rates up to 10 Hz. The SPR-response data output rate is 0.1–10 Hz in the sensorgram if one flow-cell is detected, and 0.1–4 Hz per flow-cell in multi-channel mode. Normal sensorgram noise levels are <0.3 RU RMS at data collection rate 1 Hz. The SPR-response dynamic range is 30,000 RU, within the absolute refractive index range 1.33–1.36 (RIU). In BIACORE X and BIACORE 3000, SPR-response dynamic range is 70,000 RU, within the absolute refractive index range 1.33–1.40 (RIU).

Sensor chip and opto-coupling units

One of the inventive components in the BIACORE-instrument is the sensor chip, comprising a plasmon supporting metal film deposited on a glass support, where the metal surface is coated with the bio-specific sensor chemistry. The sensor chip was made reusable and easily exchanged by the co-invented opto-coupling unit, called the opto-interface, which functions to couple the wedge-shaped light beam between the sensor chip and the stationary coupling prism in the instrument.

A stationary coupling prism provides an accurate optical angle of incidence at the surface of the separate sensor chip, enabling fast and accurate analysis. In BIACORE instruments, light and sample are simultaneously coupled to the sensor surface by docking the sensor chip with high mechanical precision between the elastic opto-interface unit and a flow-cell plateau which is part of the liquid handling module. This concept makes it possible to use a sensor chip in a routine commercial instrument while avoiding problems in delicate handling of refractive index matching smearing oils and liquids. Furthermore, this opto-interface has the advantage that the sensor-chip may be taken out of the instrument for further treatment in, e.g., sample liquid, without the need for cleaning it from opto-coupling fluid.

Sample handling system and Integrated μ-fluidic cartridge (IFC) unit

The already mentioned applications set up high functional demands on the individual sensor surface functionalising and sample handling. It should be possible to direct the sample liquid into a selected sensing area or into all areas simultaneously or in sequence. The flow path, volume, flow rate, sensor area contact time and temperature of the sample should be automatically controlled for ease-of-use, while minimising carry-over and sample consumption.

Driven by these requirements, Biacore AB developed the integrated fluid handling system providing multi-flow-cell biomolecular analysis, as described by Sjölander and Urbaniczky (1991). According to this design, the optical module together with the sensor chip and a micro-fluidic cartridge form the integrated sample handling system (see Figure 9.4b).

The flow injection analysis technique (Ruzicka and Hansen 1987) is suitable for reproducing qualitative and quantitative measurements with rapid response time. Sample and reagent volumes can be minimised by integrating and miniaturising sample loops, valves and conduits. Integration also enhances repeatability (Ruzicka and Hansen 1983).

The BIACORE 2000 liquid handling system consists of the following modules:

- two liquid delivery pumps for maintaining a constant flow of liquid over the sensor surface and for handling samples in an autosampler;
- an autosampler for pre-programmed sample handling including mixing and injection to the integrated micro fluidic cartridge, as well as sample recovery and fraction collection;
- autosampler rack base which may accommodate two microtitre plates or aluminium thermo-racks;
- microprocessors for the control of pumps, autosampler and IFC-valves.

See Figure 9.6 for a photograph of BIACORE 2000.

Fluid channels and valves are integrated in the micro-fluidic cartridge to control and direct the flow through loops and the thin-layer flow-cells. The cartridge has interfacing connectors to carrier stream liquid and auto-injector driven by external pumps in connection with waste and carrier reservoirs.

The detector flow cells are formed directly on the IFC by docking the sensor chip against the IFC flow-cell block, which consists of four grooves, rectangular in cross section. Each groove has its own liquid delivery channel and control valve in the

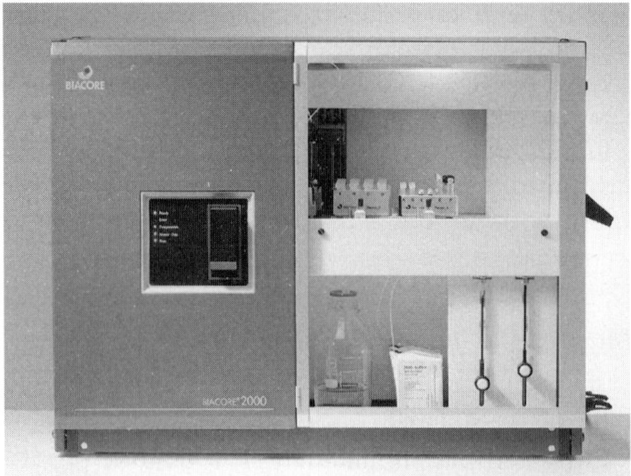

Figure 9.6 A photograph of BIACORE 2000 instrument

IFC. The block is pressed against the sensor chip, forming four separate flow cells with sensor chip surface as one wall. Effluent from the flow cells is collected in an outlet channel and directed either to waste or a recovery cup. Each flow thin-layer cell is 1.6 mm long, 500 μm wide, and 50 μm high. In BIACORE 2000/3000 the IFC allows single or multi-channel analysis in up to four flow cells.

In the BIACORE-instrument, the biomolecular interaction takes place on the bio-specific active surface of the sensor chip, which serves as one of the walls in a thin-layer flow cell. This allows for direct sensing of the interaction with SPR, where the thin-layer cell is used to transfer the analyte in the sample onto the sensing surface with a known mass-transfer rate and efficiency. The mass-transfer characteristics for a thin-layer cell are well known, and an analytical solution for the mass-transfer equation was given by Matsuda (1967). It is valid when the diffusion layer thickness next to the bio-specific active surface is sufficiently thin relative to the channel height in the thin-layer cell.

It turns out (Sjölander and Urbaniczky 1991) that if low detection limit is the primary goal, the thin layer cell should be designed to enable transfer of the analyte in the sample onto the sensing surface with (ultimately) 100 per cent adsorption efficiency. A drawback with a low flow rate is the longer time to apply the sample volume, e.g., at a flow rate of 10 mm/s it will take more than three minutes for a sample volume of 50 μl to pass the thin layer cell with a thickness of 50 μm. Flow rates lower than this would increase the adsorption efficiency further; however, the assay time then becomes impractically long. Mass transport is the important parameter and in BIACORE 3000 the flow cell height has been reduced to 20 μm that can be used for either smaller volumes or higher mass transfer rate. In practice, the maximum assay time and sample consumption set the limits.

For kinetic studies a high mass-transfer rate is important to be able to resolve high rate constants. If the mass-transfer rate is much higher than the heterogeneous reaction rate, then diffusion/convection influence can be neglected. For kinetic measurements it is important that the sample is delivered to the flow cell in a well-defined plug in the carrier stream with minimum dispersion at the sample-buffer boundaries. In the IFC the sample plug is created by switching between sample and buffer flow with help of the valves. Dispersion is minimised by the construction of the IFC, where the dead volumes between valves and the flow cell are kept as small as possible.

In the BIACORE instrument it is possible to inject a sample volume into the thin-layer cell in two ways. The normal mode, which increases the sample turn around time, is direct injection of the sample and reagent solutions into the channel connected with the thin-layer cell.

Thereby, the flow cell washing time also decreases, especially when solutions for the surface regeneration are injected, as well as when secondary reagents are used in an amplification step. The direct-inject procedure has two advantages, the total sample volume needed decreases as well as the non-specific adsorption of the analyte before reaching the sensor surface. It is very important to minimise non-specific adsorption, since the surface area of the fluidic channels is large compared with the amount of the analyte. There are, however, two questions of direct injection of the sample into the detection channel that have to be addressed: it is harder to control the sample temperature, and the flow rate for the sample injecting pump must be as precise as for the carrier stream pump.

An alternative approach is to use a sample loop, closing the loop after a selected time to control the injected volume. Loop injection is used in combination with direct injection for applications requiring large volumes. Additionally, loop injection is used for injection of one sample immediately following another which is used, e.g. for monitoring dissociation in the presence of other substances.

The injections via loop or direct-inject are performed with an air segment to prevent sample dispersion. Precision comes from time control and the low dead volume. Finally, it is not necessary to calibrate the loop volume since it is known from the flow rate and by accurate timing of the sample injection duration via IFC-valve control.

IFC-thermostatting unit

The detection limit depends on how controlled the background refractive index is. Ultimately, it is the temperature dependence for the carrier stream that determines the detection limit for an assay using the SPR-detection technique. The temperature dependence for water is $\approx -100\,\mu RIU/^\circ C$ (i.e. $-100\,RU/^\circ C$). For example, a sample temperature variation of $0.001\,^\circ C$ during the refractometric analysis corresponds to an uncertainty of $\approx 0.1\,pg\,mm^{-2}$ of analyte on the bio-specific active surface. By thermostatting the carrier stream and/or using a reference sensor surface, the temperature variation disturbance can be minimised. Next to the temperature influence, the detection limit is determined by changes in the refractive index of the bulk of the solution sample plug (so-called non-specific presence of analyte), or of the carrier solvent. By use of a reference sensor surface, this disturbance can also be minimised.

A typical operational analysis temperature range of interest is 4–$40\,^\circ C$ and the temperature equilibrium time should be below one hour for practical reasons. In the BIACORE 2000 instrument all optical units, including the sensor chip, together with the Integrated μ-Fluidic Cartridge (IFC), are mounted inside a thermally insulated box. Temperature control uses Peltier elements to maintain a constant pre-set temperature at the sensor chip surface, and a constant temperature at the units within the box. The temperature drift at the sensor chip surface is less than $\pm 3 \times 10^{-3}\,^\circ C/min$, corresponding to an SPR-response drift of $< \pm 0.3\,RU/min$.

Sensor surface chemistry

One goal for an SPR-sensor chip was to create a general-purpose chip that could be regenerated and on which the operator could tailor-make an application-guided bio-specific active surface chemistry. This challenge was first met by the Biacore sensor chip CM5, comprising a surface plasmon supporting gold film deposited on a glass support, where the metal is coated with carboxy-methylated dextran creating a flexible hydrophilic polymer to which specific ligands can be linked with a variety of established chemical methods, and through which matrix the dynamic range for specific sample binding is increased.

This sensor chip can be used for repeated measurements, 50–100 runs depending on the properties of the immobilised ligand and the regeneration conditions used. The chip can also be removed from the instrument, the immobilised biochemical layer may undergo a treatment, and be re-docked at a later stage.

Immobilisation on sensor chip surfaces

The sensor chip surface is the interface between the chemical environment and the physical surface properties needed for surface plasmon resonance to take place. The gold layer is modified with a long-chain alkanethiol (Löfås 1995) by a self-assembly process and further modified with dextran to create a hydrophilic polymeric surface. The dextran is modified to carboxydextran where the carboxyl groups can be activated with N-hydroxysuccinimide and N-ethyl-N'-(dimethylaminopropyl) carbodiimide hydrochloride in water (Johnsson and Löfås 1991). The thickness of the carboxylated dextran surface was estimated to be around 100 nm (Stenberg *et al.* 1991) and later measured by scanning force microscopy to be around 140 nm (Wigren *et al.* 1995) in physiological buffer. Under these conditions the dextran content is about 2 per cent in an aqueous environment.

By this route proteins can be coupled through amino-groups, or the sensor chip surface can be further modified with hydrazide for aldehyde coupling of periodate oxidised glycoproteins and RNA or with thiols for disulphide exchange reactions (Johnsson *et al.* 1995). Sensor chips with different polymer chain lengths and degree of carboxylation are now available, and also surfaces without dextran required for special analytical situations.

To improve the immobilisation process the residual carboxyl groups after activation are used for ionic adsorption of the protein under low ionic strength conditions at a pH under the isoelectric point of the protein. In this way low concentrations of proteins in solution can be covalently coupled to the sensor chip surface with high efficiency.

Regeneration of formed molecular complexes

To do careful experiments regeneration of formed complexes on sensor chip surfaces is important for repeated use and reproducible results. Each biomolecular interaction is unique in the combining site giving rise to specificity, affinity, kinetic properties and regeneration conditions. The sensor chip itself is very stable to low and high pH, organic solvents and combinations of different substances that can effect the chemical forces in the binding sites.

Frequently used regeneration conditions are 1 mM HCl, but often combinations of high pH and organic solvent are used. However, a new multivariate cocktail approach for identification of regeneration condition has recently been employed for BIACORE affinity biosensor. With stock solutions each representing a mixture of substance with similar chemical properties, the multivariate approach iterates towards a suitable mixture of the stock solutions optimised for effective regeneration without destroying the immobilised ligand. Due to this multivariate approach the risk of missing relevant combinations is minimised and unexpectedly good effects of EDTA as additive in regeneration cocktails containing chaotropic agents and high ion concentration were found (Andersson *et al.* 1998). Of thirteen tested antibody–antigen interactions, nine were regenerated to 90 per cent in less than 30 s and five of them were further optimised to complete regeneration with preserved activity.

Immobilisation by affinity capture

Immobilisation of water-soluble ligands on surfaces can also be achieved by strong adsorption to structures on the sensor chip that can easily be regenerated, so-called affinity capture. There are several examples of that such as adsorption through histidine tags on proteins to NTA surfaces (Nieba *et al.* 1997), or monoclonal antibodies to affinity purified polyclonal rabbit-anti-mouse antibodies. The best approach for immobilisation of nucleic acids and oligonucleotides has up to now been to use streptavidin on sensor chips to capture biotinylated nucleic acids (Nilsson 1995). Streptavidin surfaces are very hard to regenerate without destroying the protein and a way to introduce different oligonucleotides is to use a long complementary oligonucleotide that assemble the target oligonucleotide with the streptavidin immobilised oligonucleotide (Buckle 1996).

Studies of membrane associated proteins and other phospholipid anchored molecules require particular immobilisation procedures. In some early attempts to study membrane proteins from insect cells, whole membrane vesicles were immobilised to sensor chips using immobilised immunoglobulins against either avidin or biotin (Masson 1994; Masson 1995). Immobilisation of liposomes have also been done through scFv coupled to phospholipid or a fusion peptide to the antibody fragment followed by adsorption to an antigen surface (Laukkanen 1994).

For analysis of phospholipid anchored molecules, a hydrophobic surface has been used for immobilisation of liposomes. By adsorption of liposomes a phospholipid monolayer is introduced including lipid anchored molecules (Cooper 1997). Another approach has been used by MacKenzie (1997) where whole liposomes are coupled to sensor chip CM5 through an immobilised protein. In many applications related to membrane work a well-defined procedure to immobilise membrane proteins is most important. For pharmaceutical development of drug candidates the detailed analysis of kinetic properties is difficult without a better procedure for immobilisation of functional receptors.

Recently a new experimental sensor chip surface based on the dextran surface CM5 with immobilised lipophilic substances for adsorption of liposomes was introduced. Liposomes readily bind to these structures giving a possibility to create a hydrophilic environment for analysis of membrane proteins. Very few experiments have been performed to date on such surfaces, although they hold interesting potential for use in combination of membranes and membrane proteins. A summary of affinity capture ligands is shown in Table 9.1.

Table 9.1 The table shows some commonly used ligands for affinity capture of molecules. The binding should be very stable but reversible under changed chemical conditions in order to useful. Affinity capture is used in situations when the molecule of interest is not stable for regeneration or if the molecule is not purified

Affinity capture	*Surface ligand*
His-tagged proteins	Metal chelating, NTA
Monoclonal antibody	Rabbit anti-mouse antibody, affinity purified
GST fusion protein	Anti-GST antibodies
Biotin labelled molecules	Streptavidin
Phospholipids and hydrophobic molecules	Methylated self-assembled monolayer on gold, HPA surface
Liposomes	Lipophilic modified dextran surface

Biomolecular interaction analysis-BIA

Interactions are time dependent and in BIACORE instruments the interaction course is quantified in the sensorgram (Figure 9.5). Compared to a traditional endpoint assay there is much more information about the reaction in such time dependent interaction analysis. These sensorgrams are the basis for characterisation of the interactions and are interpreted in terms of specificity, concentration, affinity and kinetic properties. Kinetic determinations in particular have been difficult with other techniques, as they often require labels used in stop-flow fluorescent measurements.

When one of the components in a biospecific pair is immobilised an interaction with the analyte molecules in free solution can be detected by surface plasmon resonance. Due to the continuous flow the analyte is maintained at the original concentration when it enters the detection cell. However, diffusion to the sensor chip surface has to be considered in relation to the used concentration, the kinetics of the interactions and the molecular weight of the substance.

When BIACORE was introduced (1990) very little was known about the potential use of the technology. Most of the application work was done with monoclonal antibodies where the basic principles easily could be demonstrated. However, with time a learning curve of how to interpret results from affinity biosensors has been implemented and today the methods are accepted for characterisation of interactions. It is, however, very important to understand the underlying operational principles and to set up appropriate experiments to be able to draw conclusions about interaction parameters, since the results are not absolute. BIA-technology correlates well with other techniques such as microcalorimetry (Hensley 1996), ultracentrifugation (Johanson *et al.* 1995), ELISA for $\Delta\Delta G$ of a mutant series based on affinity determinations (Cunningham and Wells 1993), although good correlation studies have not been done for kinetics. Obviously all methods have different types of experimental procedures giving results that must be viewed as apparent or relative affinities and kinetic constants. As one component in the biospecific pair is immobilised, comparisons with other technologies to get information about correlations are therefore very important to perform.

Affinity and kinetic analysis

The most used application for BIACORE instruments has been kinetic determinations of biorecognition pairs. The technology offers potential to determine kinetic constants without any label and over a wide dynamic range (Malmqvist and Karlsson 1997). After the introduction of the system approach (Jönsson *et al.* 1991), the first attempts were done by Karlsson *et al.* (1991). The concept has been further developed by a series of publications where pros and cons have been discussed (O'Shannessy 1994; Karlsson and Ståhlberg 1995; Karlsson and Fält 1997).

By experience it has been considered that it is important to have high flow rates for kinetic analysis to minimise diffusion limitations. When the sample pulse has passed only dissociation of the molecular complex is expected as the freed analyte is transported away by the flow. However, rebinding can occur and is a problem unless a low amount of ligand is immobilised and global fit analysis is used for interpretation of results.

In a recent review by Myszka (1997) important aspects of how to avoid problems in experimental set-up are described. Typical parameters to consider are surface concentration, liquid flow, concentration ranges and avidity effects due to multivalency. It is important to use global fit analysis of all experimental data from experiments with different concentrations for interpretation of kinetic data under partial diffusion limitation. Global fit analysis and numeric integration has made it possible to separate mass transport contributions from reaction properties and also to determine diffusion coefficients. A combination of a fifty-fold increase in refractometric sensitivity from 1990 up to BIACORE 3000 launched in 1998, and global fit analysis has made it possible to measure association rate constants up to $10^8 \mathrm{M}^{-1}\mathrm{s}^{-1}$ from about 10^5 in the original BIAcore instrument.

Over the years there have mainly been two concerns about BIA-technology based methods for affinity determinations. One is that the dextran on sensor chips may affect the measured parameters and the other is the mass transport limitation of the system approach (Schuck 1996). For the first hypothetical problem it has been shown by Parsons and Stockley that surfaces with and without dextran give very similar kinetic properties for their studied protein-DNA interactions (Parsons and Stockley 1997) and the same is true for protein–protein interactions (Karlsson and Fält 1997).

The diffusion limitation can be handled by including partial mass transport limitation in the global fit analysis of interaction parameters. In work with very high association rate constants, Myszka has determined larger than $10^7 \mathrm{M}^{-1}\mathrm{s}^{-1}$ for protein interactions (Myszka, personal communication). Working with such high association rate constants requires very low concentrations both of ligand and analyte and set high demands on instrument sensitivity. For dissociation rate constants the time for buffer exchange is the limiting parameter in order to directly determine dissociation rate constants larger than $10^{-1}\mathrm{s}^{-1}$, and on the other extreme is the long time drift of the instrument.

With the specification of the flow-based system, very large differences in affinities from mM (Ohlson *et al.* 1997) to pM range have been described. Of particular interest is the determination of very high dissociation rate constants estimated to be between 1 and $10\mathrm{s}^{-1}$ for cell surface bound CD molecules (van der Merwe *et al.* 1997).

The technology has also been used for analysis of binding affinities based on kinetic determinations at different temperatures for thermodynamic analysis of antibody interactions (Zeder-Lutz *et al.* 1997). By careful thermodynamic analysis Roos *et al.* also found that the temperature dependence was larger for dissociation rate constant than of the association rate constant (Roos *et al.* 1998).

Concentration determinations with BIA

Traditional approaches such as sandwich or inhibition assays can be performed with BIA technology as well as with labelled technologies. The difference is that formed complexes are determined directly and it is not possible to distinguish biospecifically adsorbed material from unwanted binding. This puts high demands on the surface properties and on careful control of the sample matrix for unwanted contributions to the detected response. Another difference is that the formed complex is measured directly without any enhancement of signal. This is important to understand when the techniques are compared with ELISA or extremely sensitive fluorescent assays. A series of food analyte inhibition assays for the vitamins biotin and folic acid have been developed for commercial use based on BIACORE instrument system. The precision in determinations is below 10% CV dose over the quantification range 2–70 ng/ml. Other examples are assays for antibiotics that correlate well with HPLC methods (Sternesjö 1996).

The good precision is the result of the flow system where diffusion limitation is a positive contribution to assay performance. The diffusion characteristics are defined by flow cell properties, diffusion constant of the molecule and concentration. In this way a high precision is reached as molecular properties, flow control are easier to control than a critical surface modification that is important in most other types of assays. After the first attempts by Karlsson (1993) a further development of the diffusion limitation concentration determination of active molecules has been taken by Richalet-Sécordel (Christiansen 1997; Richalet-Sécordel *et al.* 1997).

Careful determination of active concentration is important to increase the reproducibility of association rate constant. Active concentration is hard to determine by other methods but with BIACORE it is possible to use one standard curve for antibodies with many different specificities as the instrument system is calibrated and not each assay. This is very important when affinities and rate constant are to be compared for protein mutants as the yield of substance varies individually with mutations.

Epitope mapping

Epitope mapping is a way to characterise binding sites in topological relation to each other. Normally, a series of monoclonal antibodies for one single antigen is used in combination with natural binding partners (Fägerstam *et al.* 1990). One antibody is bound to rabbit anti-mouse Fc, RAMFc on the sensor chip surface followed by antigen and secondary antibody. In this way the relative binding can be compared and, by running the whole chess-board of antibody combinations, a detailed description of related binding sites can be obtained (Daiss *et al.* 1995). A detailed epitope mapping of MUC1 mucin has been performed with BIACORE and other techniques by a large combined effort in different laboratories (Price *et al.* 1998).

Transcription complex

An area of great biological interest has been transcription factors (Galio *et al.* 1997) and DNA binding estradiol hormone receptors (Cheskis *et al.* 1997) and their interactions with DNA sequences. Transcription as a polymerisation process has been studied by Buckle *et al.* (1996) and the interaction with DNA bending proteins in the transcriptional complex in procaryotic systems (Muskhelishvili 1997).

Screening of small molecules in BIA-technology

With the increased sensitivity of current systems, interactions between a large immobilised protein and a small organic substance can be detected and analysed for kinetic properties. In large screening efforts by pharmaceutical companies the results from high throughput yes/no assays (HTS) must be further analysed. When the total number of interesting compounds has been reduced to 100–1000 it is possible to use the BIACORE instrumentation of today to characterise the interactions. The technology has made it possible to correlate kinetic properties in binding a receptor to biological models in the evaluation of results. This is an important issue to consider in relation to doses and time responses of future drugs.

BIACORE has been used in a series of experiments screening for inhibitors of HIV-1 protease (Markgren *et al.* 1998). As most compounds are dissolved in DMSO an internal calibration procedure has to be used in order to verify a maximal response level of 10–20 RU in the presence of a bulk refractive index change that can amount to several thousand RU. Another aspect to consider is the large amount of receptor that has to be immobilised in order to get a useful response signal from the low molecular weight substance. This also affects the reference surface that ought to have the same amount of immobilised proteins in order to be useful.

Carry over effects from very efficient binders have to be considered in a series of experiments. Careful washing procedures after analysis of high affinity binders are important in order to get reproducible results. In this secondary screen of compounds a simple plot of maximum response signal during the association phase versus a report point in the dissociation phase plotted in one diagram gives a classification of binders. It is possible to see groups of high affinity binders and high dissociation binders that could be a good start for a lead chemical structure (Figure 9.7).

Real-time Monitoring for Hit Identification

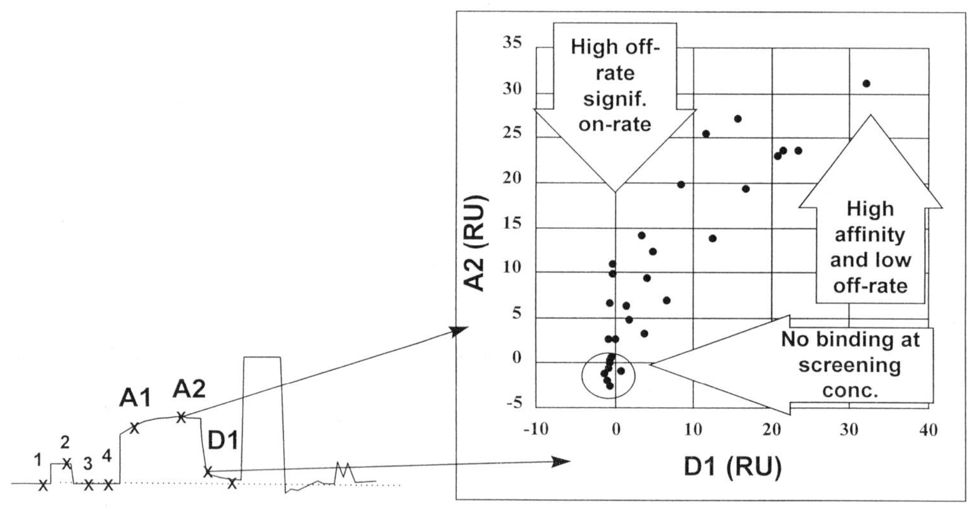

Figure 9.7 With real-time monitoring, several quality parameters in interactions between an inhibitor and an enzyme can be studied in the search for new drugs. The sensorgram illustrates injection of buffer for calibration of reference surface at number 1–3. Number 4 is the base line marker and A1, A2 and D1 are report points that characterise the interaction between an inhibitor and enzyme and the regeneration that follows for analysis of next sample. The plot of report points A2 versus D1 gives direct information about interaction behaviour for a set of drug candidates. For HIV-1 protease inhibitors the commercial drugs all have high affinity and low off-rate. In the group of high off-rate and significant on-rate drug lead candidates can be found

Combination of BIA-technology and mass spectrometry

Mass spectrometry (MS) and optical biosensors complement each other as the sensor cannot separate mass changes from different molecules whereas mass spectrometry give extremely high precision in molecular weight determinations. The increased sensitivity of MALDI and time of flight measurements have made it possible to determine mass of protein molecules with high precision. However, each molecule gives different responses dependent on the ionisation process and concentration determinations with MS have been very difficult.

There have been two approaches to combine BIA technology and MS. Both exploit binding of analyte to the sensor surface as an affinity separation step. In one approach, the MALDI matrix is applied directly on the chip which is then analysed with a MALDI-TOF instrument (Nelson *et al.* 1997a; Nelson *et al.* 1997b). The other approach is to carefully wash both the surface and the IFC followed by elution of bound material for MS measurements with the sensor chip still in the BIACORE instrument (Sönksen *et al.* 1998). Using BIACORE for affinity separation provides an easy way to immobilise and separate affinity substances at the same time as information of the interaction is obtained and the amount of analyte bound to sensor chip surface is controlled. BIACORE is thus both an analytical and a micro-preparative system that is efficient enough to make preparation of substance for a sensitive complementary technique. Antibodies, DNA-binding proteins and enzyme inhibitors have been demonstrated to be possible to analyse in these situations. In drug abuse in sports and animal production it is important that a positive immuno-assay result is followed by MS verification of the drug. Immunoassays with BIA can thus be combined with MS for such analysis.

These combined uses of high performing techniques BIA and MS can open the traditional route of using function as the handle for identification and characterisation of proteins. Over a long period of time the genome route with identification of function from genome sequences with sequence analogy and calculation of function based on structural information has been assigned extensive resources (Figure 9.8). Will biosensors technology change the way function–structure relationship is established to a procedure based on experience and detailed kinetic and affinity determinations for structural modified molecules? This is one reason why functional databases are important for future biological research and applied work.

We can modify molecular structure with genetic or synthetic methods under defined conditions. How such modification will affect biological systems is both a matter of molecular properties and concentration in different compartments. It is a wide open question how biological effects are consequences of different kinetic properties of the same molecule. Therefore we need the high quality analytical tools to study the molecular functional properties such as affinity and kinetic properties of a molecular interaction in relation to biological effects such as proliferation and differentiation.

Marketing

The introduction of BIACORE as a new technology in 1990 required several different marketing approaches. One way was to approach well-known scientists to develop applications as soon as possible, while another was to arrange symposia

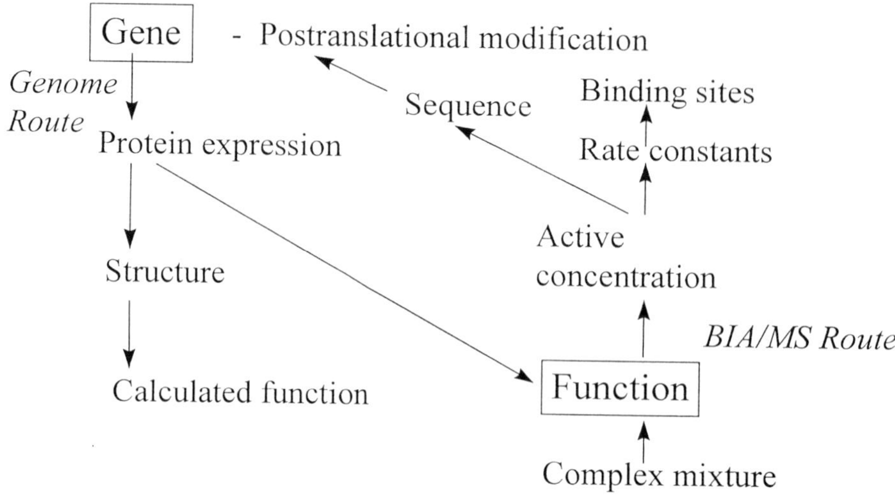

Figure 9.8 With combined approaches such as BIA–mass spectrometry, interactions form the basis for identification and characterisation. It is possible to elute what has adsorbed to immobilised ligands on a sensor chip and identify the component with MS. Another approach is to apply MALDI matrix directly on the sensor chip with the formed complex and analyse the adsorbed molecules. This route to identify genes is important when functional complexes and molecular networks and how they are disturbed by mutated proteins and inhibitors begin to be studied

about the technology. The most recent BIA-symposium was held in Edinburgh (1998), where scientific lectures, workshops and an exhibition were presented. Scientists described their work and questions in relation to technology, and during workshops Biacore personnel presented technical know-how and applications. This mixture seems to be of interest as a very open technical discussion about scientific results, and a lot of new tips and tricks to achieve the best results, were held. For many scientists it is also a way to meet and discuss future application and give input for further technical development.

From the latest BIA-symposium it can be of interest to report a few presentations that really forced the technology. Myszka and Hensley have shown the importance of working with ultracentrifugation, titration calorimetry and BIA together to optimise, compare and use the pros and cons for the three technologies in a detailed kinetic evaluation of erythropoietin ligand–receptor interactions. They characterised a 1:1 complex by ultracentrifugation, analysed the thermodynamic properties of the reaction by calorimetry and determined the kinetic and affinity by BIA. The reaction was clearly mass transport limited and k_a was determined to be $8*10^7\,M^{-1}s^{-1}$ and the affinity $3*10^{12}\,M^{-1}$. For an antibody interaction with interleukin-5 they determined the affinity to be 0.44 pM and also determined the diffusion coefficient to $7.1*10^{-7}\,cm^2 s^{-1}$ compared to D determined by ultracentrifugation to be $6.7*10^{-7}\,cm^2 s^{-1}$ in free solution. This means that BIACORE can also be used for quick determination of diffusion coefficients that together with a known molecular weight can give information about molecular shape. Diffusion coefficients have, up to now, seldom been of biological interest but with the rapidly growing interest in proteomics and compartmentalisation diffusion will probably receive greater attention in the future.

With more than 1600 instruments worldwide (2001) and a reference list of more than 1000 references, BIA-technology is now accepted as a valuable tool in biophysical characterisation of protein interactions. It is one of the few alternatives to study interactions in a general way but it also places some demands on the user. As always with science and technology, results have to be evaluated from the basic performance of the technology. With the precision and sensitivity today it is a technology that is most needed in the rapid transfer from analysis of the genome to the analysis of the proteome. There is a huge challenge in the characterisation of the molecular interactions in networks and pathways and to give kinetic properties a biological meaning.

References

Andersson, K., M. Hämäläinen and M. Malmqvist (1998) Identifying and optimising regeneration conditions for affinity biosensor assays – a multivariate cocktail approach for BIACORE. *Analytical Chemistry* 71(13), 2475–2481.

Benner, R.E. *et al.* (1979) Angular emission profiles of dye molecules excited by surface plasmon waves at a metal surface. *Optics Communications* 30, 145–149.

Buckle, M., R.M. Williams, M. Negroni and H. Buc (1996) Real time measurements of elongation by a reverse transcriptase using surface plasmon resonance. *Proc. Natl. Acad. Sci. U.S.A.* 93(2), 889–894.

Cheskis, B.J., S. Karathanasis and C.R. Lyttle (1997) Estrogen receptor ligands modulate its interaction with DNA. *The Journal of Biological Chemistry* 272(17), 11384–11391.

Christiansen, L.L.H. (1997) Theoretical analysis of protein concentration determination using biosensor technology under conditions of partial mass transport limitation. *Anal. Biochem.* 249(2), 153–164.

Cooper, M.A., D.H. Williams and Y.R. Cho (1997) Surface plasmon resonance analysis of glycopeptide antibiotic activity at a model membrane surface. *Chemical Communications* 1625–1626.

Cunningham, B. and J.A. Wells (1993) Comparison of a structural and a functional epitope. *J. Mol. Biol.* 234, 554–563.

Daiss, J.L., E.R. Scalice and D.J. Sharkey (1995) Topographical characterization of the DNA polymerase from *Thermus aquaticus*: Defining groups of inhibitor mAbs by epitope mapping and functional analysis using surface plasmon resonance. *J. Immunol. Methods* 183, 15–26.

Fägerstam, L.G., Å. Frostell, R. Karlsson, A. Larsson, M. Malmqvist and H. Butt (1990) Detection of antigen–antibody interactions by surface plasmon resonance. Application to epitope mapping. *J. Molecular Recognition* 3, 208–214.

Galio, L., S. Briquet, S. Cot, J.-G. Guillet and C. Vaquero (1997) Analysis of interactions between huGATA-3 transcription factor and three GATA regulatory elements of HIV-1 long terminal repeat, by surface plasmon resonance. *Analytical Biochemistry* 253, 70–77.

Garland, P. (1996) Optical evanescent wave methods for the study of biomolecular interactions. *Quarterly Reviews of Biophysics* 29(I), 91–117.

Hensley, P. (1996) Defining the structure and stability of macromolecular assemblies in solution: the re-emergence of analytical ultracentrifugation as a practical tool. *Structure* 4(4), 367–373.

Johanson, K., E. Appelbaum, M. Doyle, P. Hensley, B. Zhao, S.S. Abdel-Meguid, P. Young, R. Cook, S. Carr, R. Matico, D. Cusimano, E. Dul, M. Angelichio, I. Brooks, E. Winborne,

P. McDonnell, T. Morton, D. Bennett, T. Sokoloski, D. McNulty, *et al.* (1995) Binding interactions of human interleukin 5 with its receptor alpha subunit. Large scale production, structural, and functional studies of Drosophila-expressed recombinant proteins. *The Journal of Biological Chemistry* 270(16), 9459–9471.

Johnsson, B. and S. Löfås (1991) Immobilization of proteins to a carboxymethyldextran modified gold surface for biospecific interaction analysis in surface plasmon resonance. *Analytical Biochemistry* 198, 268–277.

Johnsson, B., S. Löfås, G. Lindquist, Å. Edström, R.-M. Müller Hillgren and A. Hansson (1995) Comparison of methods for immobilization to carboxymethyl dextran sensor surfaces by analysis of the specific activity of monoclonal antibodies. *Journal of Molecular Recognition* 8(1–2), 125–131.

Jönsson, U., L. Fägerstam, B. Ivarsson, B. Johnsson, R. Karlsson, K. Lundh, S. Löfås, B. Persson, H. Roos, I. Rönnberg, S. Sjölander, E. Stenberg, R. Ståhlberg, S. Urbaniczky, H. Östlin and M. Malmqvist (1991) Real-time biospecific interaction analysis using surface plasmon resonance and a sensor chip technology. *BioTechniques* 11(5), 620–627.

Jönsson, U., M. Malmqvist and I. Rönnberg (1985) Immobilization of immunoglobulin on silica surfaces. *Biochem. J.* 227, 363–378.

Jönsson, U., I. Rönnberg and M. Malmqvist (1985) Flow injection ellipsometry – an *in situ* method for the study of biomolecular adsorption and interactions at solid surfaces. *Colloids and Surfaces* 13, 333–339.

Karlsson, R., L. Fägerstam, H. Carlsson and B. Persson (1993) Analysis of active antibody concentration. Separation of affinity and concentration parameters. *J. Immunol. Methods* 166(1), 75–84.

Karlsson, R. and A. Fält (1997) Experimental design for kinetic analysis of protein–protein interactions with surface plasmon resonance biosensors. *J. Immunol. Meth.* 200, 121–133.

Karlsson, R., A. Michaelsson and L. Mattsson (1991) Kinetic analysis of monoclonal antibody–antigen interactions with a new biosensor based analytical system. *J. Immunol. Methods* 145, 229–240.

Karlsson, R. and R. Ståhlberg (1995) Surface plasmon resonance detection and multi-spot sensing for direct monitoring of interactions involving low molecular weight analytes and for determination of low affinities. *Analytical Biochemistry* 228, 274–280.

Kretschmann, E. (1978) The ATR method with focused light – application to guided waves on a grating. *Optics Communications* 26, 41–44.

Laukkanen, M.L., K. Alfthan and K. Keinänen (1994) Functional immunoliposomes harboring a biosynthetically lipid-tagged single-chain antibody. *Biochemistry* 33, 11664–11670.

Liedberg, B., C. Nylander and I. Lundström (1983) Surface plasmon resonance for gas detection and biosensing. *Sensors and Actuators* 4, 299–304.

Löfås, S. (1995) Dextran modified self-assembled monolayer surfaces for use in biointeraction analysis with surface plasmon resonance. *Pure and Applied Chemistry* 67(5), 829–834.

MacKenzie, C.R., T. Hirama, K.K. Lee, E. Altman and N.M. Young (1997) Quantitative analysis of bacterial toxin affinity and specificity for glycolipid receptors by surface plasmon resonance. *The Journal of Biological Chemistry* 272, 5533–5538.

Malmqvist, M. and R. Karlsson (1997) Biomolecular interaction analysis: affinity biosensor technologies for functional analysis of proteins. *Curr. Opin. Chem. Biol.* 1(3), 378–383.

Markgren, P.-O., M. Hämäläinen and U.H. Danielson (1998) Screening of compounds interacting with HIV-1 proteinase using optical biosensor technology. *Analytical Biochemistry* 265, 340–350.

Masson, L., A. Mazza and R. Brousseau (1994) Stable immobilization of lipid vesicles for kinetic studies using surface plasmon resonance. *Analytical Biochemistry* 218, 405–412.

Masson, L., A. Mazza, R. Brousseau and B. Tabashnik (1995) Kinetics of Bacillus thuringiensis toxin binding with brush border membrane vesicles from susceptible and resistant larvae of Plutella xylostella. *The Journal of Biological Chemistry* 270(20), 11887–11896.

Matsuda, H. (1967) Theory of the steady-state current potential curves of redox electrode reactions in hydrodynamic voltammetry. II Laminar pipe-and channel flow. *J. Electroanal. Chem.*, 15, 325–336.

Muskhelishvili, G., M. Buckle, H. Heumann, R. Kahmann and A.A. Travers (1997) FIS activates sequential steps during transcription initiation at a stable RNA promoter. *The EMBO Journal* 16(12), 3655–3665.

Myszka, D.G. (1997) Kinetic analysis of macromolecular interactions using surface plasmon resonance biosensors. *Current Opinion in Biotechnology* 8, 50–57.

Nelson, R.W., J.R. Krone and O. Jansson (1997a) Surface plasmon resonance biomolecular interaction analysis mass spectrometry. 1. Chip-based analysis. *Analytical Chemistry* 69(21), 4363–4368.

Nelson, R.W., J.R. Krone and O. Jansson (1997b) Surface plasmon resonance biomolecular interaction analysis mass spectrometry. 2. Fiber optic-based analysis. *Analytical Chemistry* 69(21), 4369–4374.

Nieba, L., S.E. Nieba-Axmann, A. Persson, M. Hämäläinen, F. Edebratt, A. Hansson, J. Lidholm, K. Magnusson, Å.F. Karlsson and A. Plückthun (1997) BIACORE analysis of histidine-tagged proteins using a chelating NTA sensor chip. *Analytical Biochemistry* 252(2), 217–228.

Nilsson, P., B. Persson, M. Uhlén and P.-Å. Nygren (1995) Real-time monitoring of DNA manipulations using biosensor technology. *Anal. Biochem.* 224, 400–408.

Oda, K. and M. Fukui (1986) Instantaneous observation of angular scan-attenuated total reflection spectra. *Optics Communications* 59, 361–365.

Ohlson, S., M. Strandh and H. Nilshans (1997) Detection and characterization of weak affinity antibody antigen recognition with biomolecular interaction analysis. *Journal of Molecular Recognition* 10, 135–138.

O'Shannessy, D.J. (1994) Determination of kinetic rate and equilibrium binding constants for macromolecular interactions: a critique of the surface plasmon resonance literature. *Curr. Opin. Biotechnol.* 5, 65–71.

Parsons, I.D. and P.G. Stockley (1997) Quantitation of the *Escherichia coli* methionine repressor-operator interaction by surface plasmon resonance is not affected by the presence of a dextran matrix. *Analytical Biochemistry* 254(1), 82–87.

Price, M.R., P.D. Rye, E. Petrakou, A. Murray, K. Brady, S. Imai, S. Haga, Y. Kiyozuka, D. Schol, M.F.A. Meulenbroek, F.G.M. Snijdewint, S. von Mensdorff-Pouilly, R.A. Verstraeten, P. Kenemans, A. Blockzjil, K. Nilsson, O. Nilsson, M. Reddish, M.R. Suresh, R.R. Koganty, *et al.* (1998) Summary report on the ISOBM TD-4 Workshop: Analysis of 56 monoclonal antibodies against the MUC1 mucin. San Diego, CA, November 17–23, 1996. *Tumor Biology* 1, 1–20.

Rauffer-Bruyère, N., J. Chatellier, E. Weiss, M.H.V. Van Regenmortel and D. Altschuh (1997) Cooperative effects of mutations in a recombinant Fab on the kinetics of antigen binding. *Molecular Immunology* 34(2), 165–173.

Richalet-Sécordel, P.M., N. Rauffer-Bruyère, L.L.H. Christiansen, B. Ofenloch-Haehnle, C. Seidel and M.H.V. Van Regenmortel (1997) Concentration measurement of unpurified proteins using biosensor technology under conditions of partial mass transport limitation. *Anal. Biochem.* 249(2), 165–173.

Roos, H., R. Karlsson, H. Nilshans and A. Persson (1998) Thermodynamic analysis of protein interactions with biosensor technology. *J. Molecular Recognition* 11, 204–210.

Ruzicka, J. and E. Hansen (1983) Flow injection analysis from test tube to integrated microconduits. *Analytical Chemistry* 55, 1040A–1053A.

Ruzicka, J. and E. Hansen (1987) *Flow Injection Analysis.* New York, Wiley.

Schuck, P. (1996) Kinetics of ligand binding to receptor immobilized in a polymer matrix, as detected with an evanescent wave biosensor. I. A computer simulation of the influence of mass transport. *Biophysical Journal* 70(3), 1230–1249.

Sjölander, S. and C. Urbaniczky (1991) Integrated fluid handling system for biomolecular interaction analysis. *Analytical Chemistry* 63, 2338–2345.

Sönksen, C.P., E. Nordhoff, Ö. Jansson, M. Malmqvist and P. Roepstorff (1998) Combining MALDI mass spectrometry and biomolecular interaction analysis using a biomolecular interaction analysis instrument. *Analytical Chemistry* 70(13), 2731–736.

Stenberg, E., B. Persson, H. Roos and C. Urbaniczky (1991) Quantitative determination of surface concentration of protein with surface plasmon resonance by using radiolabeled proteins. *J. Colloid. and Interface Science* 143, 513–526.

Sternesjö, Å., C. Mellgren and L. Björck (1996) Analysis of sulfamethazine in milk by an immunosensor assay based on surface plasmon resonance. *Immunoassays for Residue Analysis, ACS* 463–470.

Swalen, J.D. *et al.* (1980) Plasmon surface polariton dispersion by direct optical observation. *Am. J. Phys.* 48, 669–672.

van der Merwe, P.A., D.L. Bodian, S. Daenke, P. Linsley and S.J. Davis (1997) CD80 (B7–1) binds both CD28 and CTLA–4 with a low affinity and very fast kinetics. *Journal of Experimental Medicine* 185(3), 393–403.

Welford, K. (1991) Surface plasmon-polaritons and their uses. *Optical and Quantum Electronics* 23, 1–27.

Wigren, R., P. Billsten, R. Erlandsson, S. Löfås and I. Lundström (1995) Force measurements on a hydrogel with an SFM. *J. Colloid. Interf. Sc.* 174, 521–523.

Zeder-Lutz, G., E. Zuber, J. Witz and M.H.V. Van Regenmortel (1997) Thermodynamic analysis of antigen–antibody binding using biosensor measurements at different temperatures. *Anal. Biochem.* 246(1), 123–132.

10 IAsys

The resonant mirror biosensor

R.J. Davies and P.R. Edwards

Introduction

Optical evanescent biosensors greatly facilitate the study of the interaction between a wide range of biomaterials. Since 1993 the resonant mirror (RM) as embodied in IAsys, has been used for work on eukaryotic and prokaryotic cells, viruses, liposomes, antibiotics, viruses, protein-coated latex and gold sols, polysaccharide, monosaccharide, bacterial toxins, antibiotics and vitamins, and the more conventional protein and nucleic acid binding studies. In principle the user links one of the interactants to the biosensor surface, adds the other interactant(s) and follows the binding response with time. With highly specific interactions, of reasonable affinity, measurements can be made in a realistic support medium, e.g. blood plasma. The optical 'signal' is generated by the change in refractive index as the binding species displaces water from the interrogated surface. In addition to revealing details such as the strength and rate of binary complex formation and dissociation, the order of assembly of higher complexes can be elucidated. This information will complement the high resolution structural data that is becoming increasingly available.

This article will cover some of the publications that have appeared since 1997 in which the RM has made a substantial contribution. Earlier RM-based works are covered in other reviews (Davies *et al.* 1994; Yeung *et al.* 1995; Purvis *et al.* 1998). For completeness the optical phenomenon and surfaces available for use in IAsys are briefly described.

Modus operandi – light and surfaces

The origin of the optical resonance and it angular measurement have been described (Cush *et al.* 1993), however the following) in conjunction with Figure 10.1, is a more succinct account. The RM is an approximately 1 cm cube of a lead-doped glass with one surface coated in turn with a low (relative to the cube) refractive index silica spacer ($t \approx 500$ nm) and a high index oxide layer ($n_D \approx 2$, $t \approx 100$ nm). Coatings are optimised to give maximum sensitivity. Laser light ($\lambda = 670$ nm), is passed through a polariser to equalise the transverse electric and magnetic (TE and TM) intensities, then swept repeatedly across one face of the cube. At a particularly well defined angle for each polarisation, light entering the cube will propagate along the high index layer which acts as a monomode waveguide. At this 'resonance angle' an evanescent field is set up at the aqueous/high index interface and light couples out of the high index layer and interferes with light reflected from the top of the cube.

Phase sensitive optics and a detector are used to measure the light intensity which reaches a maximum at the resonant angle. This angle is determined by the refractive index of the aqueous sample, particularly within a few 100 nm of the surface. Optical modelling indicates that the field intensity falls exponentially into the sample with a decay length of about 100 nm. Thus a molecule 100 nm from the surface would give 1/eth (i.e. ≈37 per cent) of its response at the surface, at 200 nm this falls to ≈13.5 per cent. IAsys measures the resonance angle to better than 0.1 arc second (i.e. 1/36,000 of a degree). For a particular protein the response is directly proportional to its concentration at the sensor surface. A calibration has been established using $[I^{125}]$-labelled proteins: for the planar surfaces it is 600 arc sec/ng/mm^2, for the standard carboxymethyl dextran (CMD) it is 200 arc sec/ng/mm^2. The higher angular shifts on the planar surfaces simply reflect the evanescent nature of the RM. The response is related to the *mass* concentration of protein at the surface. Where a mole measure is needed, for example to determine binding stoichiometry, the signal of each binding partner should be divided by its relative molecular mass (M_r). For lipid, carbohydrate and nucleic acids no calibration exists. In part this is because, unlike protein, these adopt a wide variety of structural forms. For example even a one component monolayer of lipid adsorbed on a hydrophobic surface would be expected to give a response which depended upon its lateral packing density. This will be governed by pH, ion-binding and temperature. With more complex structures, other factors apply. For instance consider two close-packed films of monodisperse, spherical, unilamellar liposomes of different size. Although the mass per unit area of lipid on the surface is

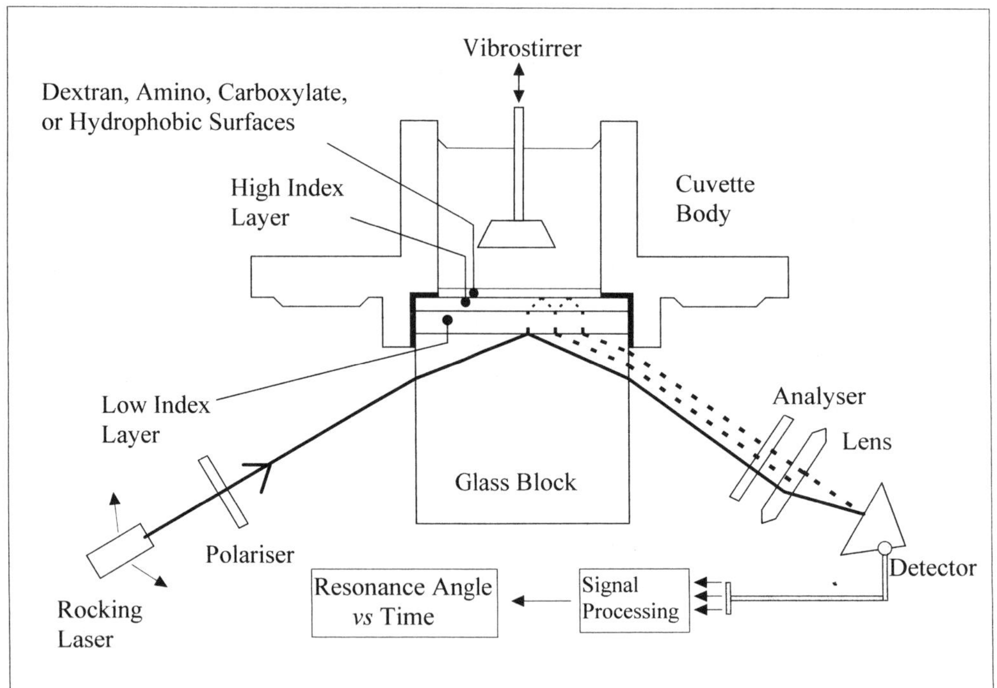

Figure 10.1 Schematic of the IAsys cuvette with integral prism showing the relative dispositions of the optical components and the vibrostirrer

independent of liposome diameter, because of the evanescent field decay the layer of larger vesicles will give a smaller signal.

To add versatility, cuvettes with several surface types are available; some guidance on which surfaces may be preferable for particular applications are supplied by the manufacturer. However, because of the plethora of biomolecules with different structures, rigid rules for surface selection are not possible. The CMD surface, with a degree of carboxylation of 0.55 per glucose residue and an estimated thickness of about 200 nm, has found widest applicability. A large area (17 mm^2) CMD cuvette is also available for 'ligand fishing'. The other surfaces are planar. Carboxylate and primary amino are more suitable for work with large species unable to enter the CMD (e.g. cells, liposomes, large viruses and highly charged molecules that may be electrostatically repelled from the CMD). The hydrophobic surface (water contact angle $\geq 104°$, critical surface tension ≈ 22 mNm^{-1}) has been designed for lipid monolayer deposition and hydrophobic 'binding' of macromolecules. The biotinylated surface is designed for work with biotinylated molecules which are attached by pre-coating it with avidin, streptavidin or neutravidin. One advantage of this surface over that of simply coupling avidin to the CMD surface is that it can be regenerated and reloaded with avidin. IAsys is available in manual and automated versions with dual channel cuvettes to enable control and test measurements to be made together. The automated version has temperature-controlled pads for microtitre plates and reagent wells enabling solutions to be held at 4 °C. Both have a dedicated suite of software to allow rapid data analysis.

Developments in kinetics

The interaction of a dissolved molecule (the ligate, L) with one immobilised on the sensor (the ligand, G) can be represented as:

$$G + L \; \underset{k_d}{\overset{k_a}{\rightleftharpoons}} \; GL \tag{10.1}$$

where k_{ass} and k_{diss} are the association and dissociation rate constants. More complex binding interactions are possible and have been considered in the context of optical biosensors (Myszka *et al.* 1997).

The RM's ability to measure both the affinity (K_D) and the component rate constants (k_{ass}, k_{diss}) of many biomolecular interactions gives it wide appeal. These constants may, years from now, become as important as K_m, k_{cat} are to the metabolic biochemist. Mechanisms of enzyme regulation may also have their parallel in the assembly of biomolecular complexes. Optical biosensors will play a large part in this arena. An outline of their measurement and more recent developments is presented below.

Data analysis

For kinetic investigations the concentration of ligate is assumed to be always in excess of the ligand and so remain effectively constant. Binding of ligate will then be described by an integrated rate equation:

$$R_t = R_{eq}\left(1 - e^{-(k_{ass}[L] + k_{diss})t}\right) \tag{10.2}$$

with the biosensor response at time $t(R_t)$ depending upon the equilibrium response (R_{eq}), ligate concentration $[L]$, and the two rate constants, k_{ass} and k_{diss}. The exponential term is often called the apparent on-rate, k_{on}.

Data at high ligate concentrations are often poorly described by equation 10.2. Incorporation of a second exponential term will give a better fit, returning two k_{on} values and two extent values (A and B), then equation 10.22 becomes:

$$R_t = A(1 - \exp^{-k_{on1}t}) + B(1 - \exp^{-k_{on2}t}) \tag{10.3}$$

Rate constants may be determined from plotting the on-rate, determined at differing ligate concentrations, against ligate concentration. The resulting graph should be a straight line with slope, k_{ass}, and intercept, k_{diss}. The first, rapid k_{on} value is used to construct this plot when the raw data are better described by the double exponential equation. The physical cause(s) of the slower exponential term remain unclear.

The determination of k_{diss} from the intercept of k_{on} vs $[L]$ often has a high extrapolation error. A more robust approach is to initiate dissociation by removing the ligate with a buffer wash. Under these conditions (i.e. $[L] = 0$) dissociation of ligate from immobilised ligand will be described by an equation of form:

$$R_t = R_0 e^{-k_{diss}t} \tag{10.4}$$

where R_o is the response at the initiation of dissociation. Many dissociation profiles, however, are not described by this simple equation. This is due to incomplete dissociation which has been attributed to rebinding of the dissociated ligate. Often the incorporation of an offset term gives an acceptable fit.

The binding affinity or equilibrium (dissociation) constant (K_D) can be obtained from k_{diss}/k_{ass} or the binding isotherm (i.e. a plot of R_{eq} vs $[L]$). In the latter case K_D can be calculated from:

$$R_{eq} = \frac{R_{max}[L]}{[L] + K_D} \tag{10.5}$$

where R_{max} is the maximum binding response obtained by saturation of the ligand.

The above outlines the usual methods for kinetic analysis using the IAsys biosensor. However, the analysis of binding data is not restricted to these approaches. For instance, initial rate analysis of chymotrypsin inhibitor 2 (CI-2) binding to immobilised chymotrypsin was used to determine k_{ass} (Edwards and Leatherbarrow 1997). A plot of initial rate against ligate concentration produced a straight line passing through the origin with a slope of $k_{ass}R_{max}$. Thus simply measuring R_{max} from a saturating concentration of CI-2 allowed the k_{ass} to be calculated and requires no assumptions about the interaction model. In addition, the analysis is rapid since binding need only be monitored for short times.

Experimental design and surface comparison.

Kinetic measurements at a surface can be influenced by diffusive and convective mass transport. For interactions with a high intrinsic k_{ass} diffusion of ligate through a stagnant layer to the ligand may be the governing factor, and the measured k_{ass} is then simply a fickian rate. The stagnant layer thickness is reduced by rapid stirring in IAsys. The effect of the diffusive transport limit may also be reduced by lowering the amount of ligand immobilised (Glasser 1993). High ligand loading reduces k_{ass} by steric hindrance and k_{diss} by increasing the rebinding probability (Edwards *et al.* 1997). Recommended practice is to immobilise low amounts of ligand for kinetic analysis. Also at analysis the inclusion of data at too high a ligate concentration produces a k_{on} vs [ligate] plot which is curved over at these higher concentrations. Inclusion of very low ligate data may cause positive curvature at the lower concentrations due to ligate depletion (Edwards *et al.* 1998). However, depletion only becomes significant at high ligand loadings (R_{max}), high affinities (low K_D) or with low molecular weight ligates. Under these conditions a second order kinetic equation may be applied (Edwards *et al.* 1998).

Most kinetic work has been performed with CMD surfaces and concerns about the influence of this hydrogel have been raised (Schuck 1996). Kinetic constants measured on planar amino, carboxylate and CMD with different systems are shown in Table 10.1. Generally, there was little difference between the kinetic constants determined on each of the surfaces for each of the binding systems, even in the human serum albumin (HSA): anti-HSA system when the immobilised protein was switched. Dissociation of the HSA antibody from the CMD matrix was slower than with the other two surfaces, but no reason has been established for this. Comparison of kinetic measurements made by fluorescence quench, fluorescence titration, BIAcore and IAsys (CMD in both biosensors) for lysozyme antibody (D1.3) and its single chain Fv fragment binding to lysozyme indicated a good agreement (Yeung *et al.* 1995). Suggesting that in this system any hydrogel/surface related artefacts were negligible. However, further work in this area is required to validate the biosensor measurement.

An alternative way to determine K_D, and well suited to the cuvette system of IAsys, is equilibrium titration (Winzor and Hall 1997). Here, sequential additions of ligate are made to the cuvette and the equilibrium response at each concentration is

Table 10.1 Kinetic constants obtained on three sensor surfaces with two binding systems

Surface	*System*	$k_{ass}(10^6)/M^{-1}s^{-1}$	$k_{diss}(10^{-3})\ s^{-1}$
Carboxylate	Lysoyzme/D1.3fv	1.15 ± 0.10	5.88 ± 1.74
Amino		1.36 ± 0.10	6.75 ± 1.72
CMD		1.21 ± 0.03	3.97 ± 0.36
Carboxylate	HSA/anti-HSA	0.23 ± 0.016	6.02 ± 0.65
Amino		0.20 ± 0.018	6.38 ± 0.54
CMD		0.28 ± 0.026	0.31 ± 0.49
Carboxylate	Anti-HSA/HSA	0.18 ± 0.013	2.43 ± 0.21
Amino		0.25 ± 0.088	7.15 ± 0.41
CMD		0.14 ± 0.010	4.52 ± 0.42

noted. This method does not require any regeneration steps, and is particularly suitable for ligands that cannot be regenerated, or where a suitable regenerant has not been identified.

Recent applications

Because of the versatility of the RM and the diverse nature of work undertaken with it, there is no obvious way to compartmentalise the published works. While protein–protein and, to a lesser extent, protein–nucleic acid studies have been the 'traditional' arena for optical biosensors, recent works have used a combination of molecular biology and peptide synthesis to precisely pinpoint the binding sites on proteins. Studies involving more than one protein (e.g. competitive binding of two or more proteins to the one immobilised) and the role of ions and small molecules upon interactions are becoming more common. This review will be divided along broadly biochemical lines, in which works pertaining to a particular aspect will be considered together.

Protein–protein interactions

The maintenance of normal human factor VIII (F8) levels in serum is controlled through complex formation with von Willebrand factor (vWf), and both are important for blood coagulation. An acidic region on the light chain of F8 (residues 1649–1689) is believed to be involved in the interaction. Using vWf coupled to a CMD surface, the binding characteristics (K_D, k_{ass}, k_{diss}) of F8, total light chain and proteolytic light chain digests producing fragments lying within and outside the acidic region of F8 light chain, were determined (Saenko and Scandella 1997). Radioiodinated F8 was used in competitive F8–vWf inhibition assays with native F8 and some of the proteolytic fragments and verified the biosensor K_D values. Light chain derivatives lacking the acidic region had greatly reduced affinities (K_D range: 560–660 nM) compared to the whole light chain (3.8 nM) and a 14 kDa fragment (1672–1794) containing all the acid residues (72 nM). Interestingly it was shown that the 14 kDa fragment could associate with a 63 kDa fragment (residues 1795–2332) and bind to vWf with an affinity of 3.7 nM, very close to that of the intact light chain. The biosensor data indicates that part of the acidic region and a domain on its C-terminal side are both required for high affinity binding.

IAsys was used to investigate different aspects of human plasma histidine rich glycoprotein (HRG). This protein has been implicated in coagulation, fibrinolysis, zinc transport, it binds to macrophages and platelets and may regulate the action of growth factor. Its central domain has a high histidine content (13 mole %) forming part of a 12 tandem repeat (GHHPH), and it has two N-terminal cystatin modules.

Purified complement C1q from human serum has associated with it two ≈30 kDa proteins having a similar sequence to the cystatin domains of HRG (Gorgani *et al.* 1998). HRG was shown to bind to C1q by ELISA and to C1q immobilised on CMD surfaces where two binding sites having affinities 7.8 nM and 37.3 nM were identified. When the binding order was reversed, in this case using biotinylated HRG immobilised to a streptavidin-loaded CMD (SL-CMD) surface, only one binding site was observed having an affinity of 9.2 nM. The authors reasoned that the tail region of hexameric C1q was involved in the interaction. Since it was known that

C1q can bind to the IgG-Fc region, and may form insoluble immune complexes (IIC) the binding of HRG to IgG was investigated. Using immobilised biotinylated hIgG on SL-CMD the HRG bound, and a single site of moderate affinity (85 nM) was identified. This site was located within the $F(ab')_2$ region as the binding affinity to immobilised $F(ab')_2$ fragments was close to that of the whole IgG. Zinc reduced the interaction between HRG and C1q but strengthened the HRG–IgG interaction. HRG markedly inhibited the formation of IIC between ovalbumin and a polyclonal anti-ovalbumin, presumably by obscuring the antibody paratope. The authors believe that a new functional role for HRG in the suppression of IIC formation and control of the classical complement pathway *via* its interaction with HRG may have been found.

The role of HRG as a plasma pH sensor was suggested by following its binding to heparin-BSA, immobilised on an amino surface, at different pHs (Borza and Morgan 1998). At pH 7.25 there was little HRG binding, than at more acidic pHs – where a positive charge would develop on the central domain of HRG. Equilibrium titration isotherms at pH 6.8 indicated a K_D for the HRG–heparin interaction of about 45 nM, Figure 10.2, fluorescence measurements gave 55 nM. As in the above HRG work, zinc was also shown to modulate binding to heparin. Since heparin is usually intracellularly localised, five other negatively charged cell-surface gly-cosaminoglycans were assessed as HRG binding sites. Of these dermatan sulphate was the most effective, and when terminally immobilised through its reducing end to an amino surface with cyanoborohydride it bound HRG at pHs below neutral.

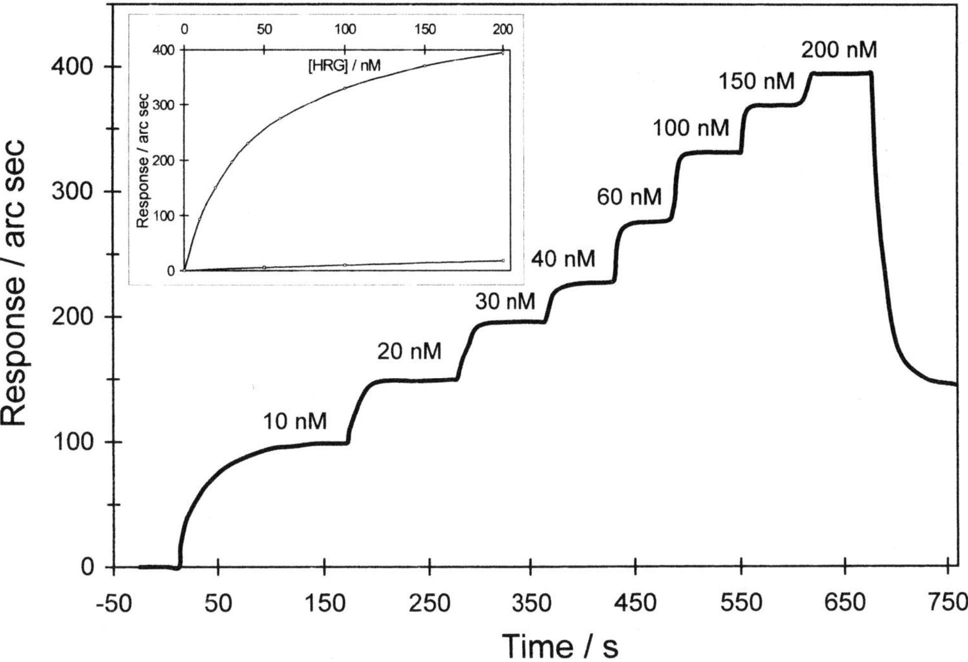

Figure 10.2 Equilibrium titration of immobilised heparin with human HRG, concentrations shown, at pH 6.8. *Inset*: isotherms of HRG bound to heparin at pH 6.8 (upper curve, converted from the titration data) and pH 7.3 (lower curve) in the absence of Zn^{2+}. Redrawn from Borza *et al.* 1998, with permission

Dermatan sulphate proteoglycan also gave similar results when coupled to an amino surface using pre-polymerised glutaraldehyde. The HRG-dermatan sulphate interaction was weaker than with heparin, and binding was not so strongly potentiated by zinc ions. The authors propose that HRG binds to circulating heparin and inhibits its anti-coagulation activity under circumstances in which the blood is acidotic. This would explain clot formation in stagnant blood flows and the interaction may be important in hypoxia and ischaemia. HRG when bound to heparin also specifically binds plasminogen, which may lead to plasminogen activation. IAsys and BIAcore measurements have indicated that plasminogen activation may be regulated by tissue factor (TF), an integral membrane protein (Fan *et al.* 1998). Purified and relipidated TF and soluble-recombinant TF were immobilised to CMD surfaces on both instruments. Plasminogen bound to TF at a site distinct from the binding site for factor VII and VIIa. Plasminogen binding to TF required kringle (lysine binding) domains 1 and 3; kringles 4 and 5 appear redundant with respect to TF binding.

Glycosaminoglycans are also important in tissue development. One group has used the RM to decipher their role in the mammary gland with the aim of understanding the aetiology of cancer. Hepatocyte growth/scatter factor (HG/SF) and acidic and basic fibroblast growth factors regulate mammary gland development through interaction with fibroblast growth factor receptors and proteoglycans (i.e. protein-linked glycosaminoglycan) such as heparan sulphate (HS) and dermatan sulphate (DS).

HS peptidoglycans were purified from cultured cell lines representative of cells forming the mammary duct; *viz.* epithelial, myoepithelial, stromal fibroblast and two malignant epithelial cell lines (Rahmoune *et al.* 1998a). They were terminally biotinylated and attached to an SL-CMD surface. Using an assay developed on streptavidin-amino RM surfaces it was shown that about 94 per cent of the HS chains were biotinylated at the peptide moiety. Three HG/SF binding sites were identified – two having a K_D of 1–3 nM but which were different since they had different rate constants; the third was of higher affinity ($K_D \approx 0.2$ nM). Each type of HS was found to have one of three binding sites for HG/SF. The origin of the sites lay in the HS 'fine structure' – the distribution of 6-O-sulphate groups on the glucosamine rather than its overall degree of sulphation. The type of HG/SF binding displayed by the malignant cells invading the stroma could enable them to capture HG/SF from stromal and basement membrane HS and so help their proliferation. Their second paper (Lyon *et al.* 1998) reported the existence of a high affinity interaction ($K_D \approx 19.7$ nM) between HG/SF and DS. While this affinity is 10–100 fold lower than the HG/SF–HS interaction (above) but similar to that for basic FGF–HS interaction (Rahmoune *et al.* 1998b) it is higher than reported for other species with DS. DS proteoglycan is located in the stromal matrix and HS proteoglycan is concentrated closer to the lumen of the mammary duct. This spatial arrangement may create a gradient of increasing affinity from the source of HG/SF (mammary fibroblasts) to the basal membrane/epithelium interface and so guide the HG/SF to its site of paracrine action.

The binding of acidic and basic fibroblast growth factors (aFGF, bFGF) to biotinylated HS proteoglycans from the cell lines used above have been measured (Rahmoune *et al.* 1998b). The ability of aFGF and bFGF to stimulated DNA synthesis was also determined. The association rate constant (k_{ass}) for bFGF to HS was about fifty times that of aFGF, which had three distinct values depending on the source of HS. Two binding sites were found for bGFG on HS and these were shown to be involved in controlling DNA synthesis.

HS has been proposed as the cell surface receptor for Herpes simplex virus, type 1(HSV-1). HSV-1 was bound to HS sulphate immobilised to CMD using EDC/NHS and a 6 carbon linker chemistry (Inoue *et al.* 1996). The virus and neutralising antibody titers by RM were similar to those obtained by a standard cytopathic method which required cultured cells and at least four days to yield a result. The same group previously followed the binding of measles virus to Vero cells, obtaining a lower limit of detection of about 103 virions/ml (Inoue and Arai 1998).

Whole cell work

Although all cells are large in relation to the evanescent field decay length, and many possess a thick, highly hydrated glycocalyx which would oppose close approach of cell and sensor surface, several publications have appeared in this area.

Helicobacter pylori (size: $\approx 0.7 \times 3\,\mu M$ with flagella located at one pole) was shown to bind a sialyl(α-2,3)-lactose polyacrylamide conjugate (20 mole% carbohydrate, total Mr = 30 kDa) immobilised on an amino surface with glutaraldehyde (Hirmo *et al.* 1998). Haemagglutinating strains gave resonant angle shifts of 600–800 arc seconds, around ten-fold higher than weakly haemagglutinating strains (Figure 10.3).

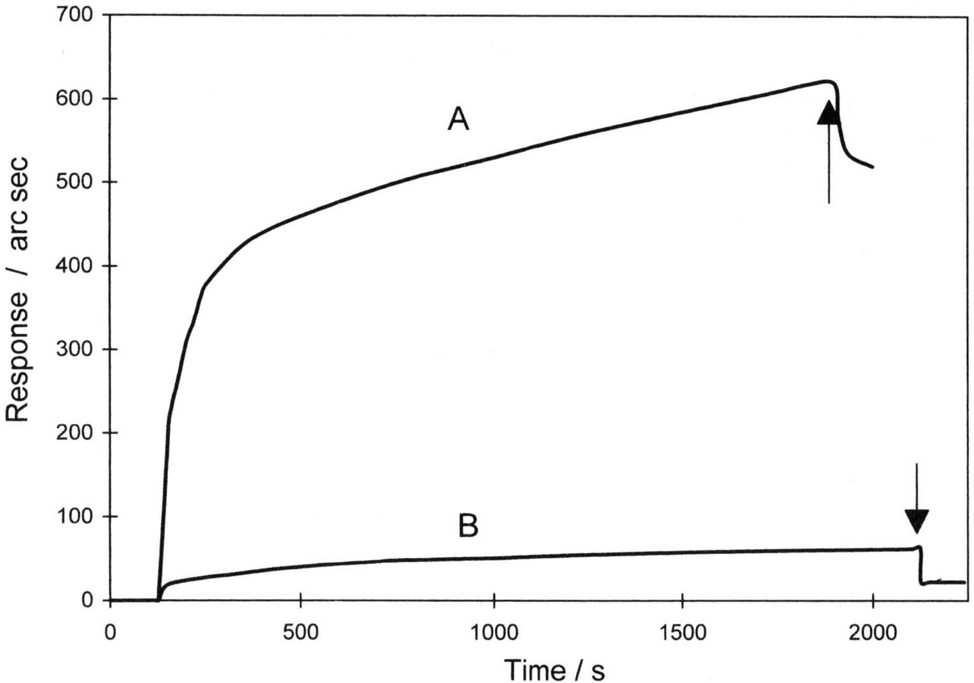

Figure 10.3 Binding of haemagglutinating (A) and weakly haemagglutinating (B) strains of *H. pylori* to sialyl(α-2,3)-lactose: polyacrylamide coupled to an amino sensor surface. The cell concentration was 10^9 colony forming units/ml. The arrow indicates a buffer wash to remove non-adherent cells. Redrawn from Hirmo *et al.* 1998, with permission

Free sialyl(α-2,3)-lactose reduced cell attachment, specifically and fully, eluting an adhesion protein from the bacterial surface which was bound to the conjugate. The RM was more rapid than standard haemagluttination and tissue culture binding assays, and information from competition-displacement assays would be useful in anti-adhesion drug development.

In an unusual study the ability of proximal tube kidney cells to bind and potentially reabsorb protein before excretion was investigated (Thakkar *et al.* 1998). The approach taken was to determine the concentration of one (of five) proteins investigated to halve the binding rate of cells to a CMD surface loaded with either albumin or retinal binding protein (RBP). An effective cell removal step was established, enabling the cuvette to be used for several inhibitory proteins. All five proteins had similar inhibition profiles indicating that they share a common receptor, even though there was, in extrema, a fifty-fold difference in protein–cell binding. Cell binding caused fairly small increases in resonance angle of ≤ 50 arc sec. Controls proved binding was not simply an artefact of cell sedimentation. The stirrer was also shown not to damage the cell membrane.

Cell specific responses of similar magnitude were also found (Morgan *et al.* 1998a). For this work a CMD surface was preferred over an amino surface because it better discriminated between the binding of HLA class 1 expressing T-cells and non-expressing cells to immobilised β2-microglobulin. Binding was carried out in a culture medium free of growth factors and antibiotics which caused non-specific binding. Having established a specific T-cell binding protocol, free β2-microglobulin at about 100 nM was shown to reduce this by half. Finally, the effect of incubating with the T-cells' two nonameric peptides – one that binds to the MHC heavy chain and one that does not, on cell binding to β2-microglobulin – loaded CMD was determined. The MHC binding peptide reduced cell binding to β2-microglobulin, while the other peptide enhanced binding. This system may be useful in the study of antigenic peptide presentation between this and other classes of HLA molecules.

Proteins acting at cell membranes

Membranes are the communications interface between the phases they encompass. They have in and on them a large number of receptors capable of either eliciting a trans-membrane response, or of acting as a transient building platform for the assembly of various protein complexes.

The influence of six peptides on the exchange of β2-microglobulin in complex with a purified HLA class 1 molecule, with β2-microglobulin bound to CMD on a sensor has been reported (Morgan *et al.* 1998b). Exchange rates depended upon the peptide, and five of these significantly enhanced the stability of the HLA:peptide:β2-microglobulin complex. Measurement of the complex stability at the cell surface by flow cytometry supported some of the biosensor findings. Analysis of the stabilising peptide sequences indicated that both primary and secondary motif residues were involved in stabilisation. Compared to other methods of determining peptide–HLA interaction, the RM has proved faster, simpler and in addition provides useful kinetic information.

Synapsins are a family of proteins involved in binding of small synaptic vesicles to the cytoskeletal F-actin in pre-synaptic terminals prior to stimulation evoked exocytosis. Binding to F-actin is regulated by calmodulin and by phosphorylation of the synapsin. One type, synapsin 1, has previously been shown to bind to calmodulin, a Ca^{2+}-binding protein. Using an RM, synapsin II (SII) was shown to bind to immobilised biotinylated calmodulin in a Ca^{2+} dependent manner (Nicol *et al.* 1997). Examination of the binding kinetics revealed that phosphorylated synapsin II (SIIPi) bound calmodulin with a $K_D \approx 59\,nM$, and in the non-phosphorylated state (SIIPo) the affinity was approximately doubled. Larger differences were apparent in the rate constants, with SIIPi binding about ten-fold faster and dissociating twenty times quicker than SIIPo. The K_D values, determined from equilibrium binding isotherms, were $54\,nM$ (SIIPi) and $29\,nM$ (SIIPo). The K_D values were internally consistent (Schuck and Minton 1996). Two non-overlapping peptide stretches in SII were suspected calmodulin binding sites. Only one, when synthesised and tested, inhibited binding of SII with calmodulin, and its K_D was $32\,nM$ – the same as that of SIIPo – suggesting that binding activity resides fully in this peptide. Non-phosphorylated Synapsin I also bound calmodulin, with similar affinity and rate constants as SIIPo, although a fruit fly synapsin II had a lower affinity ($K_D \approx 120\,nM$). The authors concluded that calmodulin–synapsin interaction is likely to be a common factor in the regulation of synaptic function.

The cytoplasmic domain of the transmembrane protein CD44 binds protein 4.1, a $78\,kDa$ cytoskeletal protein, with $K_D \approx 0.1\,\mu M$ (Nunomura *et al.* 1998). There was a sixteen-fold increase in the dissociation rate of the complex in the presence of calmodulin and Ca^{2+}, while the k_{ass} was unchanged. Neither of these alone affected binding. A sequence critical for binding of protein 4.1 to CD44 was found in the $30\,kDa$ domain of protein 4.1, a domain also involved in the binding of this protein to erythrocyte membrane band 3 protein. A clear sigmoidal relationship was obtained between the pCa^{2+} and the fractional occupancy of CD44 sites by protein 4.1. The binding of cytoskeletal Ankyrin to CD44 was modulated by protein 4.1.

Membrane cytoskeletons are dynamic structures that respond to external environmental conditions by modulating the interactions between their components. The above studies have began to dissect these interactions not simply in terms of which proteins bind together, as in the photochemical cross-linking studies carried out in the 1970s and 1980s, but also how tight they bind, how quickly they can associate and dissociate and whether species, like Ca^{2+}, soften or harden these interactions. These factors are likely to be microscopic measures of the flexibility of the individual cytoskeletal joints, and will have a bearing on the deformability of the membrane/cytoskeletal assembly.

The cytosolic second messenger Calmodulin/Ca^{2+} also binds with high affinity ($K_D \approx 48\,nM$) to a protein kinase C activating protein in the post-synaptic neurone and interferes with protein kinase C binding (Faux and Scott 1997).

TGN38 is an integral membrane protein with one tyrosine in its internalisation motif (-YQRL-). It cycles between the plasma membrane and the trans-golgi network, and a hexapeptide sequence -SDYQRL- in its cytoplasmic domain was known to bind to the $\mu 2$ subunit of the AP2 (plasma membrane) clathrin adaptor. By immobilising various thioredoxin-TGN38 (cytoplasmic domain) constructs on a CMD surface and following binding of soluble $\mu 2$ the requirement for a serine residue in the hexapeptide was shown (Stephens *et al.* 1997). The importance of serine just upstream from the internalisation motif was implicated in an earlier work. Other methods, i.e. two hybrid analysis and immunoprecipitation, were unable to demonstrate the serine requirement.

Interactions between regulatory proteins

A large number of hormones and neurotransmitters exert their action via a large group of GTP hydrolysing proteins. These heterotrimeric G-proteins have GTPase activity in their α subunit and their tightly associated $\beta\gamma$ subunits can dissociate when GTP is bound to activate downstream effectors.

Interaction between the α subunit and heterodimers $\beta\gamma$ have been investigated to determine whether heterodimer composition controls the functionality of the complex (Figler *et al.* 1997). G-proteins are involved in transmembrane signal transduction and couple the binding event at the cell membrane receptor to the production of a cytosolic 'messenger'. Both $\gamma1$ (naturally farnesylated) and $\gamma2$ (bearing a geranylgeranyl group) heterodimers ($\beta\gamma$) bound to the biotinylated α-subunit on SL-CMD surface. Both heterodimers had a similar affinity for the surface ($\approx100\,nM$), which is in the range (1–$300\,nM$) determined by other methods. RM, crystal structure and immunoprecipitation data also support the view that the nature of the γ subunit does not influence assembly of the G-protein heterotrimer.

Two other G-proteins have also been studied. Rab 5 is one of many cytosolic factors directing small vesicle movement between organelles. RM measurements indicated that rab 5 bound rabaptin -5β, a known effector which was immobilised on an amino surface. The binding was tighter when rab 5 had also bound GTPγs (a non-hydrolysable GTP analogue) than when complexed with GDP (Gournier *et al.* 1998). The membrane-associated ras oncogene product has been shown by RM to bind to phosphatidylinositol kinase γ in the presence of GTP, but not GDP (Rubio *et al.* 1997). The recombinant kinase used bore a glutathione-S-transferase tag which bound to reduced glutathione (GSH) linked to CMD surface via a 6 aminohexanoic acid spacer; without the spacer, kinase-rab 5 binding could not occur.

Protein–lipid interactions

There are two methods of studying these. The first is to put the protein on the sensor surface, the second to put the lipid on the sensor surface. Both approaches have been used.

The recently released hydrophobic surface facilitates the latter approach. The first work on this surface looked at the formation of various phospholipid monolayers, both pure and as binary mixtures from a organic lipid solution (Athanassopoulou *et al.* 1999). Using BSA as a probe, it was established that monolayers fully covered the sensor surface, with their headgroups facing the aqueous solution. This method of monolayer formation is less wasteful of materials and time than film formation from liposomes. To investigate protein binding monolayers of purified brain G_{M1} ganglioside and phosphatidylcholine alone, and as mixtures, were formed and binding of cholera toxin B subunits to G_{M1} followed. This system indicated that even a fairly water soluble lipid such as G_{M1} could form a stable monolayer. The film could be repeatedly treated with $0.1\,M$ HCl, a common regenerant in protein–protein studies on biosensors, to release the bound B sub-units and allow further bindings to take place. The affinity of the interaction was essentially identical to that measured by other methods, and the amount of subunit bound depended upon the concentration of G_{M1} in the monolayer forming solution (Figure 10.4). The

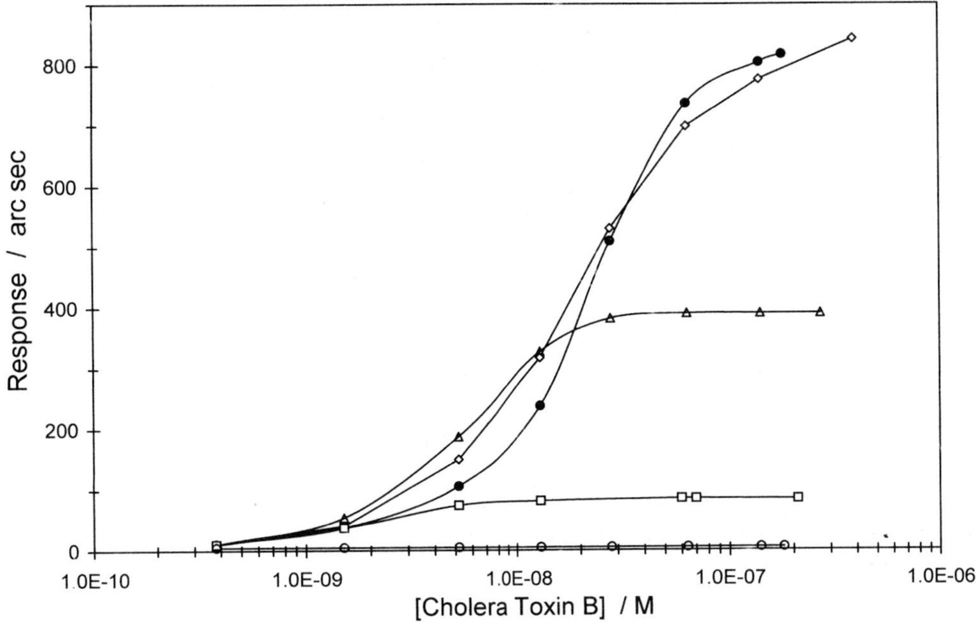

Figure 10.4 Equilibrium binding responses of cholera toxin B-subunits with lipid monolayers formed on a hydrophobic sensor surface from monolayer-forming solutions in propanol having the composition: pure bovine brain G_{M1} (●); dioleoyl phosphatidylcholine (DOPC) containing G_{M1} at 25 mole% (◇), 3.8 mole% (△), and 1.3 mole % (□). Control monolayers of DOPC and DOPC with 25 mole% phosphatidyl serine – having a negative charge like G_{M1}– (○) indicated no binding. From Athanassopoulou *et al.* 1999, with permission

surface concentration of B subunits, calculated from the maximum shift in resonance angle, was consistent with electron diffraction images of cholera toxin bound to membranes.

A more realistic bilayer membrane could be made by anchoring liposomes to the sensor surface. Work at Affinity Sensors has shown that unilamellar phospholipid liposomes (30 nm and 100 nm in diameter) incorporating around 1 mole% phosphatidylethanolamine-LC-biotin will form a monolayer on a streptavidin-loaded biotin surface. Furthermore, the small signals that such liposomes give can be amplified by loading them with metrizamide®, a membrane impermeable, iodinated solute.

Another approach is to form extended, planar bilayers at the surface (Puu and Gustafson 1997). Factorial design was used to limit the number of measurements needed to explore the effect of composition (four lipid components and one membrane protein extract) and temperature. Unilamellar liposomes ($\phi = 200$–300 nm) were made by detergent dialysis. Experiments were carried out using hydrophilic silicon wafers (for ellipsometry) and silicon nitride surfaces for the RM. Negatively charged liposomes, with or without membrane extract adsorbed reproducibly in the presence of Ca^{2+} to give RM responses of 1372 ± 185 arc seconds, about twice the response for a lipid monolayer on the hydrophobic surface, and would suggest that the bilayer is separated from the sensor surface by an aqueous space of no more than a few 10s of nm. In some measurements the resonance scan indicated deposition of a non-uniform layer, this may have been caused by Ca^{2+}-flocculated liposomes binding to the sensor surface. Washing and drying the bilayer coated surface did not remove it when the major component was dipalmitoyl-PC, but did when it was dioleoyl-PC. High levels of disaturated PCs appear to impart stability to the bilayer. One concern with this approach is that the thickness of the layer of water that exists between the sensor and bilayer surfaces would depend upon the charged lipid content of the bilayer, pH and the ionic strength of the electrolyte. In some environments the bilayer may separate so far from the surface as to allow thermal energy (kT) to bend and return it to liposomes.

One group (Yagisawa *et al.* 1998) has used SL-CMD to attach a biotinylated *myo*-inositol (1,4,5) tri-phosphate analogue: in nature the inositol would be attached to a phosphoglycerolipid in a membrane. A GST-phospholipase C-δ1 wild type fusion construct bound to the ligand with a $K_D \approx 0.5\,\mu M$ and constructs mutated in the N-terminal pleckstrin homology domain had a two- to ten-fold lower affinity (neutral mutations) while two single point, charge-reverse mutations had a sixty- to seventy-fold reduced affinity. The above mutations also decrease the hydrolytic activity of the enzymes with vesicles containing phosphatidylinositol.

Lung surfactant is a mixture of lipid and protein that rapidly forms monolayers to modulate the surface tension in the lung to enable breathing to occur. Excess lipid is returned to the cells lining the alveoli for repackaging and re-secretion. One of the surfactant proteins, surfactant protein A (SP-A) is involved in this process. SP-A is a calcium binding protein with a structure similar to that of the more familiar complement factor C1q. Light scattering and RM studies have been used to investigate the interaction of purified SP-A from various mammalian lungs with DPPC/egg PC/PG/cholesterol unilamellar liposomes of nominal diameter 200 nm (Meyboom *et al.* 1997). In the RM work SP-A was coupled to amino surfaces with BS[3], a water soluble *bis*-suberate. Binding of the liposomes was instigated by injection of sub-physiological levels Ca^{2+}, and could be reversed by addition of excess EDTA; naked amino surfaces, or heat denatured SP-A were unable to bind the liposomes. The effect of increasing the Ca^{2+} concentration in both resonant mirror and light scattering were similar (Figure 10.5). Both liposome binding to immobilised SP-A and aggregation by SP-A occurred around $20\,\mu M$. Data from both measurement methods could be fit to a Hill equation with coefficients of 8.8 and 5.5 respectively. The binding and aggregation of these liposomes in the presence of SP-A and Ca^{2+} was highly co-operative.

Protein–carbohydrate interactions

Capsular carbohydrate from group B streptococcus was conjugated to BSA and coupled to a CMD surface. This surface was used to determine the affinity of IgG antibody purified from human donors with a history of GBS infection (Feldman *et al.* 1998). The affinity, about 10 nM, was consistent with the inability of carbohydrate antigens to elicit antibody affinity maturation through somatic hypermutation. Affinity maturation, with $K_D s$ as low as pM being attained, was demonstrated with IAsys several years ago (Savelkoul and Pathak 1994).

The interaction between antigens terminated with galactose α1-3 galactose and

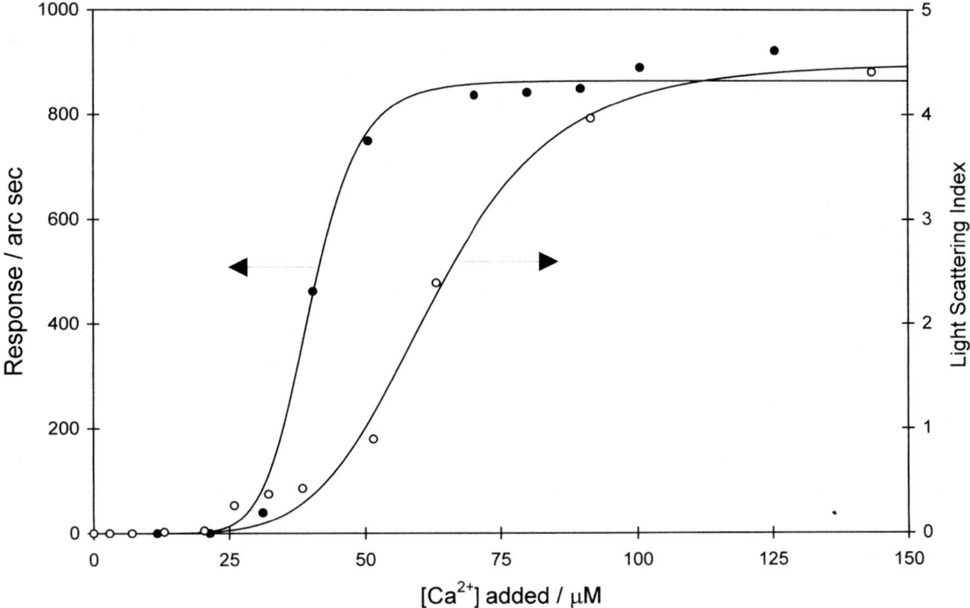

Figure 10.5 Resonant mirror and near infrared light scattering measurements of surfactant protein A-induced liposome interaction under the influence of Ca^{2+}. The lines are fitted to the data using the Hill equation. Redrawn from Meyboom *et al.* 1997, with permission

their cognate IgG and IgM antibodies, causes hyperacute rejection of xenografts (Lee *et al.* 1998). Two IgM fractions, specific to two different antigens, were purified and their binding affinity measured, using the RM, at 0.8 nM and 0.1 nM. This new finding of high affinity anti-carbohydrate antibodies has not been explained and is at odds with the belief expressed above.

Interactions involving nucleic acids

Nucleic acids are not as straightforward as their more compact, and less charged protein counterparts. Double-stranded (ds), histone-free, DNA under physiological conditions is a fairly stiff molecule with 12 per cent of its backbone ionised (Manning 1975). Single-stranded DNA is more flexible, but has a similar charge density. We might expect that homogeneous loading of a CMD surface with poly(nucleic acid) to become less likely as the degree of polymerisation approaches several hundred due to charge repulsion and steric exclusion from the matrix. Attempts to reduce the physical size of the DNA by increasing the salt concentration to screen-out charge repulsion, or even charge neutralisation by binding of di-, tri- or tetra- valent cations will be frustrated by a similar compression of the CMD matrix leading to a reduction in its 'pore' size. However, the CMD matrix has been used for work on short DNA chains (Watts *et al.* 1995). A streptavidin-loaded biotin surface can be used to immobilise terminally biotinylated poly (nucleic acids), but again the surface concentration would be expected to fall as the chains become increasingly long. The other consideration is that there must be enough room around the surface-coupled nucleic acid for the second (or third) strand to twist and hybridise. The volume required will be greater than that taken up by the 1 kT coulombic repulsion envelopes of the two strands to allow for the hybridisation of neighbouring chains. Although this is important, not just for biosensor work and other devices that aim to detect a particular nucleic acid sequence, we have not found any work directed to controlling the lateral spacing of immobilised nucleic acid. The RM and its streptavidin-loaded surface would be useful for this type of study.

Nucleic acid oligomers that form DNA triplex are used for site-specific cleavage of DNA, genome mapping and for knocking-out selected genes, thereby offering the promise of therapeutic utility. Triplex formation between a ds-23-mer and a 15-mer was investigated by isothermal titration calorimetry (ITC) and RM (Torigoe *et al.* 1997). Base and sugar chemical modifications of the 15-mer were assessed as well as replacing its poly(phosphate) backbone with a poly(phosphorothioate). No experimental details were given on the immobilisation of the nucleic acids to the RM. Affinities measured by both methods for the six triplexes were in agreement. The phosphorothioate material had a K_D (0.6 μM (ITC), 1.9 μM (RM)), an order of magnitude higher than the normal 15-mer; base and sugar modifications had a smaller effect. Kinetic measurements on the RM showed the phosphorothioate-linked polymer associated more slowly and dissociated more rapidly than the normal polymer. The phosphorothioate backbone bears a higher charge than the phosphodiester structure and this, among other factors, may destabilise triplex formation. The influence of electrostatic forces upon association of the 15-mer with the ds-23-mer was established by varying the electrolyte concentrations and measuring the rate constant for complex formation (Figure 10.6). There was linear decrease in $\log_{10} k_{ass}$ vs the $\log_{10}$ [salt], this was almost three-fold higher with the phosphorothioated 15-mer. A simple electrostatic argument would not explain triplex formation in these experiments.

Doxorubicin, an anti-tumour drug, kills cells by intercalating with DNA, but some tumour cells are resistant. Doxorubicin tolerance involves overexpression of the P-glycoprotein pump which keeps the cytosolic drug concentration below a therapeutic level. Conjugation to various macromolecules will prevent pump out and can increase the drug's efficacy. The toxicity of dextran conjugates (M_r 70, 200 and 500 kDa) towards human carcinoma cells (KB-3-1) and a resistant sub-clone (KB-V-1) with complementary RM studies on conjugate–DNA binding has been carried out (Yang *et al.* 1998). A SL-CMD surface was loaded with ≈300 arc sec of 5′-biotinylated ds-DNA (a 275-mer). Binding of the conjugates, each having on average 3–6 doxorubicins, was followed at 0–8 μM and binding rate constants determined. The k_{ass} values fell (3.4, 2.9 and 2.3 M^{-1} s^{-1}) with increasing dextran M_r. However, the decrease in k_{ass} was less than we would expect from a combination of a simple reduction in diffusion coefficient ($D = kT/6\pi\eta r_h$; η = viscosity, r_h = hydrodynamic radius (Sellen 1975)) and the reduced probability of finding doxorubicin at the surface of dextran coils of increasing surface area. Fluorescence titration measurements of the quenching of doxorubicin emission when bound to DNA in solution gave the same (within error) binding affinities for the two smaller conjugates; the 500 kDa conjugate had about twice the affinity when determined by RM as fluorescence titration.

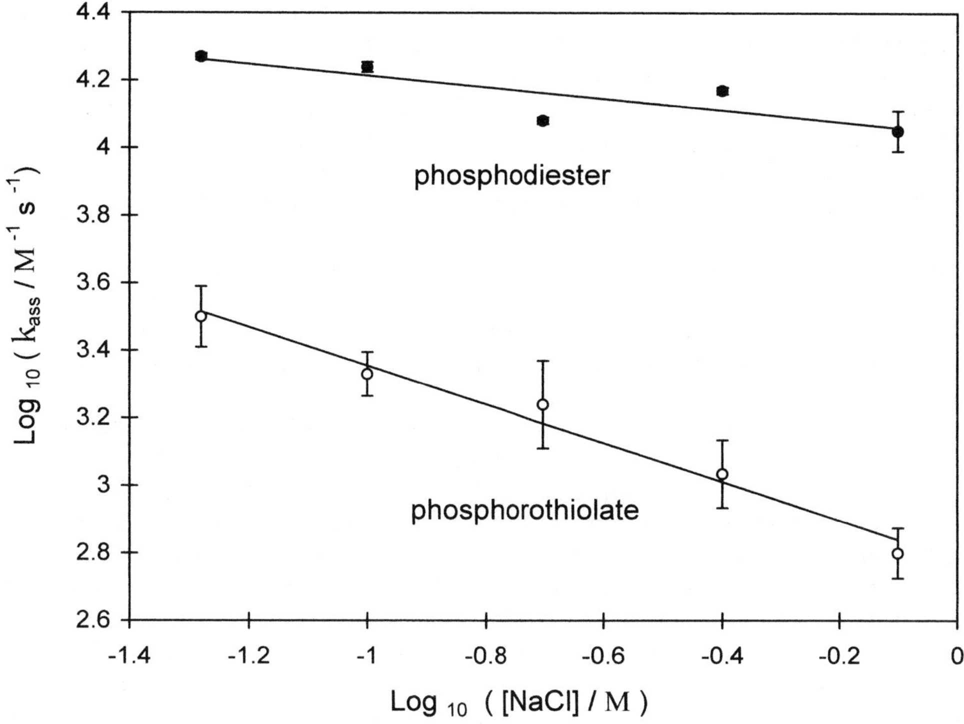

Figure 10.6 The effect of NaCl concentration of the rate of triplex formation between a 23-mer dsDNA and two 15-mer ssDNA with different backbone structures. From Torigoe *et al.* 1997, with permission

Industrial applications

Many areas in 'bioprocess monitoring' may benefit from the application of optical biosensor technology. A few papers have appeared during the past two years in which the RM has been applied. Chromatography is a commonly employed industrial step in the purification of many biomolecules from complex matrices (e.g. cell culture media). ELISA has been used for following the biosynthesis of protein products, but it is a time-consuming method requiring skilled personnel and unsuitable for determination of, for instance, when the output of a column contains the desired protein in a mixture of proteins. Such a decision needs to be taken immediately. Using a model chromatography system – a lysozyme (hen) affinity column to extract a bacterial expressed single chain lysozyme antibody (D1.3 Fv, $M_r \approx 25\,\mathrm{kDa}$) from fermentation broth supernatant – complete column loading was established by monitoring the breakthrough of excess D1.3 Fv with the RM (Bracewell *et al.* 1998). The detection system was a dual-channel RM having both biotinylated surfaces coated with neutravidin. Biotinylated hen lysozyme was loaded on the measuring surface and the control surface was biotinylated-turkey lysozyme (Figure 10.7). Specific D1.3 Fv detection was achieved, and the initial binding rate was used to determine the antibody concentration within 10s, validated later by ELISA. Several algorithms were used to automatically determine breakthrough. The system was further refined

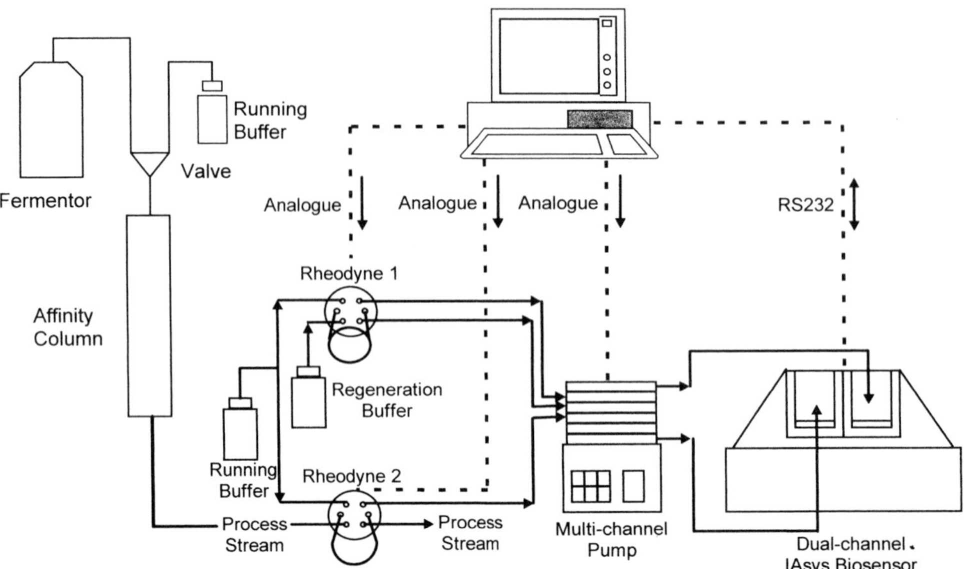

Figure 10.7 A downstream automatic bioprocess monitoring arrangement, adapted from Bracewell *et al.* 1998, with permission

to allow prediction of the breakthrough time, giving time for actions to be taken to prevent wastage of product. The same group have also used the RM for fermentation monitoring (Gill *et al.* 1998; Tsoka *et al.* 1998) and to assess the effect of impeller shear forces and air bubbles on antibody fragment activity (Harrison *et al.* 1998). IAsys has been used in industry (Genentech Inc.) to monitor the production of recombinant proteins (Cacia and Frenz 1997).

An area of growing importance is the identification of products of combinatorial synthesis with a particular binding specificity. The approach here is to immobilise the binding partner to which it is hoped to find one or more binding peptide(s), or nucleic acids, in the combinatorial library. To facilitate this a large surface area CMD cuvette whose surface acts as a micro-affinity plate to detect specific binding, determine the binding constant and allows material to be harvested from the surface for analysis by a suitable analytical method is available. It is possible, in some cases, to simply elute the bound material with SDS-PAGE sample equilibration buffer and characterise the protein by electrophoresis (IAsys Protocol 1.7). More powerful analytical tools have also be used. From a peptide library of thirty heptameric sequences electrospray mass spectrometry of material harvested from the RM surface indicated that four of them bound to an immobilised anti-β-endorphin antibody (Cao 1997). When synthesised, two were able to compete with β-endorphin for binding to the antibody, the other two may have bound to other sites on the antibody.

In the production of recombinant proteins, one of several sequences or tags (e.g. GSH, oligo-histidine, FLAG) are co-expressed to facilitate the recovery of the product. Biotin would be the ideal tag, enabling high product recovery owing to its high affinity ($K_D \approx 2\,\mathrm{fM}$) with various biotin-binding proteins. However, eluting the biotinylated protein without denaturing it would be difficult. A biotin–mimetic oligopeptide has been developed (Hengsakul and Cass 1997). When co-expressed with alkaline phosphatase its affinity for a SL-CMD surface was about 0.5 µM. The tag could be fully dissociated from streptavidin with 10 mM HCl, a far milder treatment than required to remove biotin.

Conclusion

Our review indicates that the RM has been used to investigate interactions covering four orders magnitude in size from ions and nucleotide phosphates to eukaryotic cells. In some studies the biosensor has been used in parallel with a more conventional technique and found not only to agree but to be the preferred measurement. In the early 1990s evanescent optical biosensors became available, today they have rapidly evolved in terms of surfaces and instrumentation, and their realm of application has been extended by an ever growing user base. Some have even moved from the laboratory to become bioprocess measurement tools in a more demanding industrial setting. With advances in technology they may eventually become clinical instruments, a role for which they were first envisioned but have yet to attain.

References

Athanassopoulou, N., Davies, R.J., Edwards, P.R., Yeung, D. and Maule, C.H. (1999) 'Cholera toxin and G_{MI}: a model membrane study with IAsys', *J. Biochem. Soc.* 27, 340–344.

Borza, D.B. and Morgan, W.T. (1998) 'Histidine – proline rich glycoprotein as a plasma pH sensor', *J. Biol. Chem.* 273, 5493–5499.

Bracewell, D.G., Gill, A., Hoare, M., Lowe, P.A. and Maule, C.H. (1998) 'An optical biosensor for real-time chromatography monitoring: breakthrough determination', *Biosens. Bioelec.* 13, 847–853.

Cacia, J. and Frenz, J. (1997) 'Strategies for monitoring production of recombinant proteins', *213th American Chemical Society National Meeting (Abstracts)*, BIOT: 172.

Cao, B., Urban, J., Vaisar, T, Shen, R.Y.W. and Kahn, M. (1997) 'Detecting and identifying active compounds from a combinatorial library using IAsys and electrospray mass spectrometry', in Crabb, J.W. (ed.) *Techniques in Protein Chemistry VIII*, Academic Press, San Diego, pp. 177–184.

Cush, R., Cronin, J.M., Stewart, W.J., Maule, C.H., Molloy, J. and Goddard, N.J. (1993) 'The resonant mirror: a novel optical biosensor for direct sensing of biomolecular interactions. Part 1: principle of operation and associated instrumentation', *Biosens. Bioelectron.* 8, 347–353.

Davies, R.J, Edwards, P.R., Watts, H.J., Lowe, C.R., Buckle, P.E., Yeung, D., Kinning, T.M. and Pollard-Knight, D.V. (1994) 'The resonant mirror: a versatile tool for the study of biomolecular interactions', in Crabb, J.W. (ed.) *Techniques in Protein Chemistry V*, Academic Press, San Diego, pp. 285–292.

Edwards, R. and Leatherbarrow, R.J. (1997) 'Determination of association rate constants by an optical biosensor using initial rate analysis', *Anal. Biochem.* 246, 1–6.

Edwards, P.R., Lowe, P.A. and Leatherbarrow, R.J. (1997) 'Ligand loading at the surface of an optical biosensor and its effect upon the kinetics of protein–protein interactions'. *Journal of Molecular Recognition* 10, 128–134.

Edwards, P.R., Maule, C.H., Leatherbarrow, R.J. and Winzor, D.J. (1998) 'Second order kinetic analysis of IAsys biosensor data: its use and applicability', *Anal. Biochem.* 263, 1–12.

Edwards, P.R. (1998) 'The use of optical biosensors for kinetic analysis: a critical appraisal', PhD thesis, Imperial College, London.

Fan, Z., Larson, P.J., Bognacki, J., Raghunath, P.N., Tomaszewski, J.E., Kuo, A., Canziani, G., Chaiken, I., Cines, D.B. and Higazi, A.A.R. (1998) 'Tissue factor regulates plasminogen binding and activation', *Blood* 91, 1987–1998.

Faux, M.C. and Scott, J.D. (1997) 'Regulation of the AKAP79-protein kinase C interaction by Ca^{2+}/calmodulin', *J. Biol. Chem.* 272, 17038–17044.

Feldman, R.G., Breukels, M.A., David, S. and Rijkers, G.T. (1998) 'Properties of human anti-group B streptococcal type III capsular IgG antibody', *Clin. Immunol. Immunopath.* 86, 161–169.

Figler, R.A., Lindorfer, M.A., Graber, S.G., Garrison, J.C. and Linden, J. (1997) 'Reconstitution of bovine A1 adenosine receptors and G proteins in phospholipid vesicles: βγ subunit composition influences guanine nucleotide exchange and agonist binding', *Biochemistry* 36, 16288–16299.

Gill, A., Bracewell, D.G., Maule, C.H., Lowe, P.A. and Hoare, M. (1998) 'Bioprocess monitoring: an optical biosensor for real-time bioproduct analysis', *J. Biotech.* 65, 69–80.

Glasser, R.W. (1993) 'Antigen–antibody binding and mass transport by convection and diffusion to a surface: a two-dimensional computer model of binding and dissociation kinetics'. *Anal. Biochem.* 213, 152–161.

Gorgani, N.N., Parish, C.R., Smith, S.B.E. and Altin, J.G. (1998) 'Histidine rich glycoprotein binds to human IgG and C1q and inhibits the formation of insoluble immune complexes', *Biochemistry* 36, 6653–6662.

Gournier, H., Stenmark, H., Rybin, V., Lippe, R. and Zerial, M. (1998) 'Two distinct effectors of the small GTPase rab 5 cooperate in endocytic membrane fusion', *EMBO J.* 17, 1930–1940.

Harrison, J.S., Gill, A. and Hoare, M. (1998) 'Stability of a single chain Fv antibody fragment when exposed to a high shear environment combined with air–liquid interfaces', *Biotech. Bioeng.* 59(4), 517–519.

Hengsakul, M. and Cass, A.E.G. (1997) 'Alkaline phosphatase–strep tag fusion protein binding to streptavidin: resonant mirror studies', *J. Mol. Biol.* 266, 621–632.

Hirmo, S., Artursson, E., Puu, G., Wadstrom, T. and Nilsson, B. (1998) 'Characterisation of *Helicobacter pylori* interactions with sialylglycoconjugates using a resonant mirror', *Anal. Biochem.* 257, 63–66.

Inoue, K., Arai, T. and Aoyagi, M. (1996) 'Real time observation of the binding of herpes simplex virus type 1 (HSV-1) to immobilised heparan sulphate and the neutralisation of HSV-1 by sulphonated human IgG', *J. Biochem.* 120, 233–235.

Inoue, K. and Arai, T. (1998) 'Real time observation of binding of measles virus to vero cells and neutralisation of measles virus by human immunoglobulins using optical biosensor – a real time diagnosis system for viral infections', *Kansenshogaku-Zasshi* 73(3), 273–278.

Lee, J., Cairns, T., McKane, W., Rashid, M., George, A.J.T. and Taube, D. (1998) 'Demonstration of IgM antibodies of high affinity within the anti-galα1-3gal antibody repertoire', *Transplantation* 66(8), 1117–1119.

Lyon, M., Deakin, J.A., Rahmoune, H., Fernig, D.G., Nakamura, T. and Gallagher, J.T. (1998) 'Hepatocyte growth factor/scatter factor binds with high affinity to dermatan sulphate', *J. Biol. Chem.* 273, 271–278.

Manning, G.S. (1975) 'Molecular theory of polyelectrolyte solutions with applications to electrostatic properties of polynucleotides', *Quart. Rev. Biophys.* 11(2), 179–246.

Meyboom, A., Maretzki, D., Stevens, P.A. and Hofman, K.P. (1997) 'Reversible calcium-dependent interaction of liposomes with pulmonary surfactant protein A. Analysis by resonant mirror technique and near infrared light scattering', *J. Biol. Chem.* 272, 14600–14605.

Morgan, C.L., Newman, D.J., Cohen, S.B.A, Lowe, P. and Price, C.P. (1998a) 'Real-time analysis of cell surface HLA class 1 interactions', *Biosensors and Bioelectronics* 13, 1099–1105.

Morgan, C.L., Ruprai, A.K., Solache, A., Lowdell, M., Price, C.P., Cohen, S.B.A., Parham, P., Madrigal, J.A. and Newman, D.J. (1998b) 'The influence of exogenous peptide on β2-microglobulin exchange in the HLA complex: Analysis in real-time', *Immunogenetics* 48, 98–107.

Myszka, D.G, Morton, T.A., Doyle, M.L. and Chaiken, I.M. (1997) 'Kinetic analysis of a protein antigen–antibody interaction limited by mass transport on an optical biosensor', *Biophys. Chem.* 64, 127–137.

Nicol, S., Rahman, D. and Baines, A.J. (1997) 'Ca^{2+}-dependent interaction with calmodulin is conserved in the synapsin family: identification of a high affinity site', *Biochemistry* 36, 11487–11495.

Nunomura, W., Takakuwa, Y., Tokimitsu, R., Krauss, S.W., Kawashimu, M. and Mohandas, N. (1998) 'Regulation of CD44-protein 4.1 interaction by Ca^{2+} and calmodulin. Implications for modulation of CD44–ankyrin interaction', *J. Biol. Chem.* 272, 30322–30328.

Purvis, D.R., Pollard-Knight, D. and Lowe, P.A. (1998) 'Biosensors based on evanescent waves', in Ramsay, G. (ed.) *Commercial Biosensors: Applications to Clinical, Bioprocess, and Environmental Samples*, John Wiley & Sons, Inc., New York, pp. 165–224.

Puu, G. and Gustafson, I. (1997) 'Planar lipid bilayers on solid supports from liposomes – factors of importance for kinetics and stability', *Biochim. Biophys. Acta.* 1327, 149–161.

Rahmoune, H., Rudland, P.S., Gallagher, J.T. and Fernig, D.G. (1998a) 'Hepatocyte growth factor/scatter factor has distinct classes of binding site in heparan sulphate from mammary cells', *Biochemistry* 37, 6003–6008.

Rahmoune, H., Chen, H.-L., Gallagher, J.T., Rudland, P.S. and Fernig, D.G. (1998b) 'Interaction of heparan sulphate from mammary cells with acidic fibroblast growth factor (FGF) and basic FGF', *J. Biol. Chem.* 273, 7303–7310.

Rubio, I., Buckle, P., Trutnau, H. and Wetzker, R. (1997) 'Real-time assay of the interaction of a GST fusion protein with a protein ligate using resonant mirror technique', *BioTechniques* 22, 269–271.

Saenko, E. and Scandella, D. (1997) 'The acidic region of the factor VIII light chain and the C2 domain together form the high affinity binding site for von Willebrand factor', *J. Biol. Chem.* 272, 18007–18014.

Savelkoul, H.F.J. and Pathak, S.S. (1994) 'The IAsys biosensor for affinity measurements of antibodies in immune responses', *Fusion*, 7, 10–11.

Sellen, D.B. (1975) 'Light-scattering rayleigh linewidth measurements on some dextran solutions', *Polymer* 16, 561–565.

Schuck, P. (1996) 'Kinetics of ligand binding to receptor immobilised in a polymer matrix, as detected with an evanescent wave biosensor. I. A computer simulation of the influence of mass transport', *Biophys. J.* 70, 1230–1249.

Schuck, P. and Minton, A.P. (1996) 'Kinetic analysis of biosensor data: elementary tests for self-consistency', *Trends Biochem. Sci.* 21, 458–460.

Stephens, D.J., Crump, C.M., Clarke, A.R. and Banting, G. (1997) 'Serine 331 and tyrosine 333 are both involved in the interaction between the cytosolic domains of TGN38 and the μ2 subunit of the AP2 clathrin complex', *J. Biol. Chem.* 272, 14104–14109.

Thakkar, H., Lowe, P.A., Price, C.P. and Newman, D.J. (1998) 'Measurement of the kinetics of protein uptake by proximal tubular kidney cells using an optical biosensor', *Kidney Intl.* 54, 1197–1205.

Torigoe, H., Shimizume, R., Sarai., A. and Shindo, H. (1997) 'Effect of chemical modification of oligohomopyrimidine on triplex formation: thermodynamic and kinetic studies', *Nucleic Acids Symposium Series* 37, 267–268.

Tsoka, S., Gill, A., Brookman, J.L. and Hoare, M. (1998) 'Rapid monitoring of virus-like particles using an optical biosensor: a feasibility study', *J. Biotech.* 63, 147–153.

Watts, H.J., Yeung, D. and Parkes, H. (1995) 'Real-time detection and quantification of DNA hybridisation by an optical biosensor', *Anal. Chem.* 67(23), 4283–4289.

Winzor, D.J. and Hall, D.R. (1997) 'Use of a resonant mirror biosensor to characterise the interaction of carboxypeptidase A with an elicited monoclonal antibody', *Anal. Biochem.* 244, 152–160.

Yagisawa, H., Sakuma, K., Paterson, H.F., Cheung, R.L., Allen, V., Hirata, H., Watanabe, Y., Hirata, M., Williams, R.L. and Katan, M. (1998) 'Replacements of single basic amino acids in the pleckstrin homology domain of phospholipase C-δ1 alter the ligand binding, phospholipase activity and interaction with the plasma membrane', *J. Biol. Chem.* 273, 417–424.

Yang, M., Chan, H.L., Lam, W. and Fong, W.F. (1998) 'Cytotoxicity and DNA binding characteristics of dextran-conjugated doxorubicins', *Biochim. Biophys. Acta.* 1380, 329–335.

Yeung, D., Gill, A., Maule, C.H. and Davies, R.J. (1995) 'Detection and quantification of biomolecular interactions with optical biosensors', *Trends in Anal. Chem.* 14(2), 49–56.

11 Commercial quartz crystal microbalances

Theory and applications

C.K. O'Sullivan and G.G. Guilbault

Introduction

The process of many important physical and chemical processes can be followed by observing the associated mass changes. The increased interest in using microbalances has resulted, in part, from the rapid progress in scientific instrumentation (Czanderna and Lu 1984). From a current perspective, it is difficult to realise that only thirty years ago, the ranks of experts about the piezoelectric quartz crystal microbalance were limited to those few with the resources and ability to assemble their own instruments. Today, highly sophisticated automatic, microprocessor-controlled devices are available commercially and satisfy most requirements of scientific and technological investigators. Parallel progress in vacuum science and technology has provided a means of controlling the environment required for many experiments using the QCM (Henry 1996).

Quartz crystal microbalance – theory

The theoretical foundation for the use of piezoelectricity was first pioneered by Raleigh (1885), but the first thorough investigation was by Jacques and Pierre Curie (1880). A piezoelectric quartz crystal resonator is a precisely cut slab from a natural or synthetic crystal of quartz. Quartz crystal in its perfect natural form can be seen in Figure 11.1. A quartz crystal microbalance (QCM) consists of a thin quartz disk with electrodes plated on it as can be seen in Figure 11.2. The application of an external electrical potential to a piezoelectric material produces internal mechanical stress. As the QCM is piezoelectric, an oscillating electric field applied across the device induces an acoustic wave that propagates through the crystal and meets minimum impedance when the thickness of the device is a multiple of a half wavelength of the acoustic wave. A QCM is a shear mode device in which the acoustic wave propagates in a direction perpendicular to the crystal surface (Ebersole *et al.* 1993). To make this happen, the quartz crystal plate must be cut to a specific orientation with respect to the crystal axes. These cuts belonging to the rotated Y-cut family, the AT- and BT-cuts shown in Figure 11.3 are representative. Deposition of a thin film on the crystal surface decreases the frequency in proportion to the mass of the film. A typical experimental apparatus set up is shown in Figure 11.4.

A resonant oscillation is achieved by including the crystal into an oscillation circuit where the electric and the mechanical oscillations are near to the fundamental frequency of the crystal. The fundamental frequency depends upon the thickness

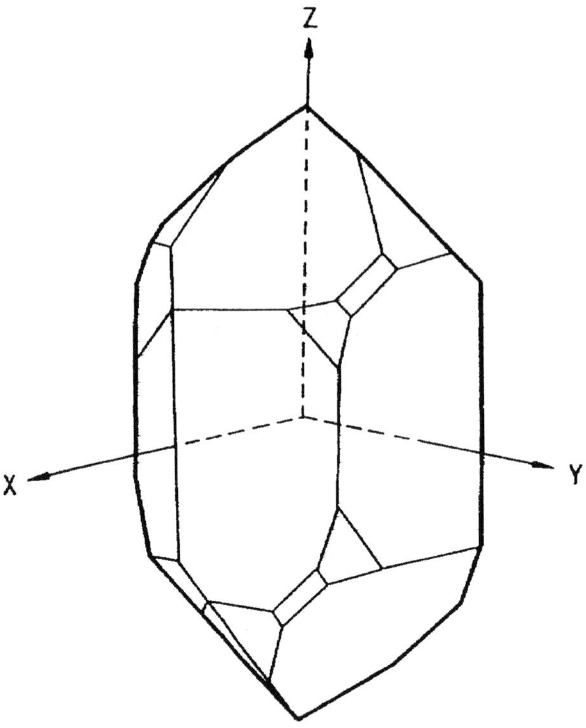

Figure 11.1 The assignment of axes to a quartz crystal

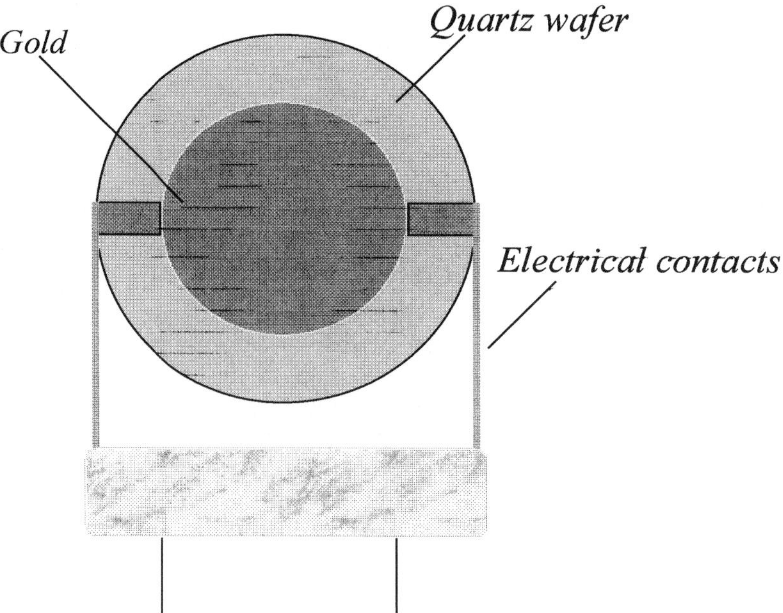

Figure 11.2 Schematic of typical piezoelectric crystal

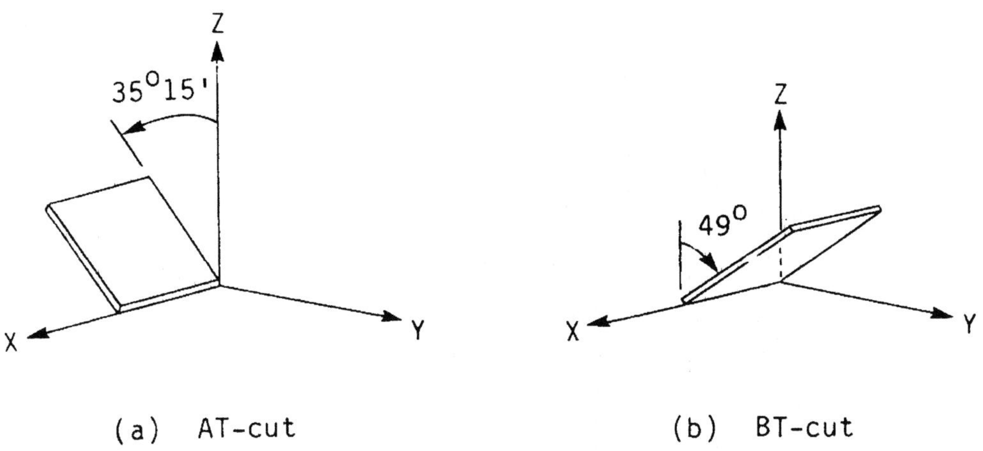

Figure 11.3 AT- and BT-quartz crystals

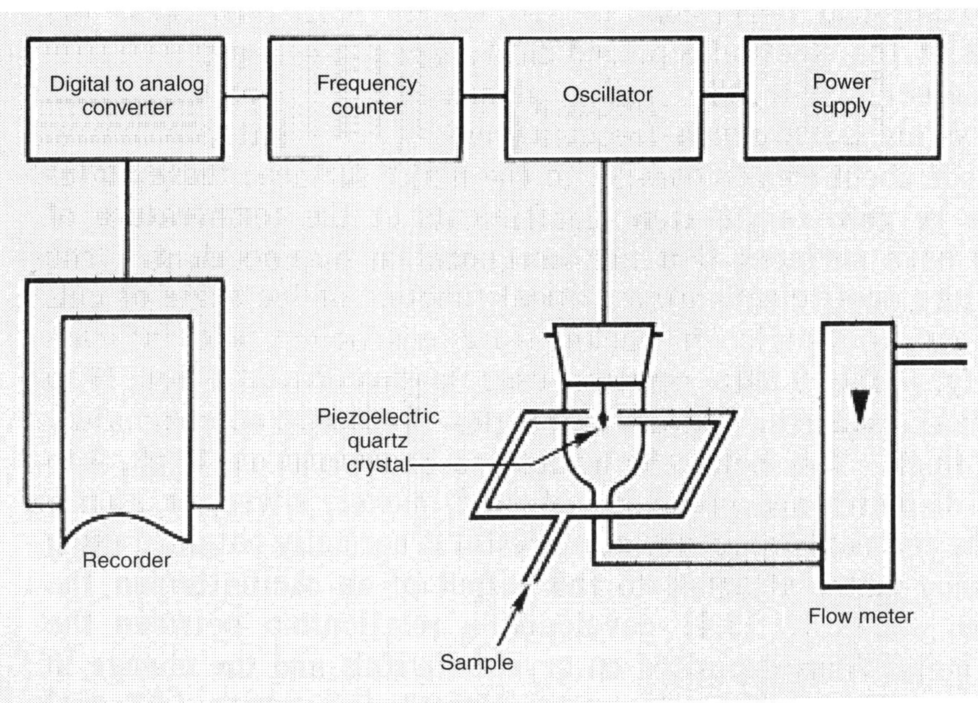

Figure 11.4 Typical experimental apparatus set-up of piezoelectric quartz crystal microbalance

of the wafer, its chemical structure, its shape and its mass. Some factors can influence the oscillation frequency, like the thickness, the density and the shear modulus of the quartz that are constant, and the physical properties of the adjacent media (density or viscosity of air or liquid). As shown by Sauerbrey, changes in the resonant frequency are simply related to the mass accumulated on the crystal by the following (1959):

$$\Delta f = \frac{-2\Delta m n f_0^2}{\eta_q \rho_q} \qquad (11.1)$$

where η_q and ρ_q are the density and viscosity of the quartz (respectively 2.648 gcm^{-3} and 2.947×10^{-11} gcm^{-1}s^{-1}), n is the overtone number, f_o is the basic oscillator frequency of the quartz and Δm is the material adsorbed on the surface per unit area. For an AT-cut crystal $\Delta f = -2.26 \times 10^{-6} f_o^2 \Delta m$.

When a crystal is dipped into a solution, the oscillating frequency depends on the solvent used. The question as to which factors determine the frequency is important for understanding the mechanism of oscillation of a crystal in solution and for its potential development as a sensor in solution (Roederer *et al.* 1983; Kanasawa *et al.* 1985; Bruckenstein and Shay 1994). When an overlayer is thick, the relationship between the f and Δm is no longer linear and corrections are necessary. The coupling of the crystal surface to a liquid drastically changes the frequency when a quartz crystal oscillates in contact with a liquid; a shear motion on the surface generates motion in the liquid near the interface. The oscillation surface generates plane-laminar flow in the liquid, which causes a decrease in the frequency proportional to $(\rho\eta)^{1/2}$ where ρ and η are the liquid density and the viscosity:

$$\Delta f = f_0^{3/2} \left(\frac{\rho\eta}{\pi\eta_q\rho_q} \right)^{1/2} \qquad (11.2)$$

The penetration depth of this sheer wave depends on $(\pi f_o \eta_q \rho_q)^{-1/2}$. For water at 20 °C this is 2500 Å for a 5 MHz crystal. Kanasawa and Gordon stress that Δf in solution is a linear function of $(\eta\rho)^{1/2}$ except for salts and high polymer solutions (1985). They, in fact, tested the linearity of the dependence of the frequency decrease on $(\eta\rho)^{1/2}$ using many solvents selected on the basis of their different viscosity, density and electrical conductivity. For a 10 MHz crystal with one face exposed to dilute aqueous solutions near room temperature Δf is ca. 2 KHz. Crystal surface roughness may increase the magnitude of the liquid phase induced frequency change by several kilohertz as observed by Bruckenstein and Shay (1994).

Two types of oscillator sensors are available – surface acoustic wave (SAW) and the piezoelectric crystal (PZ). The SAW devices can operate either on the Raleigh wave propagation principle at solid thin-film boundaries or as a bulkwave piezoelectric device (Roederer *et al.* 1983; Bastiaans 1988; Guilbault and Luong 1988). The latter follows the Sauerbrey equation (i.e. small thin films can be treated as equivalent mass changes on the crystal). SAW crystals oscillate at several hundred megahertz, instead of 9–14 MHz for the PZ. Theoretically greater sensitivity should result with the SAW, since the frequency term in the Sauerbrey equation is squared. Hence, a frequency of $(100)^2 = 10,000$ as compared to $(10)^2 = 100$ in the PZ detector.

This two-decade increase in sensitivity is, however, diminished by the requirement of the SAW detector for a very thin coating film.

The term QCM, is, in fact, not accurate as (1) in some situations the 'QCM' does not act as a microbalance (e.g. responding to viscosity) and (2) other types of quartz devices can act as microbalances. The device would be more accurately called a thickness-shear mode (TSM) resonator. As highlighted previously, the mass of a thin layer attached to the surface of a crystal can be calculated from a measured change in the resonant frequency of the device.

Commercial systems

Commercial systems are designed to reliably measure mass changes up to $\sim 100\,\mu g$, whereas the minimum detectable mass change is typically $\sim 1\,ng/cm^2$. The challenge laid down is the tailoring of the surface chemistry so that the device can be a discriminating mass detector and not simply a microbalance. As outlined above, AT-cut quartz crystals are used as QCMs due to their low temperature co-efficient at room temperature with only minimal frequency changes due to temperature in that region, as a consequence. The crystals may be rough or smooth, clear or clouded. For the liquid-phase, however, optically polished crystals are recommended as liquid can get caught in the crevices of rough crystals and sometimes result in spurious frequency changes. The crystal diameter plays an important role in stability. Crystals are generally $\frac{1}{2}$ inch in diameter, although crystals one inch or larger are available. The thickness of the crystal determines the resonant frequency, which in turn determines the mass sensitivity. Thinner devices resonate at higher frequencies and are more sensitive, but tend, however, to be very fragile. The crystal and its electrodes are incorporated into a positive-feedback oscillator circuit. A frequency counter can be employed to make measurements. For detailed research, a network analyser is deemed far superior to a frequency analyser, allowing the entire bandwidth of the resonance to be observed as well as making attenuation measurements. Another possibility is to look at phase shift measurement rather than oscillator shift measurement.

The main commercially available quartz crystal microbalances can be seen in Table 11.1. The eventual application of the QCM needs to be carefully addressed before purchasing a QCM. The fixtures for liquid and gas-phase systems are different – if a gas-phase system is used in the liquid phase, the signal will attenuate too much, and the fixture will most likely short-circuit. Liquid applications present a challenge because multiple effects occur simultaneously. The mass effects that cause a frequency shift in the gas phase engender a similar response in the liquid phase, but viscosity changes produce frequency shifts and a signal attenuation, requiring the use of an appropriate algorithm to separate the effects.

The biggest drawback of the QCM is that it is inherently non-specific and this issue is being addressed in various ways so that the device will only respond to the analyte of interest.

Applications

QCMs have traditionally been used in vacuum deposition systems and have found a plethora of other applications: thin film deposition control, estimation of stress

Table 11.1 Commercially available microbalances

Model	Model EQCN-700 Electrochemical Quartz Crystal Microbalance	Model EQCN-900 Electrochemical Quartz Crystal Microbalance	Mark Series Cryogenic Quartz Crystal Microbalances	Mark Series Thermoelectric Piezoelectric Detector
Company	Elchema 14 Elderkin St. P.O. Box 5067 Potsdam NY 13676	Elchema 14 Elderkin St. P.O. Box 5067 Potsdam NY 13676	QCM Research 2825 Laguna Canyon Rd. Box 277 Laguna Beach CA 92652	QCM Research 2825 Laguna Canyon Rd. Box 277 Laguna Beach CA 92652
Dimensions	$4.5 \times 17 \times 16.5\,$in	$6 \times 17 \times 16\,$in	$0.85 \times 0.86 \times 0.13\,$in	$1.25 \times 0.30 \times 2.95\,$in
Resolution Frequency Mass	0.1 Hz 0.1 ng	0.01 Hz 0.05 mg	0.1 Hz $4.42\,$ng/cm^2	1 Hz $442\,$ng/cm^2
Sample cell	Rotacell RTC-100 system with quartz crystal sealed to the side opening of the cell; crystal can be removed with blade for replacement; cell volumes 3.5–100 mL available	Rotacell RTC-103 system with submersible quartz crystal probes; cell volume 100–200 mL	Two crystals in a crystal pack (Au plated OFHC copper), supported by four heat-isolating struts, allowing measurement of mass change rate as a function of temperature	Measurements of mass flow can be made as it intercepts the crystal and is condensed at a specific T. Can be cooled by its own internal thermoelectric Peltier element. -80 to 80 °C active temperature range
Applications	Adsorption, electroplating, corrosion, etching, gas sensing, biosensors, ion dynamics in conductive polymers	Sensor development, *in situ* polymerisation studies, conductive polymers, batteries	Contamination in vacuum environment	Molecular flux measurement of gases; erosion and deposition rates on spacecraft
Additional information	Response time 1 ms.	Tunable reference frequency; admittance characteristics; 3 working oscillators.	3 models available; designedfor use in the gas-phase; operates at temperatures 10–398 K 3, 15, 20 and 25 MHz crystals also available.	2 models available; designed for use in the gas phase; 3, 15, 20 and 25 MHz crystals also available.

Model	Model QCA-917 Quartz Crystal Analyser	PM-700 Series	PZ-1000 Immuno-biosensorSystem	PZ-105 Gas Phase Piezoelectric Detector
Company	EG&G Princeton Applied Research, PO Box 2565 Princeton NJ 08543	Maxtek 2908 Oregon Ct. G3 Torrance CA 90503	Universal Sensors 5258 Veteran's Blvd. Suite D Metairie LA 70006	Universal Sensors 5258 Veteran's Blvd. Suite D Metairie LA 70006
Dimensions	$90 \times 230 \times 215\,mm$	$89 \times 216 \times 235\,mm$	$12 \times 23 \times 27\,cm$	$12 \times 23 \times 27\,cm$
Resolution				
Frequency	$0.1\,Hz$	$0.03\,Hz$ at $5\,MHz$	$1\,Hz$	$1\,Hz$
Mass	$1\,ng/cm^2$; $0.2\,ng$	$0.7\,ng/cm^2$	$1\,ng/cm^2$	$1\,ng/cm^2$
Sample cell	Well or dip; Teflon holder	Sensor probe made of Teflon/CPVC; oscillator housed in the probe head coupled to crystal; O-ring ensures liquid only comes in contact with one side of crystal	One side of cell constructed as static system exposed to up to $1\,mL$; other side constructed as flow system exposed to $70\,\mu L$ chamber	Gas pulled into built-in gas chamber by internal air pump
Applications	Electrochemical deposition, corrosion, adsorption, batteries, electroplating, biosensor development	Plating, etching, corrosion, polymer growth adsorption	Biosensor and non-biosensor adsorption	Biosensor, adsorption
Additional information	Non-grounded crystal that functions with any potentiostat. M270 software SoftCorr III corrosion Software; faraday cage.	Easy installation of crystal 4 models available.	PZTools software for data storage and manipulation, and for real-time display of frequency versus time 10 MHz crystals, additional flow cell.	Built-in air pump up to $100\,mL/min$; PZTools software for data storage, manipulation, real-time display of frequency versus time, and calculation of kinetic constants.

effects, etching studies, space system contamination studies and aerosol mass measurement to name but a few. These devices have, however, become more and more frequently used in the world of analysis. Various approaches can be taken once a suitable recognition layer has been coated on the crystal: to use a single sensor sensitive to a single compound or class of compounds; to use an array of sensors with different coatings combined with pattern recognition; or even to use the QCM as a chromatographic detector. The options are many but the fundamental problem is to find a suitable coating layer and a method of reproducibly applying it.

The first analytical application of piezoelectric crystals was reported by King (1964). He developed and commercialised a piezoelectric detector, which could detect moisture to 0.1 ppm and hydrocarbons such as xylene to 1 ppm. Over the following few years intensive research led to the development of many gas phase detectors for organic vapours (Ho *et al.* 1980; Guilbault 1983), environmental pollutants (Hlavay and Guilbault 1977; Edmonds and West 1980; Guilbault 1980; Guilbault 1981; Guilbault *et al.* 1981; Guilbault *et al.* 1985) and metals in solution (Nomura and Iijima 1981; Nomura and Mimatsu 1982; Nomura *et al.* 1982; Nomura and Maruyuma 1983). Chromatography detectors were also developed (Wohltjen and Dessey 1979; Konash and Bastiaans 1980). An area which has captured a lot of attention has been the use of antibodies as the crystal coating, offering inherent bioselectivity. These antibody coated crystals are referred to as QCM-based immunosensors. The first piezoelectric immunosensor was developed by Shons *et al.* (1972). The crystal was precoated with nyebar C and BSA and was used to detect BSA antibodies. The main drawbacks associated with these sensors are (1) a lack of reproducibility in antibody immobilisation on the crystal surface and (2) the viscous drag experience in the liquid phase. In order to overcome the latter problem, the resonance frequency of the crystal can be determined in air prior to the addition of sample. The crystal may then be dried, and any change in resonant frequency due to biospecific binding can be calculated. This is known as the 'dip and dry' method (Guilbault *et al.* 1992; Konig and Gratzel 1994).

A plethora of methods for antibody immobilisation on the crystal surface have been investigated. Adsorption of antibodies is a method widely used in the detection of various analytes (Muratsugu *et al.* 1993; Carter *et al.* 1995a, 1995b; Masson *et al.* 1995; Sakai *et al.* 1995; Harteveld *et al.* 1997; Yun *et al.* 1998), including HIV antibodies (Koblinger *et al.* 1992, 1994a, 1994b). Protein A is one of the most widely used pre-coatings used to aid antibody immobilisation and has been used in the detection of pesticides (Guilbault *et al.* 1992; Minunni *et al.* 1994a), bacteria (Plomer *et al.* 1992; Konig and Gratzel 1993; Jacobs *et al.* 1995; Boveniser *et al.* 1998), viruses (Konig and Gratzel 1994, 1995; Attili and Suleiman 1995) and various other analytes (Davis and Leary 1989; Prusak-Sochazewski and Luong 1990a, 1990b; Raman Suri *et al.* 1995; Nakanishi *et al.* 1996; Pei *et al.* 1998). It has been shown to give better results than other immobilisation techniques by some groups for certain systems (Raman Suri *et al.* 1995; Nakanishi *et al.* 1996), although this is not the case for all systems (Konig and Gratzel 1993; Uttenthaler *et al.* 1998). Protein G has also been used to immobilise antibodies to cocaine on the crystal surface (Attili and Suleiman 1996). Immobilisation by silanising the surface, usually with γ-aminopropyltriethoxysilane activated with glutaraldehyde, is frequently employed (Muramatsu *et al.* 1987; Minunni *et al.* 1994b; Steegborn and Skladal 1997). Polymers are also used, e.g. polyethyleneimine activated with glutaraldehyde (Prusak-Sochazewski *et al.*

1990a, 1990b; Guilbault 1981; Konig and Gratzel 1993a, 1993b, 1993c; Shao *et al.* 1993). Co-polymers of (HEMA-MMA) have been used for the detection of human IgM (Chu *et al.* 1995) and α-fetoprotein (Chu *et al.* 1996). Xia *et al.* used this polymer in their new approach to dual immunoassay for C-reactive protein and IgM using an array of 5 crystals (1996). Various other polymers have been used including electropolymerised films (Si *et al.* 1996, 1997), plasma-polymerised films (Nakanishi *et al.* 1996a, 1996b), polyamide (Mueller-Schultz 1996) and polystyrene (Yokoyama *et al.* 1995). Others have also been used (Xia *et al.* 1997). A new technique, latex piezoelectric immunoassay (LPEIA) was developed by Kurosawa *et al.* (1990). It has been used in the detection of C-reactive protein (Kurosawa *et al.* 1990), rheuma-toid factor (Ghourchian and Kamo 1994) and anti streptolysine O antibodies in which the application of initial rate LPEIA method was demonstrated (Muratsugu *et al.* 1992). Langmuir-Blodgett films have also been incorporated in the detection of ferritin (Lepesheva *et al.* 1994) and anti-fluorescein lipids (Ebato *et al.* 1994). The interaction between sulphur and gold has been exploited. Self-assembled monolay-ers (SAM) of thiols and sulphides have been used for the immobilisation of antibod-ies to the gold surface of piezoelectric crystals (Caruso *et al.* 1996; Cohen *et al.* 1996; Rickert 1996; Horacek and Skladal 1997; Storri *et al.* 1998). SAMs of cystamine (Abad *et al.* 1998) have also been used. SAMs of antibodies have been achieved using antibody-thiol complexes (Park and Kim 1998) and Traut's reagent (Neuman *et al.* 1995; Caruso *et al.* 1996) Ebersole *et al.* have shown the potential of sponta-neously formed avidin-streptavidin monolayers on gold (1990) and have also developed AMISA (Ebersole *et al.* 1988), a technique used to enhance detection limits.

Conclusion and future directions

The QCM has found a wide range of applications in areas of food, environmental and clinical analysis since its birth, due to its inherent ability to monitor analytes in real time. Piezoimmunosensors, using antibody/antigen coated crystals, offer the greatest potential, with possible applications in food, environmental and clinical analysis.

Several issues still need to be addressed, however, to produce commercially viable devices able to compete with the success of surface plasmon resonance, the technique used in the highly prosperous Pharmacia BIAcore instrument. A repro-ducible immobilisation of the biological material on the crystal surface must be achieved as well as a method to overcome problems due to non-specific binding of proteins to the antibody test surface. The re-usability of the piezoelectric crystal remains an important issue that is currently receiving a large amount of interest from a plethora of research groups. The availability of pre-coated gold crystals, which greatly reduces assay time and the requirement for skilled operators, is now viable with pre-activated crystals available from Universal Sensors Inc.

If, indeed, the QCM can accomplish the same results as the SPR method, it will offer many advantages over the BIAcore. The basic instrumentation required is far less comprehensive and thus markedly less expensive. This coupled with a predicted reduction in the cost of crystals and increase in the number of times the crystal may be regenerated promises the QCM(PZ) should supplement existing techniques in immunochemical research and technology in a most favourable manner.

Bibliography

Abad, J.M., Pariente, F., Hernandez, L., Abruna, H.D. and Lorenzo, E. (1998) Determination of organophosphorus and carbamate pesticides using a piezoelectric biosensor. *Analytical Chemistry* 70, 2848.

Attili, B.S. and Suleiman, A.A. (1995) Piezoelectric immunosensor for the determination of cortisol. *Analytical Letters* 28, 2149.

Attili, B.S. and Suleiman, A.A. (1996) A piezoelectric immunosensor for detection of cocaine. *Journal of Microchemistry* 54, 174.

Bastiaans, G.J. (1988) A surface accoustic wave sensor. U.S. Patent # 4,735,906.

Boveniser, J.S., Guilbault, G.G. and O'Sullivan, C.K. (1998) Detection of pseudomonas aeroginosa using the quartz crystal microbalance. *Analytical Letters* 31, 8.

Bruckenstein, S. and Shay, M. (1994) Dual quartz microbalance oscillator circuit. *Analytical Chemistry* 66, 1847–1855.

Carter, R.M., Jacobs, M.B., Lubrano, G.J. and Guilbault, G.G. (1995) Piezopelectric detection of ricin and affinity purified goat ant-ricin. *Analytical Letters* 28, 1379.

Carter, R.M., Mekalanos, J.J., Jacobs, M.B., Lubrano, G.J. and Guilbault, G.G. (1995) Quartz crystal microbalance detection of vibrio cholera. *Journal of Immunological Methods* 187, 121.

Caruso, F., Rodda, E. and Furlong, D.N. (1996) Piezoelectric detector. *Journal of Colloidal and Interfacial Science* 178, 104.

Chu, X., Lin, Z.H., Shen, G.L. and Yu, R.Q. (1995) Simultaneous immunoassay array and robust method. *Analyst* 120, 2829.

Chu, X., Lin, Z.H., Shen, G.L. and Yu, R.Q. (1996) *Chinese Journal of Chemistry in Chinese Universities* 17, 870.

Cohen, Y., Levi, S., Rubin, S. and Willner, I. (1996) Modified monolayer electrodes for electrochemical and pz analysis. *Journal of ElectroAnalytical Chemistry* 417, 65.

Curie, J. and Curie, P. (1880) An oscillating qwuartz crystal mass detector. *Rendu* 91, 294.

Czanderna, A.W. and Lu, C. (1984) in *Applications of Piezoelectric Quartz Crystal Microbalances*, Walsky, S.P. and Czanderna, A.W. (eds), Elsevier: New York, p. 1.

Davis, K. and Leary, T. (1989) PZ immunobiosensors for kinetic immunoassays. *Analytical Chemistry* 61, 1227.

Ebato, H., Gentry, C.A., Herron, J.N., Muller, W., Okahata, Y., Ringsdorf, H. and Suci, P.A. (1994) Investigation of specific binding of antiflouorescyl antibody and FAB to fluorescein lipids in langmuir-blodgett films using the piezocrystal microbalance. *Analytical Chemistry* 66, 1683.

Ebersole, R. and Ward, M. (1988) PZ quartz sensors for use in clinical analysis. *Journal of the American Chemical Society* 110, 8623.

Ebersole, R., Miller, J., Moran, J. and Ward, M. (1990) Analysis using pz quartz crystals. *Journal of the American Chemical Society* 112, 3239.

Ebersole, R.C., Wang, J., Ward, M.D. and Foss, R.P. (1993) PZ quartz sensors. *Analytical Chemistry* 65, 2553.

Edmonds, T.E. and West, T.S. (1980) *Analytica Chimica Acta* 117, 147.

Ghourchian, H.O. and Kamo, N. (1994) *Talanta* 41, 401.

Guilbault, G.G. (1980) Analytical uses of piezoelectric crystals. *Ion Selective Electrode Review* 2, 3.

Guilbault, G.G. (1981) Analysis of environmental pollutants using a piezoelectric crystal. *International Journal of Environmental Analytical Chemistry* 10, 89.

Guilbault, G.G. (1983) Determination of formaldehyde with an enzyme coated piezoelectric crystal. *Analytical Chemistry* 55, 1682.

Guilbault, G.G., Affolter, J., Tomita, Y. and Kolesar, E.S. (1981) Piezoelectric crystal coating for detecting organophosphorus compounds. *Analytical Chemistry* 53, 2057.

Guilbault, G.G., Hock, B. and Schmid, R. (1992) PZ Immunosensor for atrazine in drinking water. *Biosensors and Bioelectronics* 7, 411.

Guilbault, G.G., Kristoff, J. and Owen, D. (1985) Detection of organophosphorus compounds with a coated piezoelectric crystal. *Analytical Chemistry* 57, 1754.

Guilbault, G.G. and Luong, J.H. (1988) Gas phase biosensors. *Journal of Biotechnology* 9, 1.

Harteveld, J.L.N., Nieuwenhuizen, M.S. and Wils, E.R.J. (1997) Detection of staphylococcal enterotoxin B employing a PZ detector. *Biosensors and Bioelectronics* 12, 7.

Henry, C. (1996) PZ crystal sensors. *Analytical Chemistry* 68, 625A.

Hlavay, J. and Guilbault, G.G. (1977) Detection of hydrogen chloride gas in ambient air using a coated piezoelectric quartz crystal. *Analytical Chemistry* 49, 1890.

Ho, M.H., Guilbault, G.G. and Reitx, B. (1980) Continuous detection of toluene in ambient air with a coated piezoelectric crystal. *Analytical Chemistry* 52, 1489.

Horacek, J. and Skladal, P. (1997) Improved direct biosensors operating in liquid solution for the competitive label-free immunoassay of 2,4 D. *Analytica Chimica Acta* 347, 45.

Jacobs, M.B., Carter, R.M., Lubrano, G.J. and Guilbault, G.G. (1995) Piezobiosensor for listeria monocytogenes. *American Laboratory* 27, 26.

Kanasawa, K.K. and Gordon, J.G. (1985) A liquid phase piezoelectric detector. *Analytical Chemistry* 57, 1771–1775.

King, W.H. Jr. (1964) Piezoelectric sorption detector. *Analytical Chemistry* 36, 1735.

Koblinger, C., Drost, Aberl, F., Wolf, H., Koch, S. and Woias, P. (1992) PZ immunosensors for HIV viruses. *Biosensors and Bioelectronics* 7, 397.

Koblinger, C., Drost, S., Aberl, F. and Wolf, H. (1994a) Compatrison of PZ and SPR detectors. *Fresenius Journal of Analytical Chemistry* 349, 349.

Koblinger, C., Drost, S., Aberl, F., Wolf, H., Woias, P. and Koch, S. (1994b) HIV detection with a PZ immuinobiosensor. *Sensors and Actuators B* 18–19, 271.

Konash, P.L. and Bastiaans, G.J. (1980) A SAW Device for Immunoreactions. *Analytical Chemistry* 52, 1929.

Konig, B. and Gratzel, M. (1993a) Detection of viruses and bacteria with piezoimmuno-biosensors. *Analytical Letters* 26, 1567.

Konig, B. and Gratzel, M. (1993b) Human granulocytes detected with a piezoimmunosensor. *Analytical Letters* 26, 1313.

Konig, B. and Gratzel, M. (1994) A novel immunosensor for herpes virus. *Analytical Chemistry* 66, 341.

Konig, B. and Gratzel, M. (1995) *Analytica Chimica Acta* 309, 19.

Kurosawa, S., Tawara, E., Kamo, N., Ohta, F. and Hosokowa, T. (1990) *Chemical and Pharmaceutical Bulletin (Japan)* 38, 117.

Lepesheva, G.I., Turko, I.V., Ges, I.A. and Chashchin, V.L. (1994) *Biochemistry* 59, 7.

Masson, M., Yun, K., Haryama, T., Kobatake, E. and Aizawa, M. (1995) Quartz crystal microbalance for biotin. *Analytical Chemistry* 67, 2215.

Minunni, M., Skladal, P. and Mascini, M. (1994) *Analytical Letters* 11, 391.

Minunni, M., Skladal, P. and Mascini, M. (1994) A piezoelectric quartz crystal biosensor as a direct affinity sensor. *Analytical Letters* 27, 1475.

Mueller-Schultz, D. (1996) CA 112(7), 51807g, Patent.

Muramatsu, H., Dicks, J., Tamia, E. and Karube, I. (1987) A piezoelectric biosensor modified with protein A for immunoglobulins. *Analytical Chemistry* 59, 2760.

Muratsugu, M., Kurosawa, S. and Kamo, N. (1992) Detection of antistreptolysin O antibody: application of an initial rate method for latex piezoimmunoassay. *Analytical Chemistry* 64, 2483.

Muratsugu, M., Ohta, F., Miya, Y., Hosokawa, T., Kurosawa, S., Kamo, N. and Ikeda, H. (1993) Quartz crystal microbalance for the detection of microgram quantities of human serum albumin. *Analytical Chemistry* 65, 2933.

Nakanishi, K., Hiroshi, S., Karube, I., Uchida, A. and Ishida, Y. (1996b) *Analytica Chimica Acta* 325, 73.

Nakanishi, K., Masao, A., Sako, Y., Ishida, Y., Muguruma, H. and Karube, I. (1996c) Detection of red-tide causing plankton by a piezoelectric immunosensor using a novel method of immobilizing antibodies. *Analytical Letters* 29, 1247.

Nakanishi, K., Muguruma, H. and Karube, I. (1996a) A novel method of immobilizing antibodies on a quartz crystal microbalance using plasma polymerized fils for immunosensors. *Analytical Chemistry* 68, 1695.

Neuman, M.R., Perlaky, S.C., Michalski, J.P., Bass, J.B. and McCombs, C.C. (1995) EMBC, Paper 7.1.3.8.

Nomura, T. and Iijima, M. (1981) *Analytica Chimica Acta* 131, 97.

Nomura, T. and Maruyuma, M. (1983) *Analytica Chimica Acta* 143, 243.

Nomura, T. and Mimatsu, T.M. (1982) *Analytica Chimica Acta* 143, 237.

Nomura, T., Yamashita, T. and West, T.S. (1982) *Analytica Chimica Acta* 147, 365.

Park, I.S. and Kim, N. (1998) Thiolated salmonella antibody immobilization onto a PZ crystal. biosensors bioelectronics. Presented at the 5th Congress on Biosensors, Germany, 3–5th June 1998.

Pei, R.J., Hu, J.M., Hu, Y. and Zeng, Y.E. (1998) Studies of the piezoelectric immunosensor for detection of fibrin using protein A oriented immobilization of antibody. *Chinese Journal of Chemistry in Chinese Universities* 19, 363.

Plomer, M., Guilbault, G.G. and Hock, B. (1992) Development of a PZ immunosensor for detection on enterobacteria. *Enzyme and Microbial Technology* 14, 230.

Prusak-Sochazewski, E. and Luong, J. (1990a) A new approach to development of a reusable PZ biosensor. *Analytical Letters* 23, 183.

Prusak-Sochazewski, E. and Luong, J. (1990b) *Analytical Letters* 23, 40.

Prusak-Sochazewski, E., Luong, J. and Guilbault, G.G. (1990a) *Enzyme and Microbial Technology* 12, 173.

Prusak-Sochazewski, E., Luong, J. and Guilbault, G.G. (1990b) Development of a piezoelectric immunosensor for detection of salmonella. *Enzyme and Microbial Technology* 12, 185.

Raleigh, L. (1885) in Pacy, D.J. (1960) *Vacuum* 9, 261.

Raman Suri, C., Jain, P.K. and Mishra, G.C. (1995) *Journal of Biotechnology* 39, 27.

Raman Suri, C., Raje, M. and Mishra, G.C. (1994) *Biosensors and Bioelectronics* 9, 535.

Rickert, J. (1996) Quartz crystal microbalances for quantitative biosensing. *Biosensors and Bioelectronics* 11, 541.

Roederer, J.E., O'Toole, R.P., Burns, S.G. and Poter, M.D. (1983) *Analytical Chemistry* 55, 2333.

Sakai, G., Sakai, T., Uda, T., Miura, N. and Yamazoe, N. (1995) Evaluation of binding of HAS to monoclonal and polyclonal antibody by PZ immunosensing. *Sensors and Actuators B* 24–25, 134.

Sauerbrey, G.Z. (1959) Use of quartz vibration for weighing thin films on a microbalance. *Journal Physik* 155, 206–212.

Shao, B., Hu, Q., Hu, J., Zhou, X., Zhang, W., Wang, X. and Fan, X. (1993) *Fresenius Journal of Analytical Chemistry* 346, 1022.

Shons, A., Dorman, F. and Najarian, J. (1972) The piezoelectric quartz immunosensor *Journal of Biomedical Material Research* 6, 565.

Si, S., Chen, J.H., He, F.J., Nie, L.H. and Yao, S.Z. (1996) *Journal of Chinese Chemistry* 14, 3.

Si, S., Ren, F., Cheng, W. and Yao, S. (1997) Preparation of a PZ immunosensor for detection of salmonella by immobilization of antibodies. *Fresenius Journal of Analytical Chemistry* 357, 1101.

Steegborn, C. and Skladal, P. (1997) Construction and charactcrization of a PZ immunosensor for atrazine operating in solution, *Biosensors and Bioelectronics* 12, 19.

Storri, S., Santoni, T. and Mascini, M. (1998) Surface modification for development of piezoimmunosensors. Presented at the 5th Congress on Biosensors, Germany, 3–5th June 1998.

Uttenthaler, E., Koblinger, C. and Drost, S. (1998) Quartz crystal biosensor for detection of african swine fever disease. *Analytica Chimica Acta* 362, 91.

Wohltjen, H. and Dessey, R. (1979) The surface acoustic wave detector. *Analytical Chemistry* 51, 1465.

Xia, C., Guo-Li, S., Ru-Qin, Y. and Feiye, X. (1997) A polymer agglutination based PZ immunoassay for determination of human sereum albumin. *Analytical Letters* 30, 1783–1796.

Xia, C., Jian-Hui, J., Guo-Li, S. and Ru-Qin, Y. (1996) *Analytica Chimica Acta* 336, 185.

Yokoyama, K., Ikebukuro, K., Tamiya, E., Karube, I., Ichika, N. and Arikawa, Y. (1995) *Analytica Chimica Acta* 304, 139.

Yun, K., Kobatake, E., Haruyuma, T., Laukkaned, M., Keinanen, K. and Aizawa, M. (1998) Use of a quartz crystal micobalance to monitor immunoliposome-antigen interaction. *Analytical Chemistry* 70, 260.

12 The quartz crystal microbalance with dissipation monitoring (QCM-D)

Michael Rodahl, Patrik Dahlqvist,
Fredrik Höök and Bengt Kasemo

Introduction

The quartz crystal microbalance (QCM) was born as a quantitative microbalance through the careful work by Sauerbrey (1959). He demonstrated that there is proportionality between mass added to the QCM sensor and the shift in its resonant frequency.

The QCM sensor is a thin quartz crystal disk sandwiched between two electrodes (Figure 12.1). Quartz is piezoelectric. This means that when an electric field is applied between the electrodes the crystal will mechanically deform (Figure 12.2) and in an AC field the crystal can be made to oscillate. It is therefore possible to electrically both excite the crystal into oscillation and measure its resonant properties.

The strength of the device as a microbalance derives from the extremely high precision with which the resonance frequency can be measured: even a tiny mass change of the sensor causes a measurable frequency shift. Under ideal conditions the mass sensitivity is down to tens of picograms per cm². The mass sensitivity can be expressed as the so-called Sauerbrey equation:

$$\Delta m_{QCM} = \rho_f \delta_f = -\frac{t_q \rho_q}{f} \Delta f = \frac{C_{QCM}}{n} \Delta f \qquad (12.1)$$

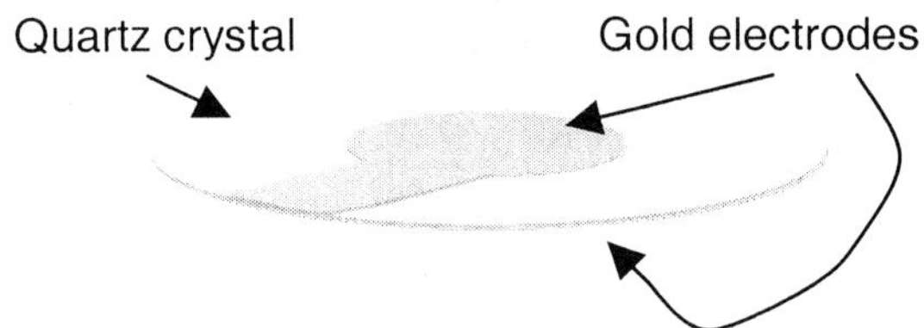

Figure 12.1 A quartz crystal with electrodes (one on each side). One of the electrodes serves as the sensing surface

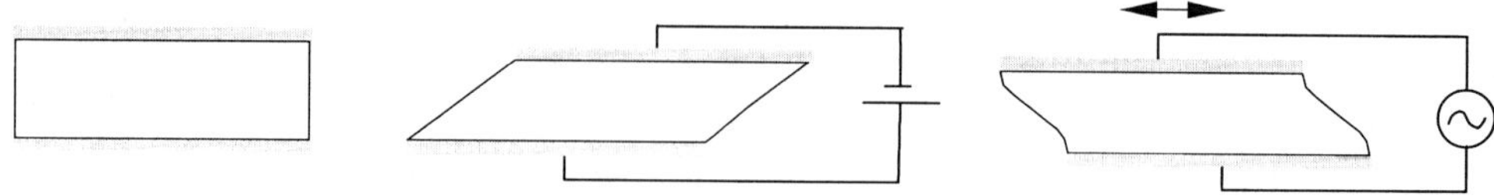

Figure 12.2 A schematic illustration of the strain induced in a quartz crystal

where ρf and δ_f are the density and thickness of the added film, ρ_q and t_q are the density and thickness of the quartz plate respectively. f is the frequency where the QCM operate, n the overtone number and C_{QCM} is the so-called mass sensitivity of the QCM. The first application of the QCM was as a film thickness monitor for film deposition in vacuum systems. Eventually, new applications were demonstrated (still in vacuum), e.g. as an oxidation rate sensor (Kasemo and Törnquist 1978; Kasemo and Törnquist 1980), adsorption sensor for gas adsorption (King Jr. 1964; Mecea 1993; Tsionsky and Gileadi 1994). Other early applications were in the direction of gas sensing in air (Guilbaut 1981).

In 1985, Kanazawa and Gordon paved the way for a whole new set of applications by demonstrating that the QCM can be operated in a stable and reproducible manner in a liquid (Kanazawa 1985). The by far most important application of this liquid phase QCM has until recently been the electrochemical QCM or EQCM (Reed, Kanazawa *et al.* 1990; Buttry and Ward 1992; Kanazawa and Melroy 1993; Takada, Diaz *et al.* 1997).

The EQCM was, however, soon challenged as the most interesting liquid phase application, by the many possible applications related to adsorption phenomena in liquids, and various processes occurring within or on adsorbed films. Among the large number of such applications we note adsorption of proteins (Muratsugu, Ohta *et al.* 1993; Aberl, Wolf *et al.* 1994; Rickert, Weiss *et al.* 1996; Caruso, Furlong *et al.* 1997; Höök, Rodahl *et al.* 1998a; Höök, Rodahl *et al.* 1998b; Höök, Rodahl *et al.* 1999), nucleotides (Niikura, Matsuno *et al.* 1998; Okahata, Kawase *et al.* 1998; Höök, Ray *et al.* 2000) and other biofilms, such as supported lipid films (Tjarnhage and Puu 1996; Keller and Kasemo 1998; Janshoff, Galla *et al.* 2000; Keller, Glasmastar *et al.* 2000), bacteria and living cells (Matsuda, Kishida *et al.* 1992; Gryte, Ward *et al.* 1993; Redepenning and Schlesinger 1993; Janshoff, Wegener *et al.* 1996; Fredriksson, Kihlman *et al.* 1998a; Fredriksson, Kihlman *et al.* 1998b; Otto, Elwing *et al.* 1999) and for blood coagulation (Vikinge, Hansson *et al.* 2000). In addition, the QCM technique has also been found valuable for investigations of polymer adsorption and/or phase-transitions (Bandey, Hillman *et al.* 1997; Forrest, Svanberg *et al.* 1998; Lucklum, Behling *et al.* 1999) and corrosion phenomena (Deakin 1989; Chandler, Ju *et al.* 1991; Eickes, Rosenmund *et al.* 2000; Wadsak, Schreiner *et al.* 2000).

In spite of the many applications of the QCM, it also became successively more evident that a single parameter measurement, i.e., frequency measurements only, had some severe limitations, especially when non-rigid films were studied. This regime can generally be referred to as the failure of the Sauerbrey relation.

The QCM beyond the Sauerbrey regime

A condition for the Sauerbrey relation to hold is that the added film perfectly couple to the shear oscillation of the sensor, as a 'dead mass'. This holds very well for, for example, metal and ceramic films of thicknesses up to a few percent of the sensor thickness. Thus, the QCM as a film thickness monitor for metals and ceramics could safely be calculated using Equation 1. However, when the QCM was used to study soft films, such as biofilms, a regime was entered where the Sauerbrey relation often failed. For many years this created confusion, misinterpretation of data and even a sceptical attitude towards the QCM for any such measurements.

It is actually quite easy to understand – qualitatively – why the Sauerbrey relation loses applicability when the applied films are soft and have appreciable thickness. Under such circumstances, the film does not follow the sensor's mechanical oscillation as a dead mass, but is deformed in the shear direction, in a way that depends on the distance from the electrode and the film's viscoelastic properties. As a pictorial analogue we can imagine jelly lying on a plate, oscillating horizontally. The jelly is periodically deformed as the plate oscillates. The frequency change caused by the jelly will not solely correspond to its static (dead) mass, but also to its elastic and viscous components. In other words, the jelly acts as a dynamic mass, whose influence on the frequency is a complex function of its viscous and elastic components.

We now realize that a thin, soft polymer film or a film of biomolecules can therefore not a priori be treated as a rigid film obeying the Sauerbrey relation. Instead we must assume that the QCM fails to measure the correct mass in such cases.

So how can we get out of this dilemma, apparently obstructing the use of the QCM for quantitative measurements with soft films of appreciable thickness, such as films composed of biomolecules? Fortunately the QCM itself offers the solution. By measuring only the frequency of the QCM we are not exploiting its full capacity by far.

If for a moment, we assume that the sensor is brought to oscillation, floating freely in space, we would see that the oscillation amplitude would decay after the driving power was removed (see Figure 12.3). The rate of decay would not be influenced if a solid film is added onto the sensor. However, if we add a soft film on top of the electrode, we would see a much faster decay of the amplitude as schematically shown in Figure 12.3.

This brings us to a very important conclusion: if we could measure not only the frequency, but also the damping (energy dissipation) of the sensor we would have a quantitative measure of the losses induced by a viscoelastic load, which can thus be used to solve our dilemma. By adding the damping measurement, D, to the traditional QCM – thus making it a QCM-D instrument (c.f. Figure 12.3) – we obtain an instrument that extends the limit of the QCM, enabling reliable measurements of

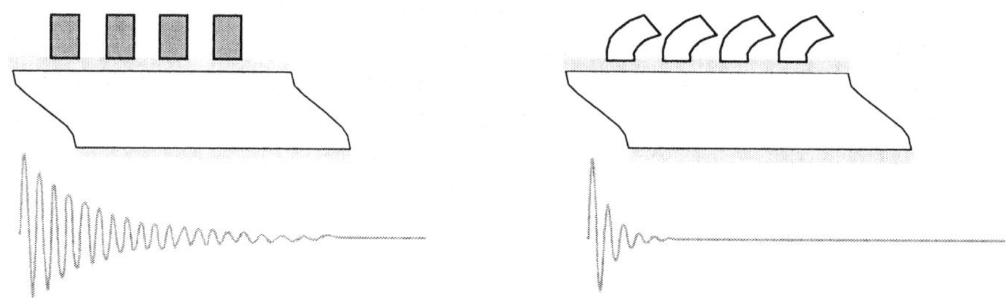

Figure 12.3 A schematic illustration of oscillating quartz crystals with a solid (left panel) and a soft film (right panel). If the driving power to the oscillating crystals would be cut off, the crystal oscillation will decay as shown below for each crystal. A soft (viscous) film that oscillates will dissipate more energy than a solid (elastic) film. Therefore, the oscillation of the crystal with the soft film will decay faster than the one with the rigid film. A QCM-D instrument works by intermittently exciting the sensor into oscillation and analysing the resulting decay curves to obtain both the resonant frequency and the dissipation (damping) factor. This can be done several times per second and at several overtones

non-rigid materials such as biomolecules and polymers. This, in turn, opens up the possibility to extract information about the viscous and elastic components of the film. The relation between an applied film's density, thickness, viscosity and shear modulus and the induced shift in resonant frequency and damping is given in the Appendix (page 313).

Application examples

The QCM-D technique offers a possibility to investigate, in real time, changes in the viscoelastic properties of adsorbed biomolecular layers, which are directly related to the conformation of the layer. Two examples of how simultaneous measurements of f and D provide such information are presented as well as a comparison with the Surface Plasmon Resonance technique (SPR) (Lofas, Malmqvist *et al.* 1991; Liedberg, Lundstrom *et al.* 1993). SPR is an optical technique offering (to a first approximation) quantitative information about the adsorbed mass from measurements of changes in the interfacial refractive index, $\Delta n \propto \Delta m$, upon replacement of water, n_{water}, for the adsorbed molecules, n_{ads}.

Example 1: DNA

It is generally so that a biosensor utilizes specific biological recognitions, such as ligand–receptor, antigen–antibody, DNA–DNA or DNA–protein interactions, where one of the involved components is immobilized on a surface. In such cases it is important that the immobilization is done in such a way that it does not influence the affinity for the substances that one wants to detect. We show here how the QCM-D technique can be used to monitor the immobilization of single-stranded DNA in a flexible conformation sufficient for subsequent hybridization with complementary DNA strands. The surface preparation used in the present case actually involves several steps and all of these steps can be studied and monitored with the QCM-D technique.

In order for surface-bound DNA strands to quickly and fully hybridize with complimentary DNA in solution, it is important that they retain their flexibility and are not sterically hindered by other molecules or the underlying surface. These problems were, in this example, solved by coupling a biotin-molecule to a single-stranded DNA that was bound to an ordered layer of streptavidin, utilizing the specific binding of biotin to streptavidin.[1] We thus obtain a layer composed of DNA strands that are unaffected by the underlying solid substrate and sufficiently spaced to avoid steric hindrance during hybridization.

This whole process is schematically illustrated in the top part of Figure 12.4, starting with the formation of the lipid bilayer from lipid vesicles to the attachment of the DNA strands and subsequent hybridization of fully complementary DNA strands (DNA_{fc}). The bottom part of Figure 12.4 shows the corresponding QCM-D response.

The QCM-D trace contains several interesting features. To start with, the initial formation of the biotin doped lipid bilayer is not a monotonic increase in coupled

---

[1]The streptavidin layer was actually a two-dimensional crystalline layer formed on a biotin doped phospholipid bilayer supported on a silicon dioxide coated QCM-D sensor.

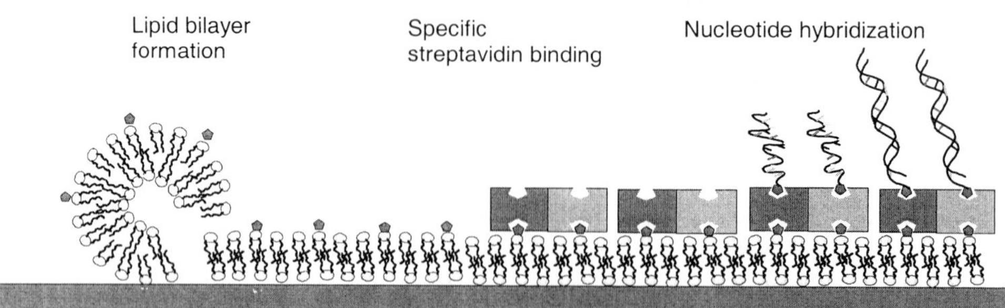

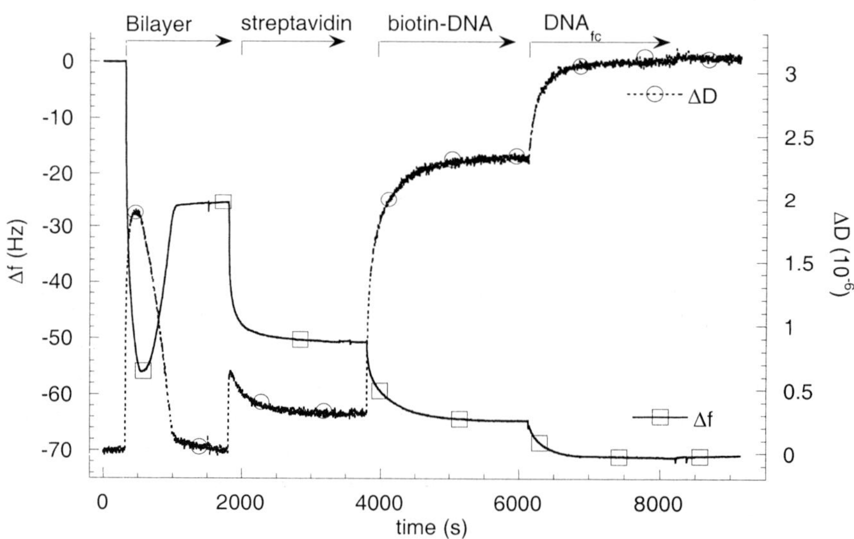

Figure 12.4 Top panel: a schematic illustration of the lipid bilayer formation followed by streptavidin binding and nucleotide hybridisation. The lipid bilayer is formed from suspended lipid vesicles that adsorb onto the sensor surface. The vesicles have incorporated biotin (pentagons) that streptavidin (boxes) subsequently binds to. DNA strands with biotin links then bind to the streptavidin layer. These DNA strands are sufficiently spaced to allow a rapid hybridisation with complementary DNA strands.

Bottom panel: corresponding Δf and ΔD versus time traces. Initially lipid vesicles adsorb inducing large f and D shift. The vesicles transforms into a lipid bilayer when the surface concentration is high enough which is seen as an increase in f (a mass loss as water leaves the oscillating system) and a decrease in D as the bilayer is more rigid than vesicles. Streptavidin is then adsorbed seen as an increase in mass (decrease in f). Crystallisation of the streptavidin is seen as a decrease in D as the formed layer becomes more rigid. Biotin-DNA and complementary DNA is then added in two steps inducing high D values confirming the DNA-strands to be flexible

mass (decrease in resonant frequency) as one might have expected, but goes through a maximum. This is in contrast to SPR data, which actually shows only a simple (monotonous) mass-uptake curve (not shown) (Keller and Kasemo 1998; Keller, Glasmastar *et al.* 2000). The simultaneously measured dissipation factor gives an important clue to the apparent discrepancy. The large peak in D that coincides with the dip in f during the bilayer-formation process shows that the film at this time is far from rigid. This indicates that the lipids vesicles do not break up and form a bilayer directly upon adsorption, but adsorb as intact vesicles that break up first when a critical coverage is reached (Keller and Kasemo 1998; Keller, Glasmastar *et al.* 2000). This view is also supported by subsequent Monte Carlo based simulation of the bilayer formation process (Zhdanov, Keller *et al.* 2000).

In contrast to SPR, which estimates mass based on the difference in refractive index between the adsorbed lipids and replaced interfacial water, the QCM techniques senses all coupled mass, including water trapped in the vesicles. The difference in mass-uptake measured by QCM and SPR is thus simply the amount of water coupled to the layer composed of unbroken vesicles.

The f-response due to streptavidin adsorption onto the biotin-doped lipid bilayer is rather straightforward and is a monotonic increase in mass with time. However, there is a small but significant peak in the measured dissipation factor. This indicates that streptavidin initially forms a mobile layer that solidifies as more and more proteins are packed on the surface.

From the large increase in D due to the adsorption of the biotin linked DNA onto the streptavidin layer it is apparent that a highly flexible layer is formed. This layer can, as seen in Figure 12.4, rapidly hybridize with complementary DNA. It is interesting to note that single-stranded DNA attached directly to a bare surface, such as gold, induces very low energy dissipation (not shown). This signals that the binding to a bare surface involves several contact points to the surfaces, which can also explain the significantly lower ability to hybridize with complementary DNA in these cases (not shown). Hence, the possibility to be able to perform combined f and D measurements can be extremely valuable when manufacturing surfaces designed for specific biorecognition (Höök, Ray *et al.* 2000).

Example 2: Mussel adhesive protein

In this example it is illustrated how the QCM-D can be used to monitor structural changes in a protein film at the same time as kinetic, viscous and elastic information is extracted. The conformation and structure of biological macromolecules play an important role for their ability to participate in reactions. One practical situation in which surface-induced structural changes of adsorbed proteins are believed to be critical is the first contact between an implant and the host organism (Ratner 1996; Kasemo 1998; Kasemo and Gold 1999). Another example is the growth of algae, mussels, etc. on marine vessels, which is a serious problem in shipping industry as well as for recreational yachting.

We have used, as a model system highly relevant for the latter case, one of the components in the 'glue' that mussels use to anchor to a solid surface, namely the *Mytilus edulis foot protein, Mefp-1* (Waite 1985; Waite 1987; Williams, Marumo *et al.* 1989). It is composed of repeating units of almost identical decapeptides that has an open flexible conformation in solution. Cross-linking using, for instance, $NaIO_4$ can, as we will show here, dramatically change this conformation as well as the mechanical properties of the protein.

From a QCM-D point of view, we use the adsorption and cross-linking of *Mefp-1* as an example to show how the additional information obtained by simultaneously measuring both f and D at several frequencies can be used to make not only accurate thickness measurements where the Sauerbrey equation fails but also extract shear viscosity and shear modulus.

Figure 12.5 shows a QCM-D measurement of the adsorption of *Mepf-1* to a hydrophobic surface (CH_3 terminated thiols self-assembled on a QCM-D sensor with gold electrodes) and subsequent cross-linking using $NaIO_4$ (Fant, Sott *et al.* 2000). Changes in f and D were simultaneously measured at both 5 MHz (the sensor's fundamental frequency) and at 15 MHz (the sensor's third overtone). According to the Sauerbrey equation, the frequency change upon a given mass change at the third overtone is three times higher than at the fundamental frequency. Therefore the response at 15 MHz has been scaled with a factor of three. It is therefore obvious that this data cannot be interpreted using the Sauerbrey equation. This is due to the fact that the adsorbed *Mepf-1* film cannot be considered rigid, as also indicated from the large increase in D during adsorption of *Mefp-1*. After cross-linking, which results in a large decrease in D, there is still a discrepancy, but it is much less pronounced. Taken together, these observations show how the film becomes more rigid upon cross-linking.

The results have therefore been analysed with the extended QCM-D model (implemented in the analysis software QTools, Q-Sense) to extract thickness, shear elastic and viscous modulus (Höök, Fant *et al.* 2000). The result is shown in Figures 12.5b and 12.5c. As a comparison, we have done the same measurement using a Biacore 2000 surface plasmon resonance instrument, as shown in Figure 12.6.

There is a striking difference in the two measurements of film mass. Whereas QCM-D measures a decrease in film mass during rinsing and cross-linking, SPR measures no change and increase in mass, respectively. Also, the apparent mass of the *Mefp-1* film before cross-linking is much larger when measured with QCM-D than SPR. However, if we consider the difference in how QCM-D and SPR measures mass (as discussed above) it is easy to understand that this discrepancy in mass is due to the water present in the *Mefp-1* layer.

We can now interpret these measurements like this. The film that forms when *Mefp-1* adsorbs contains a significant amount of water. During rinsing and cross-linking, water is expelled from the film whereby its total mass decreases (as well as its thickness) even though the dry weight of the film increases (due to the addition of $NaIO_4$). There is a small but significant increase in shear modulus and shear viscosity when the film thickness decreases during rinsing. There is however no loss of 'dry mass'. The addition of $NaIO_4$ radically increases the shear modulus and shear viscosity.

We have also conducted parallel film-thickness measurements of adsorbed *Mefp-1* before and after cross-linking with ellipsometry (Table 12.1). These results agree within 15 per cent with those of QCM-D using the extended model (Höök, Fant *et al.* 2000), but much less using the Sauerbrey relation. It can thus be concluded that for sufficiently non-rigid films, inducing large D and big variations in the frequency response at different harmonics, modelling is required in order to accurately determine the coupled mass.

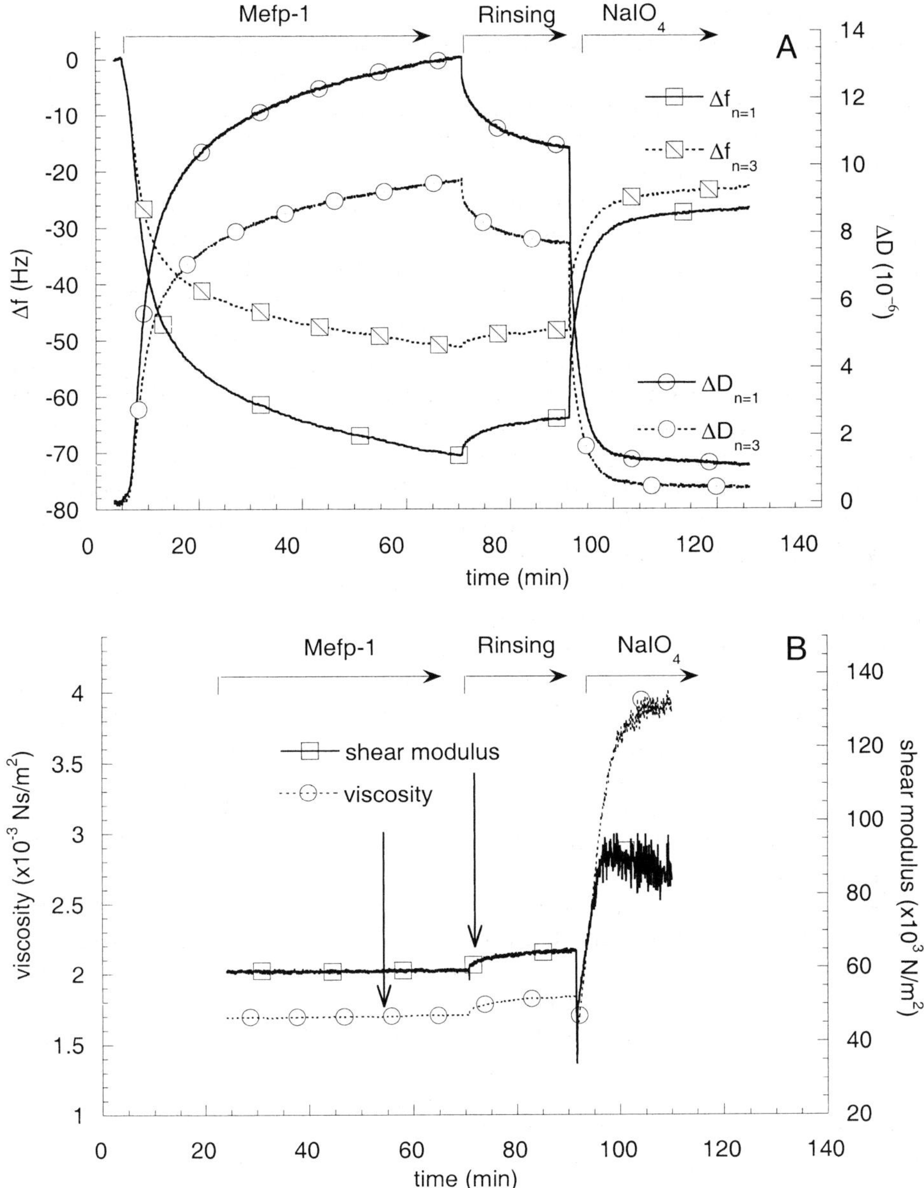

Figure 12.5 a Δf and ΔD at $n = 1$ (5 MHz) and $n = 3$ (15 MHz) versus time for exposure of a CH$_3$-terminated surface to *Mefp-1* followed by addition of NaIO$_4$. *Mefp-1* adsorbs as a non-rigid film, with a substantial amount of trapped water, as evident by the high D values and the difference between the two frequency measurements. Water is expelled from the protein layer during cross-linking as the layer is exposed to NaIO$_4$, which causes an increase in f. The cross-linking also stiffens the protein layer, which decreases D

b Top panel: The modeled changes in the shear elastic modulus (μ) and shear viscosity (η) versus time when *Mepf-1* cross-link due to addition of NaIO$_4$

c The modeled change in mass when *Mepf-1* cross-link

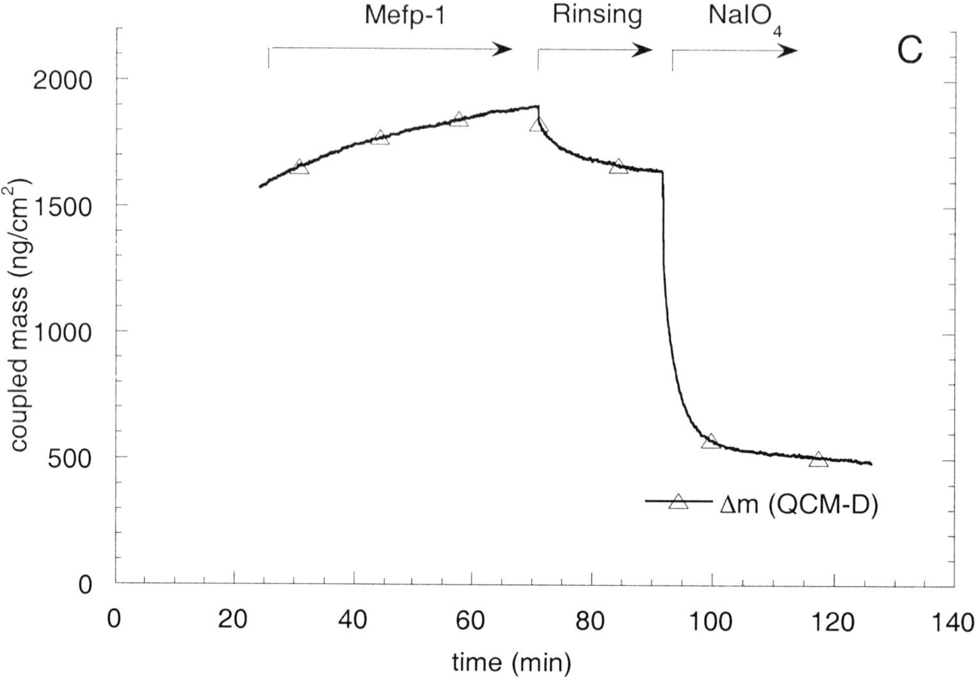

Figure 12.6 Adsorbed (optical) mass of *Mepf-1* versus time measured with SPR. Cross-linking due to addition of NaIO$_4$ is also shown

Table 12.1 Comparison between thicknesses and masses achieved with QCM, Ellipsometry, SPR, and QCM-D. Acoustic methods measure all oscillating mass, both protein and water, while optic methods only measures the protein mass resulting in the large differences between the methods. Correlation between QCM-D and Ellipsometry concerning thickness is however much better when the viscous and elastic components is considered and not possible with the classic QCM

	QCM	*Ellipsometry*	*SPR*	*QCM-D*
Mass before cross-linking (ng/cm^2)	1250	134	149	1900
Mass after cross-linking (ng/cm^2)	480	130	174	490
Thickness before cross-linking (nm)	12	23	n/a	18
Thickness after cross-linking (nm)	4	3,7	n/a	4

Concluding remarks

Traditional QCM measurements of soft adlayers (such as biofilms) may at first sight appear confusing compared to optical methods like ellipsometry and SPR. However by simultaneously measuring the dissipation factor of the QCM sensor (as well as the resonant frequency) the confusion can be cleared and we have a unique platform to obtain exciting information about soft films.

It is in this context also worthwhile to point out that modelling is, in certain cases, necessary in order to correctly estimate the total coupled mass (see Figure 12.7 and

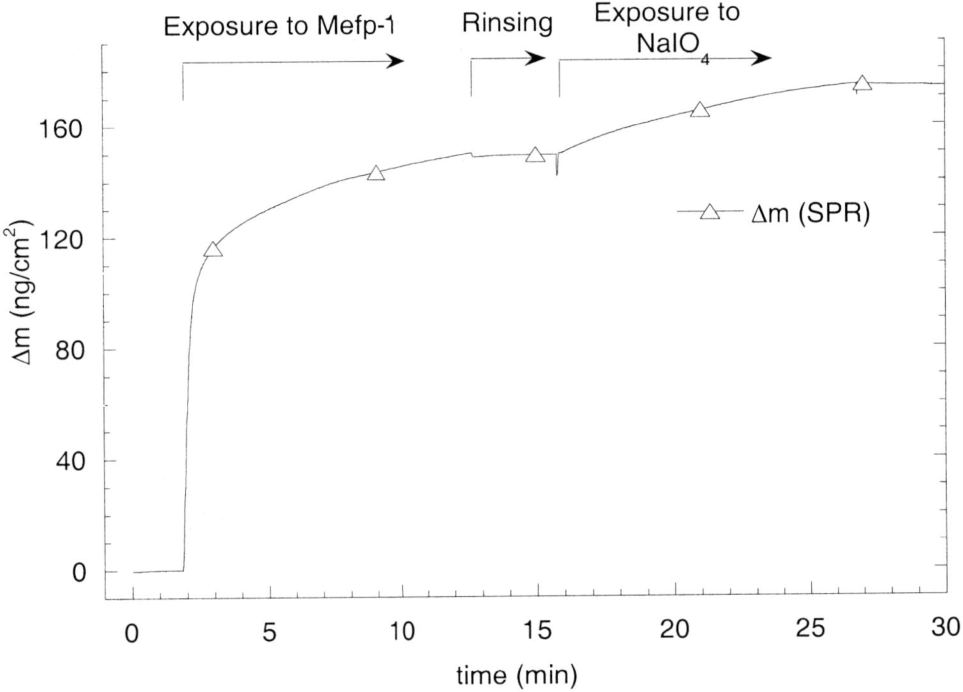

Figure 12.7 The adsorbed mass vs. time for the same sequence of events as in Figure 12.5 from SPR measurements

Table 12.1). However, in several cases, such as for adsorbed globular proteins like albumin, hemoglobin or streptavidin (see Figure 12.5) and even multi-domain proteins, such as antibodies and fibrinogen, the failure of the Sauerbrey relation is less than 10 per cent. For DNA (depending on their actual length) it is more critical, and is about 25 per cent in the example shown in Figure 12.5. Nevertheless, as soon there is a detectable change in D ($>\sim 0.2 \cdot 10^{-6}$) the viscous and the elastic component of the film can be extracted, which is a significant advantage compared to the classical QCM technique (measuring the frequency only). Regarding the latter, the QCM-D technique actually compares with and nicely complements ellipsometry, which in addition to thickness- and mass-measurements, also allow quantification of changes of the interfacial refractive index. However, while ellipsometry is restricted to highly reflective surfaces, the QCM-D technique can operate with a surface of any material as long as it can be deposited as thin ($<\sim 10\,\mu$m) film on its sensor surface.

Appendix

The details of the derivations and a full discussion of this model have been published elsewhere (Voinova, Rodahl *et al.* 1999). Here, we will only give the solution to the case where we have one film on top of the quartz crystal electrode that is submerged in a Newtonian bulk liquid such as water. (The model is easily extended to

314 *Michael Rodahl* et al.

contain slip between film and electrode, several film layers, and thin layers of Newtonian liquids.) The change in resonant frequency, Δf, and dissipation factor, ΔD, due to the presence of the viscoelastic film and bulk liquid are:

$$\Delta f = \frac{\mathrm{Im}(\beta)}{2\pi t_q \rho_q} \tag{A1}$$

and

$$\Delta D = \frac{\mathrm{Re}(\beta)}{\pi f t_q \rho_q} \tag{A2}$$

where

$$\beta = \xi_1 \frac{2\pi f \eta_1 - i\mu_1}{2\pi f} \frac{1 - \alpha e^{2\xi_1 h_1}}{1 + \alpha e^{2\xi_1 h_1}} \qquad \alpha = \frac{\dfrac{\xi_1}{\xi_2} \dfrac{2\pi f \eta_1 - i\mu_1}{2\pi f \, \eta_2} + 1}{\dfrac{\xi_1}{\xi_2} \dfrac{2\pi f \eta_1 - i\mu_1}{2\pi f \, \eta_2} - 1}$$

$$\xi_1 = \sqrt{-\frac{(2\pi f)^2 \rho_1}{\mu_1 + i2\mu f \eta_1}} \qquad \xi_1 = i\sqrt{\frac{2\pi f \rho_2}{\eta_2}}$$

and where h_1 is the thickness of the viscoelastic film, ρ_1 and ρ_2 denotes the density of the film and the bulk liquid, respectively, η_1 and η_2 denotes the viscosity of the film and the bulk liquid, respectively, μ_1 and the elastic modulus of the film.

References

Aberl, F., H. Wolf, *et al.* (1994) HIV serology using piezoelectric immunosensors. *Sensors and Actuators B* 18–19: 271.

Bandey, H.L., A.R. Hillman, *et al.* (1997) Viscoelastic characterization of electroactive polymer films at the electrode/solution interface. *Faraday Discussions 107: Acoustic Waves and Interfaces, Lester UK* 107: 105–122.

Buttry, D.A. and M.D. Ward (1992) Measurement of interfacial processes at electrode surfaces with the electrochemical quartz crystal microbalance. *Chemical Review* 92: 1355–1379.

Caruso, F., D.N. Furlong, *et al.* (1997) Characterization of ferritin adsorption onto gold. *Journal of Colloid and Interface Science* 186(1): 129–140.

Chandler, C., J.B. Ju, *et al.* (1991) Corrosion and electrodissolution studies of cobalt and copper thin-films with the quartz crystal microbalance and microelectrodes. *Corrosion* 47(3): 179–185.

Deakin, M.R. (1989) Monitoring corrosion processes in solution with the QCM. *Abstracts of Papers of the American Chemical Society* 198: 114–ANYL.

Eickes, C., J. Rosenmund, *et al.* (2000) The electrochemical quartz crystal microbalance (EQCM) in the studies of complex electrochemical reactions. *Electrochimica Acta* 45(22–23): 3623–3628.

Fant, C., K. Sott, *et al.* (2000) Adsorption behavior and enzymatically or chemically induced cross-linking of a mussel adhesive protein. *Biofouling* 16(2–4): 119–132.

Forrest, J.A., C. Svanberg, *et al.* (1998) Relaxation dynamics in ultrathin polymer films. *Physical Review E* 58(2): R1226–R1229.

Fredriksson, C., S. Kihlman, *et al.* (1998a) *In Vitro* real-time characterization of cell attachment and spreading. *Journal of Material Science: Materials in Medicine* 9: 785.

Fredriksson, C., S. Kihlman, *et al.* (1998b) The piezoelectric quartz crystal mass and dissipation sensor: a means of studying cell adhesion. *Langmuir* 14: 248–251.

Gryte, D.M., M.D. Ward, *et al.* (1993) Real-time measurement of anchourage-dependent cell adhesion using a quartz crystal microbalance. *Biotechnol. Prog.* 9: 105–108.

Guilbaut, G.G. (1981) Analysis of environmental pollutants using a piezoelectric crystal detector. *International Journal of Environmental Analytical Chemistry* 10: 89–98.

Höök, F., B. Kasemo, *et al.* (2001) Variations in coupled water, viscoelastic properties, and film thickness of a Mefp-1 protein film during adsorption and cross-linking: a quartz crystal microbalance with dissipation monitoring, ellipsometry, and surface plasmon resonance study. *Analytical Chemistry* 73(24): 5796–5804.

Höök, F., A. Ray, *et al.* (2001) Characterization of PNA and DNA immobilization and subsequent hybridization with DNA using acoustic-shear-wave attenuation measurements. *Langmuir* 17(26): 8305–8312.

Höök, F., M. Rodahl, *et al.* (1998a) Energy dissipation kinetics for protein and antibody–antigen adsorption under shear oscillation on a quartz crystal microbalance. *Langmuir* 14(4): 729–734.

Höök, F., M. Rodahl, *et al.* (1998b) Structural changes in hemoglobin during adsorption to solid surfaces: effects of pH, ionic strength, and ligand binding. *Proc. Natl. Acad. Sci.* 95(21): 12271–12276.

Höök, F., M. Rodahl, *et al.* (1999). *The dissipative QCM-D technique: Interfacial phenomena and sensor applications for proteins, biomembranes, living cells and polymers.* 1999 Joint Meeting EFTF-IEEE IFCS, IEEE.

Janshoff, A., H.J. Galla, *et al.* (2000) Piezoelectric mass-sensing devices as biosensors – an alternative to optical biosensors? *Angewandte Chemie-International Edition* 39(22): 4004–4032.

Janshoff, A., J. Wegener, *et al.* (1996) Double mode impedance analysis of epithelial cell monolayers cultured on shear wave resonators. *European-Biophysics-Journal-With-Biophysics-Letters.* 25(2): 93.

Kanazawa, K.K. (1985) Parasitic drag on a quartz resonator in liquid. *Electrochemical Society Extended Abstracts* 85(2): 652.

Kanazawa, K.K. and O.R. Melroy (1993) The quartz resonator: electrochemical applications. *IBM Journal of Research Development* 37(2): 157–171.

Kasemo, B. (1998) Biological surface science. *Current Opinion in Solid State Materials Science* 3(5): 451–459.

Kasemo, B. and J. Gold (1999) Implant surfaces and interface processes. *Advances in Dental Research* 13: 8–20.

Kasemo, B. and E. Törnquist (1978) Quartz crystal microbalance measurements of O/Ti and CO/Ti atom ratios on very thin Ti films. *Surface Science* 77: 209.

Kasemo, B. and E. Törnquist (1980) Weighing fractions of monolayers: application to the adsorption and catalytic reactions of H_2, CO, and O_2 on Pt. *Physical Review Letters* 44: 1555.

Keller, C.A., K. Glasmastar, *et al.* (2000) Formation of supported membranes from vesicles. *Physical Review Letters* 84(23): 5443–5446.

Keller, C.A. and B. Kasemo (1998) Surface specific kinetics of lipid vesicle adsorption measured with a quartz crystal microbalance. *Biophysical Journal* 75(3): 1397–1402.

King Jr., W.H. (1964) Piezoelectric sorption detector. *Analytical Chemistry* 36: 1735–1739.

Liedberg, B., I. Lundstrom, *et al.* (1993) Principles of biosensing with an extended coupling matrix and surface–plasmon resonance. *Sensors and Actuators B-Chemical* 11(1–3): 63–72.

Lofas, S., M. Malmqvist, *et al.* (1991) Bioanalysis with surface–plasmon resonance. *Sensors and Actuators B-Chemical* 5(1–4): 79–84.

Lucklum, R., C. Behling, *et al.* (1999) Role of mass accumulation and viscoelastic film properties for the response of acoustic-wave-based chemical sensors. *Analytical Chemistry* 71: 2488–2496.

Matsuda, T., A. Kishida, *et al.* (1992) Novel instrumentation monitoring *in situ* platelet adhesivity with a quartz crystal microbalance. *ASAIO-J* 38: M171–3.

Mecea, V.M. (1993) Tunable gas sensors. *Sensors and Actuators B* 15–16: 265–269.

Muratsugu, M., F. Ohta, *et al.* (1993) Quartz crystal microbalance for the detection of microgram quantities of human serum albumin: relationship between the frequency change and the mass of protein adsorbed. *Analytical Chemistry* 65(20): 2933–2937.

Niikura, K., H. Matsuno, *et al.* (1998) Direct monitoring of DNA polymerase reactions on a quartz-crystal microbalance. *Journal of the American Chemical Society* 120(33): 8537–8538.

Okahata, Y., M. Kawase, *et al.* (1998) Kinetic measurements of DNA hybridisation on an oligonucleotide-immobilized 27-MHz quartz crystal microbalance. *Anal. Chem.* 70(7): 1288–1296.

Otto, K., H. Elwing, *et al.* (1999) The role of type 1 fimbriae in adhesion of *Escherichia coli* to hydrophilic and hydrophobic surfaces. *Colloids and Surfaces B-Biointerfaces* 15(1): 99–111.

Ratner, B.D. (1996) The engineering of biomaterials exhibiting recognition and specificity. *Journal of Molecular Recognition* 9(5–6): 617–625.

Redepenning, J. and T.K. Schlesinger (1993) Osteoblast attachment monitored with a QCM. *Anal. Chem.* 65: 3378–3381.

Reed, C.E., K.K. Kanazawa, *et al.* (1990) Physical description of a viscoelastically loaded AT-cut quartz resonator. *Journal of Applied Physics* 68: 1993–2001.

Rickert, J., T. Weiss, *et al.* (1996) Self-assembled monolayers for chemical sensors: molecular recognition by immobilized supramolecular structures. *Sensors and Actuators B-Chemical* 31(1–2): 45–50.

Sauerbrey, G. (1959) Verwendung von Schwingquarzen zur Wägung dünner Schichten und zur Mikrowägung. *Zeitschrift für Physik* 155: 206–222.

Takada, K., D.J. Diaz, *et al.* (1997) Redox-active ferrocenyl dendrimers: thermodynamics and kinetics of adsorption, *in situ* electrochemical quartz crystal microbalance study of the redox process and tapping mode AFM imaging. *Journal of the American Chemical Society* 119(44): 10763–10773.

Tjarnhage, T. and G. Puu (1996) Liposome and phospholipid adsorption on a platinum surface studied in a flow cell designed for simultaneous quartz crystal microbalance and ellipsometry measurements. *Colloids and Surfaces B – Biointerfaces* 8(1–2): 39–50.

Tsionsky, V. and E. Gileadi (1994) Use of the quartz crystal microbalance for the study of adsorption from the gas phase. *Langmuir* 10(8): 2830–2835.

Vikinge, T.P., K.M. Hansson, *et al.* (2000) Comparison of surface plasmon resonance and quartz crystal microbalance in the study of whole blood and plasma coagulation. *Biosensors & Bioelectronics* 15(11–12): 605–613.

Voinova, M.V., M. Rodahl, *et al.* (1999) Viscoelastic acoustic response of layered polymer films at fluid–solid interfaces: continuum mechanics approach. *Physica Scripta* 59(5): 391–396.

Wadsak, M., M. Schreiner, *et al.* (2000) Combined *in situ* investigations of atmospheric corrosion of copper with SFM and IRAS coupled with QCM. *Surface Science* 454: 246–250.

Waite, J.H. (1985) Catechol oxidase in the bussys of the common mussel. *Marine Biol. Assoc. UK* 65: 359–371.

Waite, J.H. (1987) Nature's underwater adhesive specialist. *International Journal of Adhesion and Adhesives* 7(1): 9–14.

Williams, T., K. Marumo, *et al.* (1989) Mussel glue protein has an open conformation. *Archives of Biochemistry and Biophysics* 269: 415–422.

Zhdanov, V.P., C.A. Keller, *et al.* (2000) Simulation of adsorption kinetics of lipid vesicles. *Journal of Chemical Physics* 112(2): 900–909.

Index

Note: Figures and Tables are indicated by *italic* page numbers

2,4-D: immunoassay 223, *224*, 228

acoustic force microscopy (AFM) 192
acoustic plate mode (APM) 179–180
acoustic plate mode (APM) sensors 193–195; applications 194–195, *196*, *197*; comparison with other acoustic sensors *201*; theory 193–194
acoustic transducers 176–206; comparison of various devices 200–202
acoustic wave devices 180–183; as biomolecular sensors 184–200; sensitivity 200–201
ADP/ATP cycling *211*, 214
adsorption: immobilisation by *51*, 60
adsorption isotherms *91*
affinity capture: immobilisation by 258
affinity-conversion mechanism 114
affinity determination: BIA-technology used for 260, 264
agarose 56, 61
alanine-scanning mutagenesis 8, 15, 22
aldehydes: coupling to *65*, 69–70
alkaline phosphatase (AP) measurement: applications 222, 223, *224*, 225–226; detection limits 217, 221–222; principle 217–222
amides: formation of 66–67, 69
amines: coupling to *66*, 73
amplified flow immunoassay (AFIA) 223, *224*; applications 225–227
analyte recycling electrodes 209–22
antibodies: orientation-control techniques for 74–78
antibody-binding proteins 75–76
antibody fragments: tagging with 77
antibody specificity 13–14
antigen–antibody interactions 9–13, 74; recognition mechanisms 12–13
antigenicity 3
antigens 3; cross-reactivity 13, 14

aryl azides 80–81
aryl diazirines 81
association rate constants 90; antigen–antibody interactions 12; protein–protein interactions 25
attenuated total reflection (ATR) lens/prism 246
avidin/streptavidin capture systems 76–77, 188

barnase–barstar interaction *20*, 23, 24–25
benzophenone derivatives 81
BIAcore instrument(s) 148–149, 241–268; compared with quartz crystal microbalance 299; computer control 253; and coupling reactions 67; heterogeneous-phase measurements 27; integrated micro-fluidic cartridge *250*, 254–255; kinetic analysis by 259–260; light source 248–249; marketing of technology 263–265; and mass spectrometry 263, *264*; optical system 149, *150*, 248–249, *249*, 250–251; prism couplers in 148–149, 253; rate constants measured using 99, 104; receptor density 104; sample cell geometry 104, 109, 117, 255; sample handling system 254; and screening of small molecules 262; sensor chip 57–59, 68–69, 253, 256, 257, 258; simulation procedures 111, *113*; thermostatting unit 256; *see also* surface plasmon resonance
bidiffractive grating couplers 147, *148*
bi-enzymatic substrate recycling 209–216; electrodes *211–213*
binding constants 90; determination of 89–92, 272, 273; factors affecting 93
binding isotherms 93–98, *200*
binding kinetics 87–120; fitting software for 108, 112, 114; protein–protein interactions 28–29

binding reactions: computer simulation of 109–112, *113*; and diffusion 100–107
bioelectrocatalytic substrate recycling 216–217; sensors *218–219*
biological recognition 1–46
biomolecular interaction analysis (BIA) 241–242, 259–263; affinity separation using 260, 263; combined with mass spectrometry 263, *264*; concentration determination using 260–261; epitope mapping using 261; high-throughput screening using 261–262; kinetic analysis using 259–260; symposium 264; *see also* BIAcore instrument(s)
biotin–avidin/streptavidin system 76–77, 94, 188–189, *190–191*, 271, 307
Bizelesin (DNA bis-alkylator) 37–39
Brewster angle *130*, 155
Brewster angle reflectometry 155–156
bulk acoustic wave (BAW) devices 181–182
bulk acoustic waves 177; surface-generated 179–180
Butterworth–Van Dyke (BVD) circuit 186, *187*

calmodulin 278, 279
carbodiimide reagents 66–67, 68
carbohydrate–protein interactions 282–283
carbohydrates 56; functional groups in 72, 73
carbon electrodes 56–57
carboxylic acids: coupling to 64–69
carboxymethyl dextran (CMD) 57–59, 68–69, 271; in kinetic studies 273
CD44 transmembrane protein 279
cellulose-binding protein 77
chirped gating couplers *142*, 143
chromatographic detectors 298
cocaine: determination of 223, 225–227
combinatorial synthesis: identification of products 286
competitive assays 124, *125*; compared with displacement assays *231*, 232; immunoassays 208, 222, 223, 228, 230, 232; kinetics 111, 115–117
computer simulation: of binding reactions 109–112, *113*
conducting polymers: entrapment within 63
covalent coupling: aldehydes *65*, 69–70; amines *66*, 73; carboxylic acids 64–69; hydroxy compounds *66*, 72–73; immobilisation by *51*, 52, 64–73; thiols *66*, 70–72
cyclopropanepyrrololindole- (CPI-) containing molecules 37–39

deposition of liquid samples: electro-chemical control of 82–83; physical placement of 79–80

derivatisation of surface groups 57, 62
dermatan sulphate 275–276
detection limits: alkaline phosphatase 221, 222; biomolecules 192; phenolic compounds 217, *218–219*; virions 277
dextran-based sensor surfaces 56, 57–59, 68–69, 256, 257
dextran sulphate 59
dialysis tubing 60, 61
diffusion: and binding reactions 100–107; electrical analogue model 102–103
diffusion coefficients: determination by BIAcore 264–265
direct assay 124, *125*
direct-detection optical transducers 133–158
displacement assays 223; compared with competitive assays *231*, 232
dissociation rate constants 90; antigen–antibody interactions 12; and half-lives *12*; protein–protein interactions 25
distal sensing fibre optic transducers 169–170
disulphide formation 70–72
dithiothreitol (DTT) 71–72
DNA: A-form 33, 34, 35; alkylation of 37–39; analogue (shape/structure) recognition of 36–37; B-form 33, 34, 35; bendability/flexibility 35–36; digital (sequence readout) recognition of 39–42; functional groups within grooves 34–35; intercalating with 39, 284; structure 33–36; Watson–Crick form 34; Z-form 34
DNA interactions 33–42; and acoustic transducers 187–188, 195, *197*; and quartz crystal microbalance 307–309; and resonant mirror sensors 283–284
double mutant cycle analysis 23–25
doxorubicin 284

elastic waves: in solids 176–180
electrochemical control of deposition 82–83
electrochemical immunoassays 207–238
electrochemical quartz crystal microbalance (EQCM) 305
electropolymerisation 63, 82
ellipsometry 153–155, 241; compared with other techniques *312*; disadvantages 154–155
encounter complex 26
energetic epitope 8
entrapment: in conducting polymers 63; in hydrogels *51*, 61–63; immobilisation by *51*, 60–63; by membranes *51*, 60–61
enzymatic amplification electrodes *211–213*; immunoassays using 207–238
enzymatic cleavage: of antibodies 75; of immunoglobulins 5

enzyme-labelled immunoassays 222–223
enzyme linked immunosorbent assay
(ELISA) 207, 222
enzyme recycling electrodes 209–216;
immunoassays using 222–226
epitope mapping 261
epitopes 3, 4; continuous 6–9; discontinuous
6–9; methods for localising in proteins *8*;
structure 6–9
equilibrium titration *275*; kinetic analysis by
273–274
esters: active 67; displacement of 68–69
1-ethyl-3-(3-dimethylaminopropyl)carbodii
mide (EDC) 67
evanescent wave 131

fast-flow injection immunoassay (FIIA) 230,
231
ferrocene derivatives 227–228
fibre optic transducers 167–170; for
multichannel sensing 170
Fick's law(s) of diffusion 102, 105
fitting routines/software: for binding kinetics
108, 112, 114
flow immunoassays 223, 225–226
fluorescence 158–160
fluorescence capillary fill device (FCFD)
163–164
fluorescence intensity 160
fluorescence labelling 162
fluorescence transducers 158–170, 171;
configurations 161–162; multichannel
detection in 170, 171; optical-fibre-based
167–170; planar-waveguide-based
163–166; sensitivity 89, 161
fluorescently labelled antibodies 73
fluorophores 158; emission spectra 159, *160*;
excitation spectra 159, *160*
focused light ATR 247
Fresnel equations 129
fusion-protein approach 77

G-proteins 76, 280, 298
β-galactosidase: as label in immunoassay
223, *224*, 228
glass transducer surface 55
global fit analysis 114, 260
glycosaminoglycans 275–276
glycosidic coupling 74
gold 53–55, 149–150; self-assembled
monolayers on 53–54, 94, *95*
grating couplers: for fluorescence sensors
164–165; for SPR sensors 151; for
waveguide sensors 141–143
guided-wave modes 135–139
guided-wave optical sensors/transducers
135, 139–141

haptens 3
Helicobacter pylori 277–278
helix-turn-helix motif 36
herbicides: immunoassay of 223, 228
herpes simplex virus 277
high-throughput screening (HTS) 261–262
histidine-rich glycoprotein (HRG) 274–276
human growth hormone (hGH): interaction
with receptor *20*, 29
human thyroid stimulating hormone
(hTSH): immunoassay of 222–223, *224*
hydrogels: entrapment within 60–63
hydroxy groups: coupling involving *51*,
72–73
N-hydroxysuccinimide (NHS): coupling by
62, 67

IAsys 269–290; details 270; industrial
applications 285–287; membrane-protein
studies 278–279; protein–carbohydrate
interaction studies 282–283; protein–lipid
interaction studies 280–282;
protein–protein interaction studies
274–277; regulatory-protein interaction
studies 280; whole cell studies 277–278;
see also resonant mirror transducer
imines 69
immobilisation of biomolecules 47–120; by
adsorption *51*, 60; advantages and
disadvantages of various methods *51*; by
affinity capture 258; chemistry 49–86; by
covalent coupling *51*, 52, 64–73; by
entrapment *51*, 60–63; orientation-control
techniques 74–78, 97; properties required
for transducer surface 50–52;
spatial-control techniques 78–83; on SPR
sensor surface 57–59, 257
immobilised metal chelate chromatography
(IMAC) 77–78
immunoassays 207; coupled with enzymatic
recycling electrodes 207–238
immunoenzymometric assay (IEMA) *231*,
232
immunogenicity 3
immunoglobulins 3;
complementarity-determining regions 3;
paratopes in 5–6; properties *4*; proteolytic
cleavage of 5; structure 4–5
immunological applications: Love wave
sensor 198–200
immunological specificity 13–14
immunosensor: TSM resonator as 189, 192
induced-fit (antigen–antibody recognition)
model 12–13
inkjet deposition 79–80
interdigital transducers (IDTs) 182, *183*
interference of light 131–133

interleukin-2 ligand–receptor interactions
    114, *115*
ion-selective sensors 62–63

Jablonski diagram *159*

kinetic analysis 87–120; BIA-technology
    used for 259–260, 264; RM sensor used for
    271–274
kinetic rate constants 90; determination of
    92, 272; mass-transport effects on
    measurement 98–100

lactate/pyruvate determination *212*, 214,
    215
Langmuir equation/isotherm/model 90, 126
latex piezoelectric immunoassay (LPEIA)
    299
leaky surface (acoustic) waves 178
light-directed immobilisation 80–82
lipid (bi)layers, supported 88, 281; study of
    interfacial properties 189
lipid–protein interactions 280–282
lithography 79
lock-and-key model 12, 13
Love wave 178, *179*
Love wave sensors 196–200; applications
    198–200; comparison with other acoustic
    sensors *201*; theory 197–198
lung surfactant 282

Mach–Zehnder interferometers (MZIs)
    144–146, 147–148, 171
marker enzymes: detection in
    electrochemical immunoassays 217,
    220–222
mass spectrometry: and BIA-technology
    263, *264*
mass transport: measurement of rate
    constants affected by 98–100; in
    simulations 109
membrane cytoskeletons 279
membrane proteins 278–279
membranes: entrapment behind *51*, 60–61
micro-contact printing method 79
mimotope 9
mixed anhydrides 69
mixed monolayers 54, 94, *96*
molecular recognition interface: creation on
    transducer 49–50
multichannel sensors: fluorescence sensors
    170, 171; SPR sensors 150–151
mussel adhesive protein: QCM-D studies
    309–310, *311*
mutagenesis methods: protein–protein
    interactions studied by 22
mutational analysis 25

NANP-peptide system 93–94, *95*, 99–100,
    105, 107, 109
nitrophenol: coupling by 67
non-specific binding 51–52; suppression of
    88, 93, 97, 128
nucleic-acid interactions 33–42; studies
    187–188, 195, *197*, 283–284

optical-fibre-based fluorescence transducers
    167–170
optical transducers 123–175; direct detection
    transducers 133–158; dynamic measuring
    range 25; fluorescence transducers
    158–170; general considerations 124–133;
    see also surface plasmon resonance

paratopes 4; structure 5–6
peptide nucleic acids (PNAs): recognition of
    DNA by 39–41
pesticides: detection of 298
Pharmacia BIAcore instrument(s) *see*
    BIAcore instrument(s)
phenolic derivatives: immunoassay of *211*,
    214, 215, *218–219*
photochemical immobilisation techniques
    80–82
photo-deprotection methods 82
photon excited photons 246; *see also* surface
    plasmon resonance (SPR)
photoresist methods 78–79
piezoelectric crystals *292*; compared with
    SAW devices 294–295; materials 180–181,
    291, 304
piezoelectric detectors 298
piezoelectric immunosensors 298
planar waveguides: compared with surface
    plasmon modes 140; fluorescence
    transducers based on 163–166;
    light-coupling methods for 137–138
plane acoustic waves 177
plasminogen activation 276
plasmon resonance reflectance curves 248
polarimetric transducers 147
polarised light: waveguiding modes for 137
poly(L-lysine) 59–60
polymers: in sensors 55–56; surface-modified
    56; surface-tethered 57–60
poly(vinyl alcohol) 61–62
prism couplers: for SPR sensors 148–151,
    246; for waveguide sensors 143–144
Protein A (*or* G) capture systems 76, 298
protein–carbohydrate interactions 282–283
protein complexes: association mechanism
    for 26–27
protein–DNA interactions 36–37
protein–lipid interactions 280–282
protein–nucleic-acid interactions 36–37

protein–protein interactions 10, 19–32; affinity vs activity 29; heterogeneous-phase measurements 27–28; IAsys studies 274–277; kinetic parameters *20*, *22*, *25*
protein–protein interfaces: chemical characteristics 21; conformation changes upon complex formation 22; double mutant cycle analysis of 23–25; mutational induced structural rearrangement in 25; secondary structure of binding site 21–22; size and shape *20*, 21; structural characterisation of 19–22; surface complementarity 21; thermodynamic characterisation of *20*, 22–23
proteins: antibody-binding 75–76; regulatory 280; tagged 77–78
proteoglycans 276
pyrrole–imidazole polyamides: recognition of DNA by 41–42

quartz crystal microbalance (QCM) 185, 291–316; applications 187–193, 295, 298–299; commercial systems 295, *296–297*; compared with other techniques 299, *312*; liquid-phase applications 305; for soft films 305–307; theory 185–187, 291–295, 304–305; *see also* thickness-shear mode (TSM) resonators
quartz crystal microbalance with dissipation monitoring (QCM-D) 306–307; applications 307–310, *311*; compared with other techniques *312*; mathematical discussion 313–314

rab 5 binding 280
random-walk model of diffusion 100–101
rapid mixing model 89, 90, 112
Rayleigh wave 177–178
receptor density 104, 112; in NANP-peptide system 99, 105, 107, 109
recombinant proteins: monitoring production of 286–287
redox label immunoassay *208*, 227–228, *229–231*
reflected bulk wave (RBW) 179
reflection of light 129–131
reflectometric interference spectroscopy (RIfS) 156–157, 171
reflectometric techniques 153–158; sensitivity enhancement 157–158
refraction of light 131
refractive index: and biological molecules 133–135
refractometric sensors 133, 135, 141–152; mass labels for 152; referencing for 153; sensitivity 89, 152; temperature-control

requirements 152–153; *see also* surface plasmon resonance
resonance angle 149, 269–270
resonant mirror (RM) transducer 143–144, 269–290; applications 274–287; kinetic investigations 271–273; principle *144*, 269–271; *see also* IAsys

sandwich assay 124, *125*; applications 152, 222
Sauerbrey equation 185, 294, 304; for immersion in liquids 186, 294
Scatchard plots 90, *91*; bent/curved 94, 96, 97
Schiff bases 69, 74
screening assays 261–262
selectivity 127–128
self-assembled monolayers (SAMs) 53, 54–55, 93–94, *95*, *96*, 299
semi-permeable membranes 60–61
sensitivity 126–127; acoustic wave devices 200–201; guided-wave transducers 140; piezoelectric crystals 294; reflectometry 158; TSM/QCM 185, *201*; of various techniques 89, 158
sensor surfaces *see* transducer surfaces
shear-horizontal (SH) waves 178, *179*, 184
shear-horizontal surface acoustic wave (SH-SAW) sensors 193–200; *see also* acoustic plate mode sensors; Love wave sensors
signal amplification: by bi-enzymatic substrate recycling 209–216; by bioelectrocatalytic substrate recycling 216–217, *218–219*
silanised transducer surface 55, 298
silica sol-gel glasses 63
silica surfaces 55
site-directed mutagenesis 8, 15, 22
small-molecule–DNA interactions 37–39
Snell's law 131
spacer units 94, *96*
Stokes shift 158, 159
Stoneley wave 178
streptavidin: interaction with biotin 76, 94
structure–activity relationships 15
structure–function relationships 14–15
substrate patterning 78–79
substrate recycling 207, 209–217, 222
surface acoustic wave (SAW) devices 182–183; compared with piezoelectric crystals 294–295
surface acoustic waves (SAWs) 177–179
surface functional groups: transformation of 57, *58*
surface-generated bulk acoustic waves 179–180

surface plasmon resonance modes *136*, 138–139; compared with waveguide modes 140–141

surface plasmon resonance (SPR) 148–152, 241–268; advantages 149–150; and antigen–antibody interactions 12; application ranges 244; combined with planar waveguides 151–152; compared with other techniques *312*; plasmon resonance reflectance curves 248; principles *136*, 242; and protein–protein interactions 27; time-resolved analysis 248

surface plasmon resonance (SPR) instruments: configurations 245–256; control software 252–253; detector calibration 251–252; illumination unit 248–250; imaging system 250–251; multi-site refractometry 247–248; multichannel sensors 150–151; optics 247, *249*; photo-detector-array unit 251; regeneration of formed complexes 257; sample handling system 253–256; sensor surface chemistry 57–59, 68–69, 253, 256–258; thermostatting unit 256; *see also* BIAcore instrument

surface skimming bulk wave (SSBW) 179, *180*

surface-tethered polymers 57–60

synapsins 278–279

tagged proteins 77–78

TGN38 membrane protein 279

thickness-shear mode (TSM) resonator 181–182, 185–193; applications 187–193; in biotin–avidin/streptavidin interactions 188–189, *190–191*; combined with other techniques 192–193; comparison with other acoustic sensors *201*; as DNA sensor 187–188; as immunosensor 189, 192; interfacial properties studied by 189; theory 185–187, 291–295; in viscous fluid 186, 294, 305; *see also* quartz crystal microbalance

thioether formation 72

thiol groups, specifically located 74–75

thiols: coupling to *65*, 70–72, 74; monolayers 53–54

total internal reflection 131, 246; in waveguides 135, *136*

transcription complex 261

transducer materials 52–57, 149–150

transducer surfaces: creation of molecular recognition interface on 49–50; desirable properties 50–52, 87–88; functionality switching 57

transducer technology 121–238

transition state: protein association reaction 26

transmembrane proteins 279

travelling waves 128–129

two-compartment model 112, 114

V-number 168

waveguide interferometers 144–148; polarimetry-based 147

waveguide sensor geometries 141–148

waveguides: end-fire coupling to 137, *138*; grating coupling to 138, *138*, 141–143; optics 135–138; prism coupling to 138, *138*, 143–144

waveguiding modes 136–137, *136*; compared with SPR modes 140–141; for polarised light 137

whole cell studies 277–278